Laboratory Manual
for

FUNDAMENTALS OF HVACR
Second Edition

Carter Stanfield
Athens Technical College

David Skaves
Maine Maritime Academy

PEARSON

Boston   Columbus   Indianapolis   New York   San Francisco   Upper Saddle River

Amsterdam   Cape Town   Dubai   London   Madrid   Milan   Munich   Paris   Montreal   Toronto

Delhi   Mexico City   Sao Paulo   Sydney   Hong Kong   Seoul   Singapore   Taipei   Tokyo

Editorial Director: Vernon R. Anthony
Editorial Assistant: Nancy Kesterson
Director of Marketing: David Gesell
Executive Marketing Manager: Derril Trakalo
Senior Marketing Coordinator: Alicia Wozniak
Marketing Assistant: Les Roberts
Senior Managing Editor: JoEllen Gohr
Associate Managing Editor: Alexandrina
    Benedicto Wolf

Production Project Manager: Maren L. Miller
Senior Operations Supervisor: Pat Tonneman
Operations Specialist: Deidra Skahill
Senior Art Director: Diane Y. Ernsberger
Textbook Cover Designer: Bryan Huber
Manual Cover Designer: Integra
Cover Art: Shutterstock, Sergey Mironov
Printer/Binder: Edwards Brothers Malloy
Cover Printer: Edwards Brothers Malloy

Credits and acknowledgments borrowed from other sources and reproduced, with permission, in this textbook appear on the appropriate page within text.

10 9 8 7 6 5 4 3 2 1

PEARSON

ISBN 10: 0-13-287974-3
ISBN 13: 978-0-13-287974-3

INTRODUCTION

This Laboratory Manual is designed to accompany the *Fundamentals of HVACR*, second edition. Each lab worksheet title reflects the topic of coverage to ensure that the content of the exercise is clear. Some labs can be used in more than one section. The italicized labs in the Contents are not present in that section, but you can use the lab referenced in parenthesis if that better fits your curriculum.

There are approximately 200 additional labs available online in the **MyHVACLab** (www.myhvaclab.com) supplement. They can be found in folders for each unit.

All of the labs are mapped to the course objectives in **Class Master**. Class Master can be accessed through the MyLab, as well as through the instructor resources folder.

CONTENTS

Section 5

Section 6

Section 7

Section 8

LAB 2.1 HVACR JOBS

LABORATORY OBJECTIVE
You will learn about job opportunities in the HVACR field.

FUNDAMENTALS OF HVACR TEXT REFERENCE
Unit 2 Being a Professional HVACR Technician

REQUIRED TOOLS AND EQUIPMENT
Computer with Internet access

PROCEDURE
Do online research to learn about the job prospects for the field of Air Conditioning Technology.
Here are a few suggested web sites:

1. Department of Labor - http://www.bls.gov/oco/ocos192.htm

2. AHRI - http://www.ahrinet.org/

3. HVAC Excellence - http://HVACexcellence.org/

Based on your research, list three different HVACR jobs.

- HVAC System Designer
- Commercial and residential & Custom Replacements
- Precision Tune Up Specialist

Based on your research, list three employment advantages of the HVACR field.

- 2nd fastest growing profession in the US
- Top pay after a few years + Good Retirement + Benefits
- The skills that we learn on the way (a good trade)

Based on your research, what type of preparation is required for the HVACR field?

HVAC Mechanics and technicians are often Required to take state
Certification exams in order to legally work in the HVAC field.
Your state regulations and previous HVAC experience will determine
which type of training and certifications you should obtain
- Education & Training
- EPA 608

LAB 2.2 HVACR ORGANIZATIONS

LABORATORY OBJECTIVE

You will learn about professional organizations in the HVACR field.

FUNDAMENTALS OF HVACR TEXT REFERENCE

Unit 2 Being a Professional HVACR Technician

REQUIRED TOOLS AND EQUIPMENT

Computer with Internet access

PROCEDURE

Do online research to learn about HVACR trade organizations. Here are some suggested web sites:

AHRI   www.ahrinet.org

RSES   www.RSES.org
   Register for FREE subscription to RSES Journal at www.rses.org/studentjournal.aspx

NATE   www.natex.org   Arlington, VA

HVAC EXCELLENCE   www.HVACexcellence.org

SKILLS USA   http://www.skillsusa.org

ASHRAE   http://www.ashrae.org/

Give the purpose and/or mission of each organization.

Education, Training & Skills

How do you become a member of each organization?

Visit each of the sites and join there
Organizations online.

Where is each organization located?

NATE - Arlington, VA
RSES - Rolling Meadows, IL
AHRI - Arlington, VA
HVAC EXCELLENCE - Washington, DC
SKILLS USA - Leesburg, VA
ASHRAE - Atlanta, GA

LAB 3.1 MATERIAL SAFETY DATA SHEETS (MSDS)

LABORATORY OBJECTIVE

The student will demonstrate an understanding of the information provided on Material Safety Data Sheets for one of the refrigerant types to be used in the refrigeration laboratory.

LABORATORY NOTES

This lab exercise should always be the first one performed at the beginning of the course. Students should be introduced to MSD Sheets on the first day of lab.

Material Safety Data Sheets are required by law and have important information listed in specific areas so that they are easily read by emergency personnel.

You should read MSD Sheets on any material before you use it so you know how to use it properly and safely as well as knowing what to do if there is an accident involving the material.

FUNDAMENTALS OF HVACR TEXT REFERENCE

Unit 3 Safety

REQUIRED TOOLS AND EQUIPMENT

Refrigerant Material Safety Data Sheet

SAFETY REQUIREMENTS

A. None

PROCEDURE

STEP 1. Review a copy of a refrigerant MSD Sheet supplied by the Lab Instructor and answer the following questions:

A. What are the First Aid Measures for inhalation?

Remove source of contamination or move victim to fresh air. If breathing is difficult, trained personnel should administer emergency. Do not allow to move about unnecessarily, symptoms of pulmonary edema can be delayed up to 48 hours after exposure. Quickly transport victim to an ER.

B. What are the First Aid Measures for refrigerant exposure to the eyes?

Immediately flush eyes with large amounts of water for at least 15 minutes (in case of frostbite water should be lukewarm, not hot) lifting eyelids occasionally to facilitate irrigation. Get medical attention if symptoms persist.

C. What are the First Aid Measures for refrigerant exposure to the skin?

Promptly flush skin with water until all chemical is removed. If there is evidence of frostbite, bathe (do not rub) with lukewarm (not hot) water. If water is not available, cover with a clean, soft cloth or similar covering. Get medical attention if symptoms persist.

D. What are the First Aid Measures for refrigerant ingestion?

Ingestion is unlikely because of the physical properties and is not expected to be hazardous. Do not induce vomiting unless instructed to do so by a physician

E. What type of protective clothing is worn for the eyes and face?

For Normal conditions, wear safety glasses. Where there is reasonable probability of liquid contact, wear chemical safety goggles.

F. What type of protective clothing is worn for the hands, arms, and body? *General work clothing and gloves (leather) should provide adequate protection.*

QUESTIONS

To get help in answering some of the following questions, refer to the *Fundamentals of HVACR* text Unit 3.

(Circle the letter that indicates the correct answer.)

1. If an area on a MSD Sheet does not apply to a product:
   - A. it should be left blank.
   - B. it should be filed in with the refrigerant type.
   - C. there should be a line through it.
   - D. it should be marked as non-applicable.
2. The section of a MSD Sheet that provides the properties of the material, such as boiling point, vapor pressure, etc. is:
   - A. Section I.
   - B. Section II.
   - C. Section III.
   - D. Section IV.
3. If you do not follow the instructions for proper use and handling of a material as listed on a MSDS:
   - A. you could be injured.
   - B. a customer could be injured.
   - C. you could be fired.
   - D. All of the above are correct.
4. Section I of the MSDS provides:
   - A. the manufacturer's name and address.
   - B. the fire and explosion hazard data.
   - C. hazardous ingredients / identity information.
   - D. All of the above.
5. The HMIS information tells health care workers a relative number according to how significantly the material will affect health, how reactive it is, and its flammability.
   - A. True.
   - B. False.

4

LAB 3.2 HVACR PPE

LABORATORY OBJECTIVE

You will learn how and when to use the types of personal protective equipment (PPE) used in HVACR.

LABORATORY NOTES

We will identify the types of PPE used in the HVACR field, discuss when it should be worn, and demonstrate its use.

***FUNDAMENTALS OF HVACR* TEXT REFERENCE**

Unit 3 Safety

REQUIRED MATERIALS PROVIDED BY STUDENT

Shop safety manual
Safety glasses
General mechanics gloves

REQUIRED MATERIALS PROVIDED BY SCHOOL

Welding/Brazing dark safety glasses
Leather gloves for brazing
Arc-blast rated gloves
Respirator
Self contained breathing apparatus SCBA
Examples of proper and improper apparel
Examples of proper and improper footwear

PROCEDURE

The instructor will discuss and demonstrate the use of HVACR PPE. You will be prepared to discuss and demonstrate proper PPE, including:

- Appropriate shop clothing
- Appropriate shop footwear
- When safety glasses should be worn
- When welding glasses are required
- When to use general shop gloves, welding gloves, or arc blast gloves
- When to use a respirator
- When to use SCBA equipment
- How to find out the correct PPE for any type of job

Name _____

Date _____

Instructor's OK ☐

LAB 3.3 LOCATE SAFETY EQUIPMENT

LABORATORY OBJECTIVE
You will learn the location of safety equipment in the shop.

LABORATORY NOTES
We will use the shop safety manual to locate and identify all the shop safety equipment.

FUNDAMENTALS OF HVACR **TEXT REFERENCE**
Unit 3 Safety

REQUIRED MATERIALS PROVIDED BY STUDENT
Shop safety manual
Safety glasses

REQUIRED MATERIALS PROVIDED BY SCHOOL
MSDS Book
First Aid Kit
Electrical Lock Out Kit
Fire extinguishers
Eye Wash Station
Shop Vent Fan
Main Electric Switchgear
Main Gas Valve Shutoff
Emergency Exit Plan

PROCEDURE
Be prepared to show the instructor the following shop safety equipment:

- MSDS Book

- First Aid Kit

- Electrical Lock Out Kit

- Fire extinguishers

- Eye Wash Stations

- Shop Vent Fan

- Main Electric Switchgear

- Main Gas Valve Shutoff

- Emergency Exit Plan

6

LAB 3.4 LADDER SAFETY

LABORATORY OBJECTIVE
You will demonstrate how to safety set up and climb both step ladders and extension ladders.

LABORATORY NOTES
The instructor will demonstrate the correct way to set up a step ladder and an extension ladder. They will then discuss and demonstrate the correct way to climb and work on a ladder. You will demonstrate your ability to safely set up both a step ladder and an extension ladder. You will then demonstrate your ability to safely climb a ladder.

***FUNDAMENTALS OF HVACR* TEXT REFERENCE**
Unit 3 Safety

REQUIRED MATERIALS PROVIDED BY STUDENT
Shop safety manual
Safety glasses

REQUIRED MATERIALS PROVIDED BY SCHOOL
Six-foot step ladder
Extension ladder

PROCEDURE
The instructor will demonstrate how to set up and climb a typical extension ladder and a typical extension ladder. You will then demonstrate your ability to:

- Safely set up a six-foot step ladder.
- Safely climb a six-foot step ladder.
- Safely set up an extension ladder.
- Safely climb an extension ladder.
- Discuss what NOT to do on a step ladder.
- Discuss what NOT to do on an extension ladder.

LAB 4.1 USING HAND TOOLS

LABORATORY OBJECTIVE
You will demonstrate your ability to safely perform the following tasks:

- use screwdrivers and nut drivers to secure air conditioning unit panels
- tighten and loosen an assembly using wrenches
- change a hacksaw blade
- cut a piece of metal with a hack saw
- change a drill bit in a power drill
- drill a hole in a piece of metal

LABORATORY NOTES
You will perform a variety of manipulative tasks using hand tools. These include removing and replacing panels on units, disassembling and reassembling a bolted assembly, changing a hack saw blade, cutting metal with a hack saw, and drilling a hole in a piece of metal.

FUNDAMENTALS OF HVACR TEXT REFERENCE
Unit 4 Hand and Power Tools

REQUIRED MATERIALS PROVIDED BY STUDENT
Safety glasses
Flat blade screwdriver
Phillips screwdriver
¼" nut driver
5/16" nut driver
OR 6-in-1 with all the above
Adjustable jaw wrench

REQUIRED MATERIALS PROVIDED BY SCHOOL
Air conditioning units with panels
Metal for cutting and drilling
Hack saw
Power drill
Extension cord
Drill b
its
Fixed wrenches
Assembly to tighten

PROCEDURE

Safety Glasses are required for all tasks!

Panels: Determine which tool you need: flat blade, Phillips, ¼" nut driver, or 5/16" nut driver. If the

8

fasteners can use more than one type, the order of preference is nut driver, Phillips, then flat blade.

Bolted Assembly

Disassemble and reassemble the bolted assembly assigned by the instructor. Care should be taken to use tools appropriately. Make certain that wrenches fit properly. Pliers are NOT a substitute for a wrench where a wrench is needed!

Hack Saw Blade

To change the hacksaw blade: loosen the knurled wheel on the handle end. The blade should now remove easily. Pay attention to the direction of the teeth when replacing the blade; they should face forward. Many blades have an arrow that should point towards the front of the saw, away from the handle. After replacing the blade, tighten the knurled wheel by turning it clockwise.

Hack Saw

Mount the metal to be cut securely in a vise. If it is round, it should go in the jaws designed for holding tubing and round objects. Place the blade on the metal and use approximately one forward stroke per second with a moderate down pressure.

Drill Bit

Make sure the drill is unplugged! The chuck is the part of the drill that holds the drill bit. Two types of chucks are used: keyed chucks and hand chucks. Keyed chucks have holes in the base of the chuck to insert the chuck key. If the drill uses a chuck key, use the chuck key to loosen the chuck. If it is a hand chuck, loosen by hand. Insert the drill bit and tighten. The bit should be in the center of the chuck.

Drilling

NEVER USE A POWER DRILL WITHOUT SAFETY GLASSES!

Mount the metal to be drilled in a vise. Making a dimple in the metal where the hole will be helps keep the drill bit from wandering when the hole is started. Use a punch to make a dimple in the metal where the hole will be drilled by striking the punch with a hammer. Place the point of the drill bit in the dimple before stating the drill. Start the drill using a moderate amount of pressure. Forcing the bit to drill too quickly will only dull the bit and slow down the entire process.

LAB 5.1 FASTENER IDENTIFICATION

LABORATORY OBJECTIVE
You will identify the following types of fasteners and describe their use:

- sheet metal screws
- wood screws
- nuts and bolts
- nails

LABORATORY NOTES
You will receive an assortment of fasteners. You will identify each fastener, what it is used for, and the tool(s) required to use it.

FUNDAMENTALS OF HVACR TEXT REFERENCE
Unit 5 Fasteners

REQUIRED MATERIALS PROVIDED BY STUDENT
Safety glasses

REQUIRED MATERIALS PROVIDED BY SCHOOL
Assorted fasteners

PROCEDURE
Examine the assortment of fasteners provided by the instructor; identify the type of fastener, what the fastener is used for, and what tools are used with the fastener. Complete the chart on the next page as you identify the fasteners.

#	FASTENER NAME	USE	TOOL		#	FASTENER NAME	USE	TOOL
1					11			
2					12			
3					13			
4					14			
5					15			
6					16			
7					17			
8					18			
9					19			
10					20			

Name _____

Date _____

Instructor's OK ☐

LAB 7.1 BALLOON – GAS VOLUME VS. TEMPERATURE

LABORATORY OBJECTIVE

The purpose of this lab is to demonstrate the change in gas volume accompanying a change in temperature.

LABORATORY NOTES

You will compare the volume of a fully inflated mylar balloon at room temperature to the volume of the same balloon after it has been cooled below freezing.

FUNDAMENTALS OF HVACR TEXT REFERENCE

Unit 7 Properties of Matter

REQUIRED TOOLS AND MATERIALS PROVIDED BY SCHOOL

1 ultralow temp freezer, -40°
1 mylar balloon

PROCEDURE

Observe the volume of a fully inflated mylar balloon.

Place a fully inflated mylar balloon in the freezer and wait 5 minutes for the gas in the balloon to cool.

Remove the balloon from the freezer and observe its reaction.

Note: This step must be done quickly because the gas in the balloon heats up quickly.

LAB 7.2 WEIGHING SOLIDS, LIQUIDS, AND GASES

LABORATORY OBJECTIVE

The purpose of this lab is to learn to weigh solids, liquids, and gases using the shop scales.

LABORATORY NOTES

You will weigh a solid, a liquid, and a gas using both traditional and metric weights. For the liquid and the gas you will use the scale's zero setting to compensate for the weight of the container.

FUNDAMENTALS OF HVACR TEXT REFERENCE

Unit 6 Measurements
Unit 7 Properties of Matter

REQUIRED TOOLS AND MATERIALS PROVIDED BY SCHOOL

1 brick
1 quart of oil
1 empty oil quart container
1 empty recovery cylinder
Vacuum pump

PROCEDURE

Solids

- Zero the scale with nothing on it.
- Place the brick on the scale.
- Use the scale units key to see the weight in pounds, pounds and ounces, and kilograms.
- Record below.

Item	Pounds	Pounds & Ounces	Kilograms	grams
Brick				

Liquids

- Place an empty container on the scale.
- Zero the scale with the empty oil container on the scale.
- Replace the empty container with the oil you wish to weigh.
- Use the units key to see the weight in pounds, pounds and ounces, and kilograms.
- Record below.

Item	Pounds	Pounds & Ounces	Kilograms	grams
Oil				

Gas

- With the instructor's help, evacuate an empty refrigerant recovery cylinder.
- Zero the scale with the recovery cylinder on it.
- Open the valve and let air enter the cylinder.
- Record the weight of the air below.

Item	Pounds	Pounds & Ounces	Kilograms	grams
Air				

Summary

1. What is the difference between pounds and pounds and ounces?

2. Why is it necessary to zero the scale with the container on it when weighing a liquid or a gas?

3. What does the increase in the scale reading before and after letting air into the cylinder tell you about air?

14

LAB 7.3 MEASURING DENSITY, SPECIFIC VOLUME, AND SPECIFIC GRAVITY

LABORATORY OBJECTIVE
Demonstrate an understanding of density, specific volume, and specific gravity.

LABORATORY NOTES
You will measure the weight and volume of a solid, a liquid, and a gas. Using those measurements you will calculate the density, specific volume, and specific gravity of each.

FUNDAMENTALS OF HVACR TEXT REFERENCE
Unit 7 Properties of Matter

REQUIRED TOOLS AND MATERIALS SUPPLIED BY STUDENT
Standard ruler
Calculator

REQUIRED TOOLDS AND MATERIALS SUPPLIED BY SCHOOL
Metric ruler
Digital Scale
Vacuum Pump
Bicycle air pump
1 quart of oil
1 empty oil quart container
1 empty recovery cylinder

Procedure is on following pages.

PROCEDURE

Testing Solids – A Brick

Measure the volume and weight of the brick and complete the table below.

Dimension	Inches	Centimeters
Width		
Length		
Height		
Volume (WxLxH)		

Weight in Ounces	Weight in grams

Use the figures from the table and the formulas below to calculate the density, specific volume, and specific gravity of the brick

Brick Characteristics, English Units

Density = _____ ÷ _____ = _____ ounces per cubic inch
        ounces   cubic inches

Specific Volume = _____ ÷ _____ = _____ cubic inches per ounce
         cubic inches   ounces

Specific Gravity = _____ ÷ 0.58 = _____ times as heavy as water
         Density   (Density of water in ounces per cubic inch)

Brick Characteristics, Metric Units

Density = _____ ÷ _____ = _____
       grams ÷ cubic centimeters = grams per cubic centimeter

Specific Volume _____ ÷ _____ = _____
        cubic centimeters   grams   cubic centimeters per gram

Specific Gravity = Density without the units

Note: Since a gram is defined as the weight of 1 cubic centimeter of water, the specific gravity is the same thing as the density when using grams and cubic centimeters.

Testing Liquids – Oil

- Place the empty container on the scale and zero the scale.
- Replace the empty container with the full quart of oil.
- Record the weight of the full quart of oil in both ounces and grams.

Weight in Ounces	Weight in grams

Note: A quart IS a volume measurement, so we already know the volume. However we need to convert the quart to cubic inches. There are approximately 30 quarts in a cubic foot, and there are 1728 cubic inches in a cubic foot. Dividing 1728 by 30, 1 quart would be 57.6 cubic inches. Look on the quart container for the metric volume. You will see 946 milliliters. Since a milliliter is defined as 1 cubic centimeter in volume, 946 cubic centimeters is the metric volume. Use these figures to calculate the Density, Specific Volume and Specific Gravity.

Oil Characteristics, English Units

Density = _____ ÷ _____ = _____ ounces per cubic inch
         ounces   cubic inches

Specific Volume = _____ ÷ _____ = _____ cubic inches per ounce
            cubic inches   ounces

Specific Gravity = _____ ÷ 0.58 = _____ times as heavy as water
         Density   Density of Water

Oil Characteristics, Metric Units

Density = _____ ÷ _____ = _____ grams per cubic centimeter
      grams cubic centimeters

Specific Volume = _____ ÷ _____ = _____ cubic centimeters per gram
        Cubic centimeters   grams

Specific Gravity = Density

Note: Since a gram is defined as the weight of 1 cubic centimeter of water, the specific gravity is the same thing as the density when using grams and cubic centimeters.

Testing Gases

You will need an instructor's help for this portion of the experiment. Evacuate the recovery cylinder marked "Air Only." Set the scale to read ounces, place the cylinder on a digital scale, and zero the scale. Open the valve on the cylinder and let air in. You should see a small weight increase on the scale. Record this weight. Use the tire pump to increase the pressure in the tank to 20 psig. Record this weight. Use the formulas below to calculate the density, specific volume, and specific gravity. Note: the recovery cylinder volume is 1320 cubic inches.

Gas Density at 0 psig

Density = _____ ÷ _____ = ounces per cubic inch
        ounces   cubic inches

Specific Volume = _____ ÷ _____ = _____ cubic inches per ounce
             cubic inches   ounces

Specific Gravity _____ ÷ 0.58　　　= _____
            Density   (Density of water)

Gas Density at 20 psig

Density = _____ ÷ _____　　=　　ounces per cubic inch
        ounces   cubic inches

Specific Volume _____ ÷ _____ = _____ cubic inches per ounce
            cubic inches   ounces

Specific Gravity _____ ÷ 0.58　　　= _____
            Density   (Density of water)

SUMMARY

1. In general, which state has the highest density, solid, liquid, or gas?

2. In general, which state has the lowest density, solid, liquid, or gas?

3. In general, which state has the highest specific volume, solid, liquid, or gas?

4. In general, which state has the lowest specific volume, solid, liquid, or gas?

5. Which two states have the most similar density?

6. Which two states have the most similar specific volume?

7. Which state's density and specific volume is most subject to change?

8. What advantage do metric units have when performing specific gravity measurements?

9. What effect does increased pressure have on the density and specific volume of a gas?

LAB 8.1 ENERGY CONVERSION

LABORATORY OBJECTIVE

You will demonstrate that energy can be converted from one form to another and that electricity generated from a rotating magnetic device is not free.

LABORATORY NOTES

We will use a hand cranked electric generator, which is connected to several electric light bulbs. The lights will be controlled by switches. You will turn the generator by hand, producing electricity. You will control the lights with switches to demonstrate the effect of electrical loads on the generator.

***FUNDAMENTALS OF HVACR* TEXT REFERENCE**

Unit 8 Types of Energy and Their Properties

REQUIRED TOOLS AND MATERIALS PROVIDED BY SCHOOL

1 hand-cranked generator
3 electric spot switches
3 lights

PROCEDURE

1. Turn the generator with all the switches off. This is the amount of energy required to overcome the mechanical inefficiencies in the machine.
2. While you are turning the generator, turn on the smallest light.
3. Notice that the generator is now harder to turn.
4. Turn on the other lights, one at a time.
5. Note the change each time another load is added.
6. Now turn the lights off one at a time.
7. Note the change each time another load is removed.

SUMMARY

What types of energy are represented in this experiment?

LAB 9.1 MEASURING TEMPERATURE

LABORATORY OBJECTIVE
You will demonstrate your ability to measure the temperature of refrigerant lines, air, and objects.

FUNDAMENTALS OF HVACR **TEXT REFERENCE**
Unit 9 Temperature Measurement and Conversion

REQUIRED TOOLS AND EQUIPMENT SUPPLIED BY STUDENT
Safety glasses

REQUIRED TOOLS AND EQUIPMENT SUPPLIED BY SCHOOL
Fluke 16 multimeter
Temperature clamp
Thermocouple bead
Infrared thermometer
Ice
Operating air conditioner
Operating furnace
Shop heater
Shop air conditioner

LABORATORY NOTES
You will use a thermocouple bead and a Fluke 16 multimeter, a thermocouple clamp and a Fluke 16 multimeter, and an infrared thermometer to measure several temperatures. You will give the temperature readings in Fahrenheit, Celsius, Rankin, and Kelvin.

PROCEDURE
Read the following temperatures using the indicated instrument. Take readings in both Fahrenheit and Celsius, and then convert them to Rankin and Kelvin. Record your readings on the chart below.

Temperature	Using	Fahrenheit	Rankin	Celsius	Kelvin
Ice	Bead				
Ice	Infrared				
Shop	Bead				
Shop Wall	Infrared				
Skin	Bead				
Skin	Infrared				
Cold Air	Bead				

LAB 10.1 BOILING POINT VS. PRESSURE

LABORATORY OBJECTIVE
The purpose of this lab is to demonstrate the effect of pressure on the boiling point of a liquid.

LABORATORY NOTES
We know that the boiling point of water at atmospheric pressure is 212 degrees Fahrenheit. We will use a vacuum pump to reduce the pressure on a flask of water and observe the effect of reduced pressure. A vacuum pump reduces the pressure of a closed container by removing gas from the container.

FUNDAMENTALS OF HVACR TEXT REFERENCE
Unit 10 Thermodynamics – The Study of Heat

REQUIRED TOOLS AND EQUIPMENT
Flask with rubber stopper
Vacuum pump
Vacuum gauge that reads in inches of mercury vacuum

SAFETY REQUIREMENTS
None.

PROCEDURE
Step 1
Obtain an empty flask and rubber stopper.

1. Fill the flask half full with water.
2. Stopper the top of the flask.
3. Connect the flask to the vacuum pump.
4. Connect the vacuum gauge to the vacuum pump.

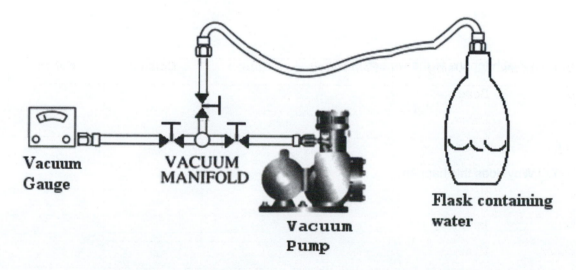

Figure 10-1-1

Step 2

Operate the vacuum pump and then answer the following questions

1. What happens to the pressure in the flask?

2. What happens to the water in the flask?

3. Explain why this happens.

Step 3

Turn off the vacuum pump and then answer the following questions.

4. What happens to the pressure in the flask?

5. Why does this happen?

6. What happens to the water?

7. Why does this happen?

LAB 10.2 PRESSURE-TEMPERATURE RELATIONSHIP OF SATURATED MIXTURES

LABORATORY OBJECTIVE
You will demonstrate the relationship of pressure and temperature for saturated gas-liquid mixes.

LABORATORY NOTES
You will observe the behavior of saturated gas by comparing the pressure of two cylinders containing different weights a saturated mix at the same temperature. Then we will change the temperature and observe the effect on pressure.

You will be using two recovery cylinders with refrigerant. Both cylinders should be the same type of refrigerant, but they should contain different amounts of refrigerant. Both cylinders should be the same temperature and both **should contain some liquid and some vapor.**

FUNDAMENTALS OF HVACR TEXT REFERENCE
Unit 10 Thermodynamics – The Study of Heat

REQUIRED TOOLS AND EQUIPMENT
Two refrigerant recovery cylinders with the same refrigerant type
Freezer or a bucket of ice
Heat gun or a sink with hot water
Gauge manifold
Infrared thermometer
Digital scale

SAFETY REQUIREMENTS
A. Wear safety goggles and gloves when working with refrigerants. Liquid refrigerant can cause frostbite when in contact with eyes and skin.

PROCEDURE
Step 1
Measure the temperature, pressure, and weight of the first refrigerant recovery cylinder as follows.

1. Place one of the refrigerant recovery cylinders on top of a digital scale and then connect the refrigerant recovery cylinder to the high side of the gauge manifold as shown in Figure 10-2-1.

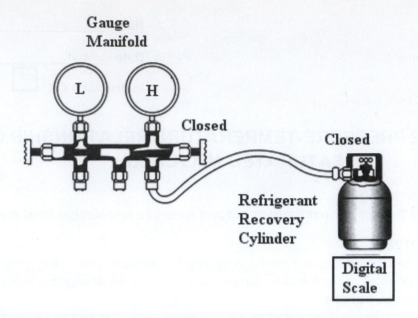

Figure 10-2-1

2. Slowly open the refrigerant recovery cylinder valve as shown in Figure 10-2-2.

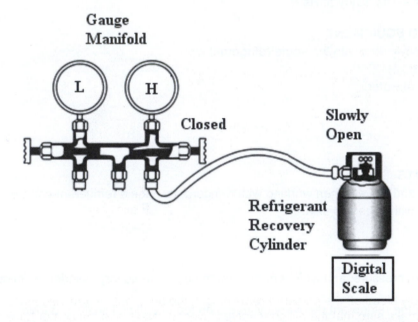

Figure 10-2-2

3. Record the pressure from the reading on the high side gauge.

Recovery Cylinder #1 pressure:

4. Measure the cylinder temperature using the infrared thermometer.

   Recovery Cylinder #1 temperature:

5. Record the cylinder weight from the reading on the digital scale.

   Recovery Cylinder #1 weight:

6. After recording the cylinder pressure and temperature, close the cylinder valve as shown in Figure 10-2-3.

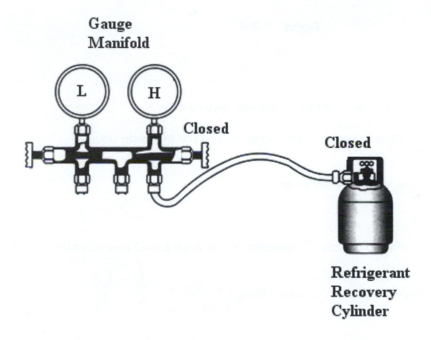

Figure 10-2-3

7. You must now bleed off the pressure remaining in the hose before you disconnect it. To do this, make sure that the cylinder valve is closed. Next take the center hose of the gauge manifold and direct it away from any person including yourself. Slowly open the high side gauge manifold valve as shown in Figure 10-2-4 and the pressure in the line will bleed off. Be careful as the hose may whip around slightly as it drains.

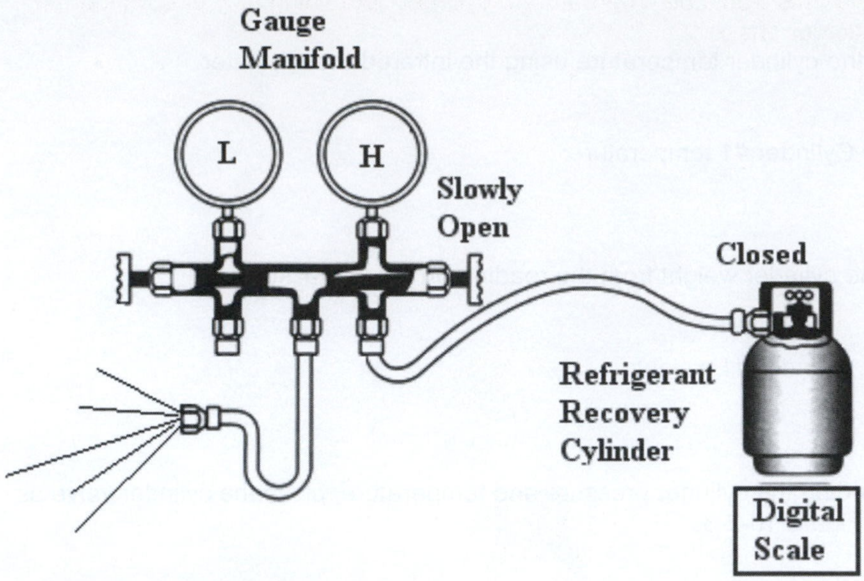

Figure 10-2-4

Step 2
Repeat step 1 using the second refrigerant recovery cylinder.

1. Record the pressure from the reading on the high side gauge.

    Recovery Cylinder #2 pressure:

2. Measure the cylinder temperature using the infrared thermometer.

    Recovery Cylinder #2 temperature:

3. Record the cylinder weight from the reading on the digital scale.

    Recovery Cylinder #2 weight:

Step 3
Referring to the measurements you recorded in steps 1 and 2, answer the following questions.

1. How does the pressure of the cylinder that contains the most refrigerant compare to the cylinder with the least refrigerant?

2. Do you think that this would be different for a cylinder containing only gas without any liquid? Explain your answer.

Step 4
Place one of the cylinders in a bucket of ice or in the freezer.

**Recovery
Cylinder
packed in ice**

Figure 10-2-5

1. Allow the cylinder to cool for 30 minutes and then measure its pressure and temperature as you did in step 1.
2. Record the pressure from the reading on the high side gauge.

    Cooled Recovery Cylinder pressure:

3. Measure the cylinder temperature using the infrared thermometer.

    Cooled Recovery Cylinder temperature:

Step 5
Calculate the expected pressure after cooling the cylinder. For this calculation use the starting pressure, starting temperature, and ending temperature. Refer to the Ideal Gas Laws from Unit 11 in the *Fundamentals of HVACR* text.

(Remember that all pressures and temperatures must be converted to absolute readings BEFORE working the problem.)

SHOW ALL OF YOUR CALCULATIONS IN THE SPACE PROVIDED.

1. How does the actual pressure compare to the pressure predicted by the gas law?

2. Use a saturation chart such as the one found in Unit 10 from the HVACR text and determine the pressure that corresponds to the temperature of the refrigerant in the cylinder.

    Saturated pressure from chart:

3. How does the pressure found from the chart compare to the actual pressure?

4. Which was more accurate in predicting the results, the ideal gas law or using the saturation chart?

Step 6
Place the recovery cylinder into a sink and run warm water over the cylinder.

CAUTION: The refrigerant cylinder temperature should NEVER exceed 125 °F.

Figure 10-2-6

1. Allow the cylinder to cool for 15 minutes and then measure its pressure and temperature as you did in step 1.
2. Record the pressure from the reading on the high side gauge.

   Heated Recovery Cylinder pressure:

3. Measure the cylinder temperature using the infrared thermometer.

   Heated Recovery Cylinder temperature:

Step 7

Calculate the expected pressure after cooling the cylinder. For this calculation use the starting pressure, starting temperature, and ending temperature. Refer to the Ideal Gas Laws from Unit 11 in the *Fundamentals of HVACR* text.

(Remember that all pressures and temperatures must be converted to absolute readings BEFORE working the problem.)

SHOW ALL OF YOUR CALCULATIONS IN THE SPACE PROVIDED BELOW.

1. How does the actual pressure compare to the pressure predicted by the gas law?

2. Use a saturation chart such as the one found in Unit 10 from the HVACR text and determine the pressure that corresponds to the temperature of the refrigerant in the cylinder.

   Saturated pressure from chart:

3. How does the pressure found from the chart compare to the actual pressure?

4. Which was more accurate in predicting the results, the ideal gas law or using the saturation chart?

5. What conclusions can you draw about the behavior of saturated liquid-gas mixtures to temperature changes?

LAB 10.3 SENSIBLE AND LATENT HEAT

LABORATORY OBJECTIVE
You will demonstrate sensible and latent heat properties.

LABORATORY NOTES
We will demonstrate sensible and latent heat properties by adding heat to water while we monitor the amount of energy used, the temperature of the water, and the weight of the water. Energy input will be determined by measuring the wattage used, temperature change will be measured with a thermocouple, and weight change will be measured with a digital scale.

***FUNDAMENTALS OF HVACR* TEXT REFERENCE**
Unit 10 Thermodynamics – The Study of Heat

REQUIRED TOOLS AND EQUIPMENT
Temperature sensor with temperature clamp and thermocouple clamp
Electric cook pot
Wattmeter

SAFETY REQUIREMENTS
A. Wear safety goggles and gloves when working with high temperature liquids.
B. The steam that is generated can easily burn your skin. Be careful not to come in contact with the hot steam.

PROCEDURE
Step 1
Obtain an electric cook pot, and digital scale.

1. Place the cook pot on top of the digital scale and pour approximately 3 pounds of water into the pot.
2. Zero the digital scale with the pot and the water.
3. Measure the temperature of the water.

Water Temperature _____

4. Connect the wattmeter to the cook pot circuit to measure the power that will be used during the heating of the water.

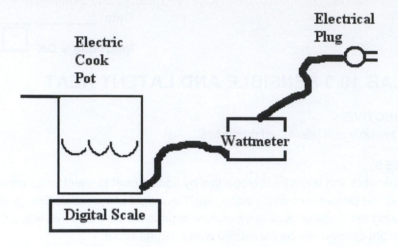

Electric
Cook
Pot

Electrical
Plug

Wattmeter

Digital Scale

Figure 10-3-1

5. Plug in the cook pot and record the time and wattage.

    Time started:

    Wattage:

Step 2
Monitor the time, temperature, and wattage as follows.

1. When the temperature reaches approximately 150 °F, record the time, the exact temperature and the wattage.

    Time _____

    Temperature_____

    Wattage_____

2. When water begins to boil, record the time, the exact temperature and the wattage.

    Time _____

    Temperature_____

    Wattage_____

3. When the water reaches a rolling boil record the time, the exact temperature, and the weight on the scale. Let the water boil for three minutes and then record these values again in part 4.

Time_____

Temperature_____

Weight_____

Wattage_____

4. Record the values after three minutes of boiling.

Time_____

Temperature_____

Weight_____

Wattage_____

Step 3
Calculate the amount of energy in BTUs required to heat the water to 150 °F.

1. Since it takes 1 BTU to heat 1 pound of water 1 °F at atmospheric pressure, you can use the formula:

(weight of the water in pounds) x (1 BTU/lb-°F) x (ending temperature – beginning temperature)= BTU required to heat the water.

SHOW ALL OF YOUR CALCULATIONS IN THE SPACE PROVIDED ON THE NEXT PAGE.

TOTAL AMOUNT OF HEAT REQUIRED (BTU)_____

2. Now calculate the measured amount of electrical heat input. One Watt is equivalent to 3.41 BTU per hour or divide this by sixty minutes in one hour for 0.057 BTU per minute. With this we can convert our measured Wattage into BTU using the formula:

(measured wattage) x (minutes to 150ºF) x (0.057 BTU/min-Watt) = BTU input

SHOW ALL OF YOUR CALCULATIONS IN THE SPACE PROVIDED BELOW.

TOTAL AMOUNT OF HEAT INPUT (BTU)_____

3. How does the amount of heat required to heat the water in BTUs compare to the measured amount of heat input? If they are not equal, then explain why.

Step 4

Calculate the amount of energy in BTUs required to boil the water for three minutes.

1. Since it takes 970 BTU to boil 1 pound of water at atmospheric pressure, you can use the formula:

(pounds of water beginning – pounds of water ending for the three minute period) x (970 BTU/lb) = BTU required to boil the water for three minutes.

SHOW ALL OF YOUR CALCULATIONS IN THE SPACE PROVIDED BELOW.

TOTAL AMOUNT OF HEAT REQUIRED (BTU)_____

2. Now calculate the measured amount of electrical heat input. One Watt is equivalent to 3.41 BTU per hour or divide this by sixty minutes in one hour for 0.057 BTU per minute. With this we can convert our measured Wattage into BTU using the formula:

(measured wattage) x (three minutes) x (0.057 BTU/min-Watt) = BTU input

SHOW ALL OF YOUR CALCULATIONS IN THE SPACE PROVIDED ON THE NEXT PAGE.

TOTAL AMOUNT OF HEAT INPUT (BTU)_____

3. How does the amount of heat required to boil the water in BTUs compare to the measured amount of heat input? If they are not equal, then explain why.

4. What happened to the temperature of the water as it was boiling?

5. What do you think the temperature of the steam would be?

6. Does it take more energy to heat the water or to boil the water?

LAB 10.4 DETERMINING REFRIGERANT CONDITION

LABORATORY OBJECTIVE
The purpose of this lab is to demonstrate your ability to determine if refrigerant is saturated, superheated, or subcooled.

LABORATORY NOTES
You will measure the pressure and temperature of several refrigeration system components. You will determine if the refrigerant in the component is saturated, superheated, or subcooled by comparing the pressure and temperature to a saturation chart for that particular refrigerant.

***FUNDAMENTALS OF HVACR* TEXT REFERENCE**
Unit 10 Thermodynamics – The Study of Heat

REQUIRED TOOLS AND EQUIPMENT
Instrumental Refrigerant Trainer
Pressure Temperature Chart

SAFETY REQUIREMENTS
None

PROCEDURE
Step 1
Identify the components on the refrigeration trainer.

Step 2
Line up the components and then start the refrigeration trainer using the trainer's instruction manual if necessary.

Step 3
Allow the refrigerant trainer to run until the temperatures and pressures have begun to stabilize and then complete the chart below. You will need to identify for each situation whether the refrigerant is subcooled, saturated, or superheated. Not all refrigeration trainers are instrumented the same, so if there is no reading available on the refrigerant trainer for the specific condition, then leave that space blank.

LOCATION	PRESSURE	TEMPERATURE FROM CHART	ACTUAL TEMPERATURE	CONDITION
Compressor IN				
Compressor OUT				
Condenser IN				
Condenser OUT				
Evaporator IN				
Evaporator OUT				

LAB 10.5 MEASURING WATER SOURCE HEAT PUMP SYSTEM CAPACITY

LABORATORY OBJECTIVE

You will demonstrate your ability to determine the system capacity of an operating water source heat pump by measuring its water flow and temperature rise.

LABORATORY NOTES

In this lab we will calculate the capacity of an operating water source heat pump system by measuring the water flow rate and the water temperature difference. A BTU is defined as a 1 degree temperature rise in 1 pound of water at atmospheric pressure. We can calculate the actual capacity accurately by multiplying the water flow in pounds per hour times the temperature rise in degrees Fahrenheit.

FUNDAMENTALS OF HVACR TEXT REFERENCE

Units 10 Thermodynamics – The Study of Heat

REQUIRED TOOLS AND EQUIPMENT

Temperature sensor with temperature clamp and thermocouple clamp
Water source heat pump
1 gallon bucket
Timer or watch that can measure seconds

SAFETY REQUIREMENTS

Always familiarize yourself with the equipment and operating manuals prior to starting up any system.

PROCEDURE

Step 1

Trace out the system and make sure that you understand the operation of the heat pump.

> *Start the water loop through the heat pump before starting the unit.*

1. After water flow has been established, you may start the heat pump and run it in the cooling mode for 15 minutes.
2. Measure the flow rate of the water. This can be done by allowing the water leaving the heat pump to flow into the one gallon bucket. Time the duration in seconds from the time the bucket is initially empty until it is full. Repeat this procedure for three different sets of readings.

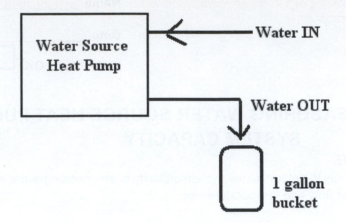

Figure 10-5-1

a) First measured time in seconds:

b) Second measured time in seconds:

c) Third measured time in seconds:

3. The average time in seconds is equal to the three readings added together and then divided by three.

    Average time in seconds = (Time 1 + Time 2 + Time 3) / 3

    Average time in seconds =

4. To determine how many gallons per minute is flowing first convert the seconds to minutes. Take the average time in seconds and divide that value by 60 (this is because there are 60 seconds in one minute).

    Gallons per minute = Total seconds / 60

    Gallons per minute (GPM) =

5. Convert gallons per minute (GPM) to pounds per minute by multiplying GPM by 8.34 (this is because there are 8.34 pounds of water in one gallon).

    Pounds per minute = GPM x 8.34

    Pounds per minute =

6. Measure the temperature of the water going in to the heat pump and the temperature of the water leaving the heat pump.

    Temperature of the water IN:

    Temperature of the water OUT:

7. Calculate the temperature difference that is equal to the temperature OUT minus the temperature IN ($°F_{OUT} - °F_{IN}$).

    Temperature Difference (ΔT) =

8. Calculate the system capacity. To do this multiply pounds per minute by the temperature difference. This will then be multiplied by the specific heat of water at atmospheric pressure which is simply 1 (one BTU for every °F for every lb of water). The calculated value will be in BTU per minute.

    (lb/min x ($°F_{OUT} - °F_{IN}$) x 1 BTU/lb-°F) = BTU/min

    System capacity in BTU/min =

9. Calculate the value for tons of refrigeration.

    Remember that one ton of refrigeration is the equivalent of melting one ton (2,000 lbs) of ice in 24 hours. Remember that the latent heat of fusion (ice to water) is 144 BTU/lb. 2,000 lbs of ice multiplied by 144 BTU/lb is equal to 288,000 BTU for one ton in 24 hours. Divide this by 24 hours per day and you have 12,000 BTU/hr. Divide this by 60 minutes per hour and you have 200 BTU/min.

    Therefore a ton of system capacity is equal to:

    1 ton = 288,000 BTU/day

    1 ton = 12,000 BTU/hr

    1 ton = 200 Btu/min

    To find capacity in tons, divide the system capacity in BTU/min by 200 BTU/min-ton.

    System capacity in tons = (BTU/min) / 200 BTU/min-ton

    System capacity in tons =

10. How does the system capacity that you calculated compare to the nameplate rating of the heat pump? Is the heat pump operating at full capacity?

Step 2
Reverse the heat pump so that now it is in the heating mode and run it in this mode for 15 minutes before recording your next measurements.

1. Measure the flow rate of the water. Repeat this procedure for three different sets of readings.
   a) First measured time in seconds:
   b) Second measured time in seconds:
   c) Third measured time in seconds:

2. The average time in seconds is equal to the three readings added together and then divided by three.

   Average time in seconds = (Time 1 + Time 2 + Time 3) / 3

   Average time in seconds =

3. To determine how many gallons per minute is flowing first convert the seconds to minutes. Take the average time in seconds and divide that value by 60 (this is because there are 60 seconds in one minute).

   Gallons per minute = Total seconds / 60

   Gallons per minute (GPM) =

4. Convert gallons per minute (GPM) to pounds per minute by multiplying GPM by 8.34 (this is because there are 8.34 pounds of water in one gallon).

   Pounds per minute = GPM x 8.34

   Pounds per minute =

5. Measure the temperature of the water going in to the heat pump and the temperature of the water leaving the heat pump.

   Temperature of the water IN:

   Temperature of the water OUT:

6. Calculate the temperature difference that is equal to the temperature OUT minus the temperature IN ($°F_{OUT}$ - $°F_{IN}$).

   (Note: The temperature OUT minus the temperature IN will have a negative value. This is because in the heating mode the heat pump is absorbing heat from the water rather than rejecting heat. You do not need to use a negative value because you are just calculating the temperature difference so you may ignore the negative sign for this experiment.)

   Temperature Difference (ΔT) =

42

7. Calculate the system capacity. To do this, multiply pounds per minute by the temperature difference. This will then be multiplied by the specific heat of water at atmospheric pressure which is simply 1 (one BTU for every °F for every lb of water). The calculated value will be in BTU per minute.

$$(\text{lb/min} \times (°F_{OUT} - °F_{IN}) \times 1 \text{ BTU/lb-}°F) = \text{BTU/min}$$

System capacity in BTU/min =

8. Calculate the value for tons of refrigeration.

Remember that one ton of refrigeration is the equivalent of melting one ton (2,000 lbs) of ice in 24 hours. Remember that the latent heat of fusion (ice to water) is 144 BTU/lb. 2,000 lbs of ice multiplied by 144 BTU/lb is equal to 288,000 BTU for one ton in 24 hours. Divide this by 24 hours per day and you have 12,000 BTU/hr. Divide this by 60 minutes per hour and you have 200 BTU/min.

Therefore a ton of system capacity is equal to:

1 ton = 288,000 BTU/day

1 ton = 12,000 BTU/hr

1 ton = 200 Btu/min

To find capacity in tons, divide the system capacity in BTU/min by 200 BTU/min-ton.

System capacity in tons = (BTU/min) / 200 BTU/min-ton

System capacity in tons =

9. How does the system capacity that you calculated compare to the nameplate rating of the heat pump? Is the heat pump operating at full capacity?

10. How does the system capacity when operating the heat pump in the cooling mode differ from running the heat pump in the heating mode. Explain your answer.

LAB 10.6 SENSIBLE AND LATENT HEAT

LABORATORY OBJECTIVE
We will demonstrate sensible and latent heat properties.

LABORATORY NOTES
We will demonstrate sensible and latent heat properties by adding heat to water while we monitor the amount of energy used, the temperature of the water, and the weight of the water. Energy input will be determined by measuring the wattage used, temperature change will be measured with a thermocouple, and weight change will be measured with a digital scale.

FUNDAMENTALS OF HVACR TEXT REFERENCE
Unit 10 Thermodynamics – The Study of Heat

REQUIRED TOOLS AND MATERIALS PROVIDED BY SCHOOL
Digital scale
Electric hot plate
Saucepan
WattsUp Pro
Fluke 16
Thermocouple

PROCEDURE

1. Put the electric hotplate and saucepan on the digital scale.
2. Put 3 pounds of water in the pot.
3. Zero the scale with the pot and water still on it.
4. Measure the temperature of the water.
5. Plug the cook pot into the Watts-Up and plug the Watts-Up into an electric outlet.
6. Check the time.
7. Monitor the temperature of the water and the time.
8. Record the wattage in the Watts-Up during this time.
9. When the temperature reaches around 150° F record the time and the exact temperature.
10. Leave the hot plate on and continue to monitor the time and temperature.
11. Note the time & temperature of the water when the water begins a "rolling boil."
12. Zero the scale.
13. Let the water boil at a rolling boil for exactly 1 minute.
14. Record the wattage during this period.
15. Record the weight at the end of a minute. If you miss it, go for two minutes and check the weight again.

Results Summary

Sensible Heat

During the first part of the experiment the weight should remain the same because we are only changing the temperature of the water. The wattage should be steady throughout because the amount of energy required to operate the cook pot should be the same. We can show that this electrical energy went into changing the temperature of the water by comparing the energy used to operate the cook-pot to the energy represented by the temperature rise of the water.

(Watts x 3.41 x minutes)/60 = BTU energy input
weight x water temperature rise = BTU energy output

Latent Heat of Evaporation

Once you start to see "steam" or "fog" over the water some of the water is evaporating. This represents latent heat of evaporation. Since water will evaporate at temperatures below the boiling point, some water will evaporate before the water reaches a rolling boil. This reduces the weight slightly. Once the water reaches a rolling boil the wattage used and the temperature of the water should stay the same. The weight change in the water is the only measurable indication that the heat is being transferred into the water. We can show this by comparing the energy used to operate the hotplate to the energy represented by the weight change of the water.

Zero the scale while noting the time. Check the weight after a minute. The weight showing on the scale will represent the CHANGE in the water weight. Now calculate the energy input and output using the formulas below.

(Watts x 3.41 x minutes)/60 = BTU energy input
weight change of water (in ounces) x 60 = BTU energy output

45

LAB 10.7 SATURATED PRESSURE-TEMPERATURE RELATIONSHIPS

LABORATORY OBJECTIVE
You will demonstrate the relationship of pressure and temperature for saturated gas-liquid mixes.

FUNDAMENTALS OF HVACR TEXT REFERENCE
Unit 10 Thermodynamics – The Study of Heat

LABORATORY NOTES
You will observe the behavior of saturated gas by comparing the pressure of two cylinders containing different weights a saturated mix at the same temperature. Then we will change the temperature and observe the effect on pressure.

REQUIRED TOOLS AND MATERIALS PROVIDED BY STUDENT
Safety glasses
Gloves

REQUIRED TOOLS AND MATERIALS PROVIDED BY SCHOOL
Two cylinders of refrigerant. Both cylinders should be the same type of refrigerant, but they should contain different amounts of refrigerant. Both cylinders should be the same temperature and both should contain some liquid and some vapor.
Freezer
Heat Gun or sink with hot water.

PROCEDURE
Preliminary
1. With the aid of the instructor, measure the pressure and temperature of each cylinder.
2. How does the pressure of the cylinder which contains the most refrigerant compare to the cylinder with the least refrigerant?

3. How is this different from the cylinders which contained only gas?

Temperature Drop
1. Place one cylinder in ice or in a freezer.

2. Measure its pressure and temperature after 30 minutes. What happens?

3. Use the beginning pressure, the beginning temperature, the ending temperature, and the gas law to predict what the new pressure should be.

4. How does the actual pressure compare to the pressure predicted by the gas law?

5. Use a saturation chart to look up the pressure that corresponds to the temperature of the refrigerant in the cylinder.

6. How does this compare to the actual pressure? Which was more accurate in predicting the results?

Temperature Rise
1. Now, run warm water over the cylinder.

2. Measure its pressure and temperature after 15 minutes.

3. What happens?

4. Use the beginning pressure, the beginning temperature, the ending temperature, and the gas law to predict what the new pressure should be.

5. How does the actual pressure compare to the pressure predicted by the gas law?

6. Use a saturation chart to look up the pressure that corresponds to the temperature of the refrigerant in the cylinder.

7. How does this compare to the actual pressure?

8. Which was more accurate in predicting the results?

9. What conclusions can you draw about the behavior of saturated liquid-gas mixtures to temperature change?

10. How are saturated mixtures different in their response to temperature change than ideal gases?

LAB 10.8 SATURATED, SUPERHEATED, OR SUBCOOLED

LABORATORY OBJECTIVE
The purpose of this lab is to demonstrate your ability to determine if refrigerant is saturated, superheated, or subcooled.

LABORATORY NOTES
You will measure the pressure and temperature of several refrigeration system components. You will determine if the refrigerant in the component is saturated, superheated, or subcooled by comparing the pressure and temperature to a saturation chart for that particular refrigerant.

REQUIRED MATERIALS PROVIDED BY SCHOOL
Refrigerant Trainer
Pressure-Temperature Chart

FUNDAMENTALS OF HVACR TEXT REFERENCE
Unit 10 Thermodynamics – The Study of Heat

PROCEDURE
Turn on the Refrigerant Trainer and let it operate until the bottom of the compressor is warm. Use the temperatures and pressures on the screen to complete the chart below.

Location	Pressure	Temperature from Chart	Actual Temperature	Condition
Compressor IN				
Compressor OUT				
Condenser OUT				
Evaporator IN				

LAB 11.1 QUANTITY-PRESSURE RELATIONSHIP OF IDEAL GASES

LABORATORY OBJECTIVE

The student will demonstrate that the pressure of an ideal gas contained in a cylinder depends upon the amount of gas in the cylinder.

LABORATORY NOTES

We will compare the pressure of a cylinder with two different weights of the same gas at the same temperature. Using the same cylinder will insure that the volume remains constant. Temperature will be verified using an infrared thermometer.

FUNDAMENTALS OF HVACR TEXT REFERENCE

Unit 11 Pressure and Vacuum

REQUIRED TOOLS AND EQUIPMENT:

An empty refillable refrigerant recovery cylinder
Vacuum pump
Digital scale
Cylinder of nitrogen gas with regulator
Infrared thermometer
Gauge manifold

SAFETY REQUIREMENTS

A. Wear safety glasses and gloves whenever handling gas cylinders and regulators.
B. Familiarize yourself with the proper procedures for operating gas cylinder regulators.

PROCEDURE

Step 1

Locate an empty refillable refrigerant recovery cylinder, a nitrogen cylinder with regulator, a vacuum pump, a gauge manifold, and a digital refrigerant charging scale.

1. Place the refrigerant recovery cylinder on the digital scale and connect the vacuum pump using the gauge manifold. Also connect the gauge manifold to the nitrogen cylinder but keep the nitrogen cylinder valves closed as shown in Figure 11-1-1.

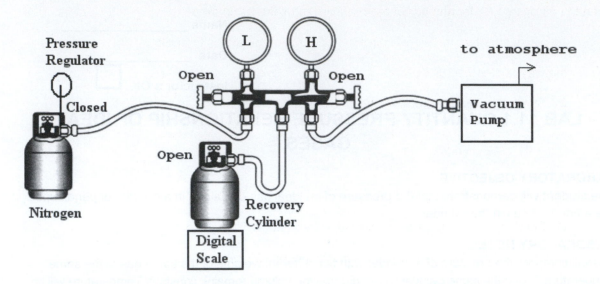

Figure 11-1-1

2. Run the vacuum pump to evacuate the cylinder and gauge manifold hoses to insure that the cylinder and hoses are empty and free from all gas. Close the high side valve (H) on the gauge manifold before shutting of the vacuum pump or else air will leak back into the lines and cylinder.

3. Close the low side valve (L) on the gauge manifold.

4. Zero the digital scale.

5. The valve arrangement should now be as shown in Figure 11-1-2.

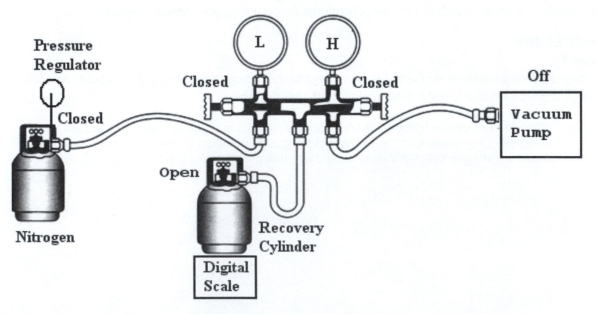

Figure 11-1-2

Step 2

Open the nitrogen cylinder and adjust the pressure regulator as follows:

A. Make sure that the nitrogen cylinder pressure regulator is turned all the way out (counterclockwise).

B. Slowly open the cylinder valve fully open to backseat it. The tank pressure should register on the regulator high pressure gauge. The pressure in the tank can be in excess of 2,000 psi. <u>Do not stand in front of the regulator "T" handle</u>.

C. Slowly turn the regulator "T" handle inward (clockwise) until the regulator adjusted pressure reaches approximately 50 psig.

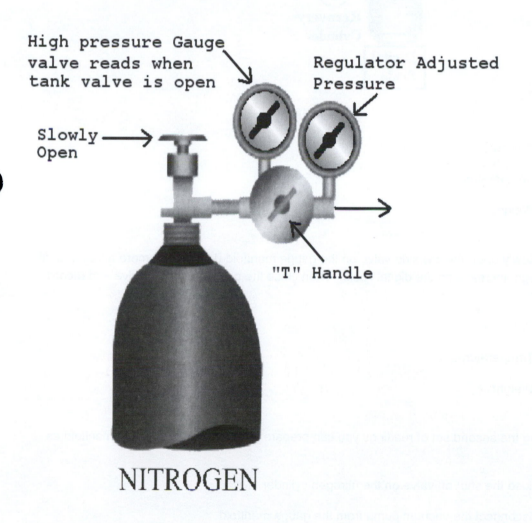

High pressure Gauge
valve reads when
tank valve is open

Regulator Adjusted
Pressure

Slowly
Open

"T" Handle

NITROGEN

Figure 11-1-3

Step 3

Slowly open the low side valve on the gauge manifold (L) until you see a weight increase on the digital scale. Then close the gauge manifold valve and record your readings.

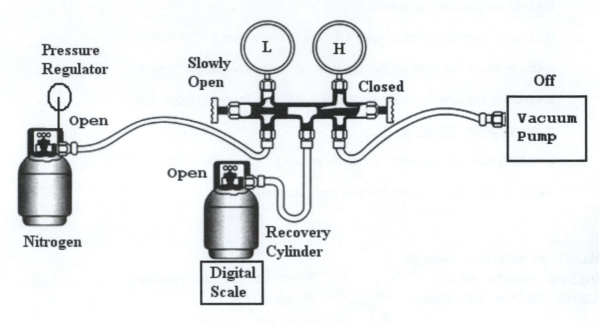

Figure 11-1-4

Gas Pressure:

Gas Temperature:

Gas Weight:

Step 4

Once again slowly open the low side valve on the gauge manifold (L) and add more nitrogen until you see a weight increase on the digital scale. Then close the gauge manifold valve and record your readings.

Gas Pressure:

Gas Temperature:

Gas Weight:

Step 5

After recording the second set of readings you can prepare to disconnect the gauge manifold as follows:

A. Close the shut off valve on the nitrogen cylinder.

B. Disconnect the vacuum pump from the gauge manifold.

C. Open the low side valve on the gauge manifold (L).

D. Slowly open the high side valve on the gauge manifold to bleed the gas pressure from the recovery cylinder and all hoses.

52

Once the pressure has been bled, you may back all the way off on the "T" handle (counterclockwise) for the pressure regulator on the Nitrogen cylinder.

● After all of the pressure has been bled off, the hoses may be disconnected.

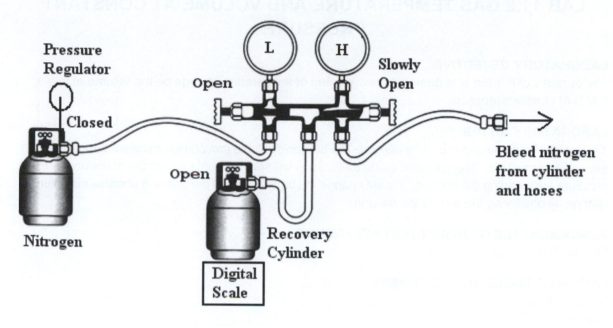

Figure 11-1-5

●

QUESTIONS

1. How does the pressure of the cylinder change when the amount of nitrogen changes?

2. What conclusions can you draw about the behavior of an ideal gas?

●

LAB 11.2 GAS TEMPERATURE AND VOLUME AT CONSTANT PRESSURE

LABORATORY OBJECTIVE
The purpose of this lab is to demonstrate the effect of temperature change on the volume of a gas that is at constant pressure.

LABORATORY NOTES
The gas will be contained in a mylar balloon. The balloon's volume can change because the balloon can expand and contract. The pressure inside the balloon will stay the same as will the atmospheric pressure surrounding the balloon. We will change the temperature of the gas and observe the volume change by observing the size of the balloon.

FUNDAMENTALS OF HVACR TEXT REFERENCE
Unit 11 Pressure and Vacuum

REQUIRED TOOLS AND EQUIPMENT
Mylar balloon
Freezer
Hair dryer

SAFETY REQUIREMENTS
A. Wear safety glasses and gloves whenever handling gas cylinders and regulators.
B. Familiarize yourself with the proper procedures for operating gas cylinder regulators.

PROCEDURE
Step 1
Inflate the mylar balloon at room temperature until it is full, but not taut.

Step 2
Place the balloon in the freezer for five minutes.

Step 3
Remove the balloon and compare its shape and size to its original shape and size.

Step 4
Heat the balloon with a hair dryer.

Step 5
Compare its shape and size to its original shape and size.

QUESTIONS

1. What happened to the balloon when the temperature dropped?

2. What happened to the balloon when the temperature rose?

3. What does this tell us about the volume of a gas compared to its temperature?

LAB 11.3 GAS PRESSURE-TEMPERATURE RELATIONSHIP AT A CONSTANT VOLUME

LABORATORY OBJECTIVE
The purpose of this lab is to demonstrate the effect of gas temperature changes on gas pressure when the volume remains constant.

LABORATORY NOTES
The gas will be contained in a recovery cylinder. Since the cylinder is a fixed volume, the gas volume will not change. We will observe the pressure and temperature of the cylinder filled with nitrogen at room temperature. Then we will cool the cylinder in ice and observe the effect of reduced temperature on the gas pressure. Next, we will heat the cylinder and observe the effect of increased temperature on gas pressure. We will use the ideal gas law to verify our results.

***FUNDAMENTALS OF HVACR* TEXT REFERENCE**
Unit 11 Pressure and Vacuum

REQUIRED TOOLS AND EQUIPMENT
Refrigerant recovery cylinder containing nitrogen
Mop bucket
Ice
Infrared thermometer
Gauge manifold

SAFETY REQUIREMENTS
Wear safety glasses and gloves whenever handling gas cylinders and regulators.

PROCEDURE
Step 1
Obtain a refrigerant recovery cylinder containing nitrogen gas.

1. Measure and record the pressure and temperature of the cylinder.

   Pressure:

   Temperature:

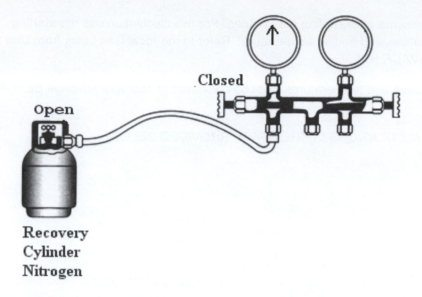

Figure 11-3-1

Step 2

Place the recovery cylinder into the mop bucket and pack it with ice.

    A.   Wait 15 minutes and then record the pressure and temperature of the cylinder.

    Pressure:

    Temperature:

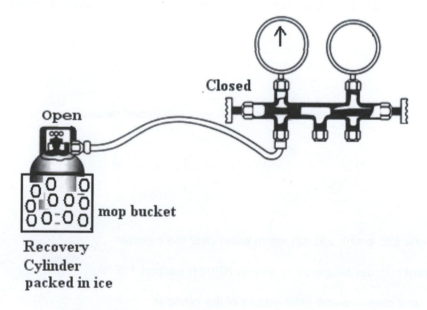

Figure 11-3-2

Step 3

Calculate the expected pressure after cooling the cylinder. For this calculation use the starting pressure, starting temperature, and ending temperature. Refer to the Ideal Gas Laws from Unit 11 in the *Fundamentals of HVACR* text.

(Remember that all pressures and temperatures must be converted to absolute readings BEFORE working the problem.)

SHOW ALL OF YOUR CALCULATIONS IN THE SPACE PROVIDED BELOW.

Calculated Pressure:

Measured Pressure:

     A.   What happened to the pressure in the cylinder when the temperature dropped?

     B.   How do your measured results compare to your calculated results?

Step 4

Place the recovery cylinder into a sink and run warm water over the cylinder.

CAUTION: The refrigerant cylinder temperature should NEVER exceed 125 °F.

     A.   Record the new pressure and temperature of the cylinder.

     Pressure:

     Temperature:

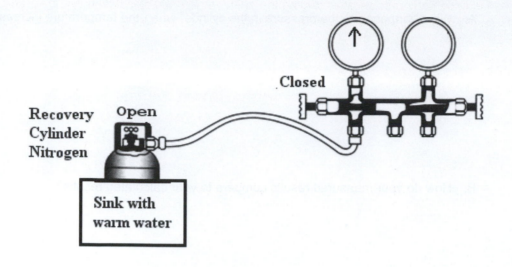

Figure 11-3-3

Step 5

Calculate the expected pressure after warming the cylinder. For this calculation use the starting pressure, starting temperature, and ending temperature. Refer to the Ideal Gas Laws from Unit 11 in the *Fundamentals of HVACR* text.

(Remember that all pressures and temperatures must be converted to absolute readings BEFORE working the problem.)

SHOW ALL OF YOUR CALCULATIONS IN THE SPACE PROVIDED BELOW.

Calculated Pressure:

Measured Pressure:

A. What happened to the pressure in the cylinder when the temperature increased?

B. How do your measured results compare to your calculated results?

C. Based upon the results from cooling and then heating the cylinder, what conclusions can you draw about the pressure of a gas compared to its temperature at constant volume?

LAB 11.4 GAS PRESSURE-TEMPERATURE-VOLUME RELATIONSHIP

LABORATORY OBJECTIVE

The purpose of this lab is to demonstrate the combined effect of gas volume and pressure changes on gas temperature.

LABORATORY NOTES

First, we will observe the temperature change in a gas that is being compressed. During compression the volume will be reduced and the pressure will be increased. Second, we will observe the temperature change in a gas as it is expanded. During expansion the volume will increase and the pressure will decrease.

***FUNDAMENTALS OF HVACR* TEXT REFERENCE**

Unit 11 Pressure and Vacuum

REQUIRED TOOLS AND EQUIPMENT

Empty refrigerant recovery cylinder
Bicycle pump
Temperature sensor with thermocouple probe
Infrared thermometer
Gauge manifold

SAFETY REQUIREMENTS

Wear safety glasses and gloves whenever handling gas cylinders and regulators.

PROCEDURE

Step 1

Obtain an empty refrigerant recovery cylinder.

1. Measure and record the pressure and temperature of the cylinder.
   (The pressure should be atmospheric)

   Pressure:

   Temperature:

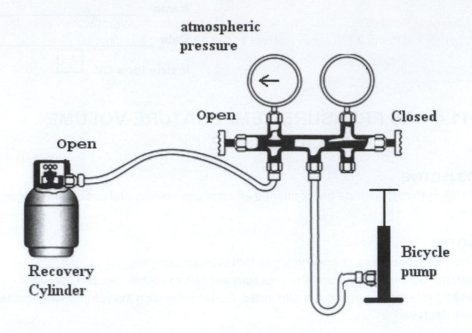

Figure 11-4-1

Step 2

Use the bicycle pump to raise the pressure of the cylinder to 40 psig then close the valve on the cylinder and record the pressure and temperature of the cylinder.

Pressure:

Temperature:

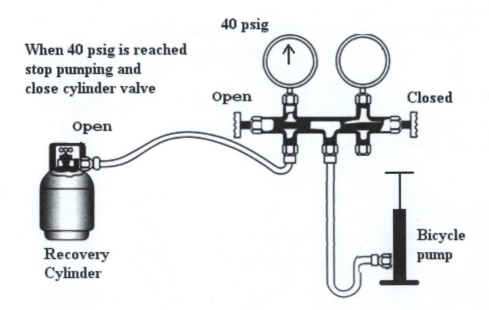

Figure 11-4-2

Step 3

Wait for the cylinder to cool to room temperature. Remove the gauge manifold and then open the gas valve on the cylinder and measure the temperature of the air leaving the cylinder (the air is expanding).

Temperature:

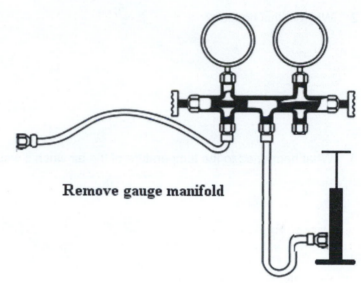

Remove gauge manifold

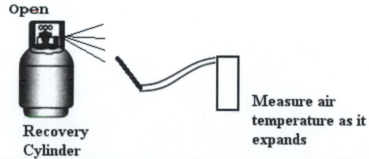

Open

Recovery Cylinder

Measure air temperature as it expands

Figure 11-4-3

1. What happened to the temperature of the air when it was compressed with the bicycle pump?

2. Why did this happen?

3. What happened to the temperature of the air when it was expanded?

4. Why did this happen?

5. How could these physical relationships of pressure-volume-temperature be useful in an air conditioning system?

LAB 11.5 AIR COMPRESSION & EXPANSION

LABORATORY OBJECTIVE
The purpose of this lab is to demonstrate the combined effect of gas volume and pressure changes on gas temperature.

LABORATORY NOTES
First, we will observe the temperature change in a gas that is being compressed. During compression the volume will be reduced and the pressure will be increased.

Second, we will observe the temperature change in a gas as it is expanded. During expansion the volume will increase and the pressure will decrease.

FUNDAMENTALS OF HVACR TEXT REFERENCE
Unit 11 Pressure and Vacuum

REQUIRED TOOLS AND MATERIALS PROVIDED BY STUDENT
Safety glasses

REQUIRED TOOLS AND MATERIALS PROVIDED BY SCHOOL
Empty refrigerant recovery cylinder
Infrared thermometer
Fluke 16 with thermocouple probe
Bicycle air pump with built in pressure gauge

Procedure
1. Measure and record the temperature and pressure of the cylinder.
2. Use the bicycle pump to increase the cylinder pressure to 40 psig.
3. Measure and record the temperature of cylinder.
4. Wait for the cylinder to cool off to room temperature.
5. Open the gas valve on the cylinder and measure the temperature of the air leaving the cylinder.

Summary
1. What happened to the temperature of the air when it was compressed?

2. Why did this happen?

3. What happened to the temperature of the air when it was expanded?

4. Why did this happen?

5. How could this be useful in an air conditioning system?

LAB 11.6 IDEAL GAS LAWS

LABORATORY OBJECTIVE
The purpose of this lab is to demonstrate the effect of gas temperature changes on gas pressure when the volume remains constant.

LABORATORY NOTES
The gas will be contained in a recovery cylinder. Since the cylinder is a fixed volume, the gas volume will not change. We will observe the pressure and temperature of the cylinder filled with nitrogen at room temperature. Then we will cool the cylinder in ice and observe the effect of reduced temperature on the gas pressure. Next, we will heat the cylinder and observe the effect of increased temperature on gas pressure. We will use the ideal gas law to verify our results.

***FUNDAMENTALS OF HVACR* TEXT REFERENCE**
Unit 11 Pressure and Vacuum

REQUIRED MATERIALS PROVIDED BY STUDENT
Safety glasses

REQUIRED MATERIALS PROVIDED BY SCHOOL
Refrigerant Recovery Cylinder containing nitrogen infrared thermometer
Refrigerant gauges
Mop bucket
Ice

PROCEDURE
1. Measure and record the temperature of the cylinder.
2. Measure and record the pressure of cylinder.
3. Use the mop bucket to pack the cylinder in ice.
4. Wait 15 minutes for the cylinder to cool off.
5. Measure and record the new cylinder temperature and pressure.
6. Calculate the expected pressure using the starting pressure, starting temperature, and ending temperature. Remember that all pressures and temperatures must be converted to absolute readings BEFORE working the problem. Review the Ideal Gas Laws from Unit 11 in the *Fundamentals of HVACR* text if you need help.
7. Place the cylinder in the sink and run warm water over the cylinder.
8. Measure and record the new temperature and pressure.
9. Calculate the expected pressure using the original starting pressure, original starting temperature, and ending temperature.

Summary

1. What happened to the pressure in the cylinder when the temperature dropped?

2. How did your results compare to the calculated results?

3. What happened to the pressure in the cylinder when the temperature rose?

4. How did your results compare to the calculated results?

5. What does this tell us about the pressure of a gas compared to its temperature?

LAB 14.1 IDENTIFYING REFRIGERATION SYSTEM COMPONENTS

LABORATORY OBJECTIVE
The purpose of this lab is to demonstrate your ability to recognize each of the four main refrigeration system components on refrigeration systems.

LABORATORY NOTES
You will be assigned four typical refrigeration systems.

The units should be labeled SYSTEM 1, SYSTEM 2, SYSTEM 3, and SYSTEM 4.

You will locate the four major refrigeration system components in each system and complete a data sheet for each unit.

You will be prepared to show the instructor each of these components and discuss the basic characteristics of each of these components.

***FUNDAMENTALS OF HVACR* TEXT REFERENCE**
Unit 14 The Refrigeration Cycle

REQUIRED TOOLS AND EQUIPMENT
Packaged refrigeration unit
Split system
Small commercial refrigeration unit
Any other unit of the instructor's choosing
Six-in-one screwdriver for removing system panels

SAFETY REQUIREMENTS
Be careful of sharp edges when removing sheet metal panels.

PROCEDURE
Step 1
Examine each unit and complete the appropriate data sheet. Remove any panels necessary to gain access to the unit components.

Datasheet System 1

1. Point out the four main refrigeration components to the Instructor.

   Instructor Check_____

2. What is the name and model number of this piece of equipment?

3. Is this a packaged unit or a split system?

4. This equipment would be installed where?

5. What style is the compressor: hermetic, semi-hermetic, or open?

6. Is the condenser air cooled or water cooled?

7. For air-cooled condensers: Is the condenser induced draft or forced draft?

8. For water-cooled condensers: Is this a waste water system or a recirculation type system?

9. Does the evaporator cool air or water?

10. What type of metering device does this system have: expansion valve, capillary tube, orifice, other?

11. Does this system use a liquid receiver? Point it out.

Instructor Check_____

12. Does this system have a suction accumulator? Point it out. What type of metering device does this systems have: expansion valve, capillary tube, orifice, other?

Instructor Check_____

Datasheet System 2

1. Point out the four main refrigeration components to the Instructor.

Instructor Check_____

2. What is the name and model number of this piece of equipment?

3. Is this a packaged unit or a split system?

4. This equipment would be installed where?

5. What style is the compressor: hermetic, semi-hermetic, or open?

6. Is the condenser air cooled or water cooled?

7. For air cooled condensers: Is the condenser induced draft or forced draft?

8. For water cooled condensers: Is this a waste water system or a recirculation type system?

9. Does the evaporator cool air or water?

10. What type of metering device does this system have: expansion valve, capillary tube, orifice, other?

11. Does this system use a liquid receiver? Point it out.

   Instructor Check_____

12. Does this system have a suction accumulator? Point it out. What type of metering device does this systems have: expansion valve, capillary tube, orifice, other?

   Instructor Check_____

Datasheet System 3

1. Point out the four main refrigeration components to the Instructor.

   Instructor Check_____

2. What is the name and model number of this piece of equipment?

3. Is this a packaged unit or a split system?

4. This equipment would be installed where?

5. What style is the compressor: hermetic, semi-hermetic, or open?

6. Is the condenser air cooled or water cooled?

7. For air cooled condensers: Is the condenser induced draft or forced draft?

8. For water cooled condensers: Is this a waste water system or a recirculation type system?

9. Does the evaporator cool air or water?

10. What type of metering device does this system have: expansion valve, capillary tube, orifice, other?

11. Does this system use a liquid receiver? Point it out.

    Instructor Check_____

12. Does this system have a suction accumulator? Point it out. What type of metering device does this systems have: expansion valve, capillary tube, orifice, other?

    Instructor Check_____

1. Point out the four main refrigeration components to the Instructor.

   Instructor Check_____

2. What is the name and model number of this piece of equipment?

3. Is this a packaged unit or a split system?

4. This equipment would be installed where?

5. What style is the compressor: hermetic, semi-hermetic, or open?

6. Is the condenser air cooled or water cooled?

7. For air cooled condensers: Is the condenser induced draft or forced draft?

8. For water cooled condensers: Is this a waste water system or a recirculation type system?

9. Does the evaporator cool air or water?

10. What type of metering device does this system have: expansion valve, capillary tube, orifice, other?

11. Does this system use a liquid receiver? Point it out.

Instructor Check_____

12. Does this system have a suction accumulator? Point it out. What type of metering device does this systems have: expansion valve, capillary tube, orifice, other?

Instructor Check_____

LAB 14.2 IDENTIFYING REFRIGERATION SYSTEM COMPONENTS

LABORATORY OBJECTIVE
The purpose of this lab is to demonstrate your ability to recognize each of the four main refrigeration system components on refrigeration systems.

LABORATORY NOTES
You will be assigned four typical refrigeration systems. You will locate the four major refrigeration system components in each system and complete the data sheet below. You will be prepared to show the instructor each of these components and discuss the basic characteristics of each of these components.

***FUNDAMENTALS OF HVACR* TEXT REFERENCE**
Unit 14 The Refrigeration Cycle

REQUIRED TOOLS AND MATERIALS PROVIDED BY STUDENT
Safety glasses

REQUIRED TOOLS AND MATERIALS PROVIDED BY SCHOOL
Four refrigeration systems, including:
* a packaged unit
* a split system
* a small commercial refrigeration unit
* any other unit of the instructor's choosing A six-in-1 screwdriver for removing system panels

PROCEDURE
Examine the four systems assigned by the instructor and complete the datasheets below. You may need to remove panels to see some of the system components. You will be required to point out the four main refrigeration cycle components on each system. You should also be prepared to discuss where each unit would be installed.

REFRIGERATION SYSTEM COMPONENTS				
	System 1	**System 2**	**System 3**	**System 4**
Type of System	Packaged Split System	Packaged Split System	Packaged Split System	Packaged Split System
Compressor	Open Hermetic Semi-hermetic	Open Hermetic Semi-hermetic	Open Hermetic Semi-hermetic	Open Hermetic Semi-hermetic
Condenser	Air Cooled Water Cooled Evaporative	Air Cooled Water Cooled Evaporative	Air Cooled Water Cooled Evaporative	Air Cooled Water Cooled Evaporative
Evaporator	Cools Air Cools Water	Cools Air Cools Water	Cools Air Cools Water	Cools Air Cools Water
Expansion Device	Orifice Capillary Tube Expansion Valve	Orifice Capillary Tube Expansion Valve	Orifice Capillary Tube Expansion Valve	Orifice Capillary Tube Expansion Valve
Refrigerant Storage	Receiver Accumulator Neither	Receiver Accumulator Neither	Receiver Accumulator Neither	Receiver Accumulator Neither

LAB 14.3 TRAINER REFRIGERATION SYSTEM CHARACTERISTICS

LABORATORY OBJECTIVE
The purpose of this lab is to demonstrate your ability to discuss the function of the four main refrigeration system components and describe the refrigerant characteristics throughout the system on the refrigeration trainer.

LABORATORY NOTES
You will be assigned a refrigeration trainer. You should be prepared to show the instructor each of the four main refrigeration components in order and discuss their function in the system. You will also discuss the characteristics of the refrigerant as it travels through the system using pressure and temperature measurements to help you determine the condition of the refrigerant throughout the trainer.

FUNDAMENTALS OF HVACR TEXT REFERENCE
Unit 14 The Refrigeration Cycle

REQUIRED MATERIALS PROVIDED BY STUDENT
Safety glasses

REQUIRED MATERIALS PROVIDED BY SCHOOL
Refrigeration trainer with clear tubes in each coil.
Fluke 16
Temperature clamp

PROCEDURE
Run the trainer and observe its operation. Note the pressure gauges that tell you what the refrigerant pressure is at every point in the system. You can measure the temperature at each location using a Fluke meter. You may also touch components to get a feel for their temperature.

CAUTION: Some components are HOT! Proceed carefully!

If you hold your hand just above a warm component you can tell if it is extremely hot or not. If you can feel radiant heat coming from it, it may be too hot to touch safely. Use the charts below to organize your data. However, the job is not complete until you TELL the instructor how the system works and respond to questions.

Completing the Refrigerant Characteristics Data Table
Record the pressure and temperature at each system location indicated on the chart. Compare the pressure and temperature using a pressure-temperature chart to determine the condition: saturated, superheated, or subcooled. Next, write in the state: gas, liquid, or mixed. Use the pressure, temperature, condition, and state to determine if the refrigerant is at a high heat content or a low heat content.

Completing the Refrigerant Changes Data Table
Look at the data to determine how the refrigerant changed as it went through each component and circle the description that best describes the changes in the conditions through each component.

Trainer Refrigeration System Characteristics

Use the following table to record system operating data for the trainer.

Location	Temperature	Pressure	Condition	State	Heat Content
Compressor in					
Compressor out					
Discharge line					
Condenser in					
Condenser center					
Condenser out					
Liquid line					
Meter device in					
Meter device out					
Evaporator in					
Evaporator center					
Evaporator out					
Suction line					

Trainer Refrigerant Changes Through the Components

Circle the description that best describes the changes that occur through each refrigeration system component.

No change to describe very slight or no significant change in a condition.

Small to describe a measurable change in a condition such as 40°F to 50°F.

Large to describe a very significant change, such as 120°F to 40°F.

Type of Change	Compressor	Condenser	Meter Device	Evaporator
Change in Pressure	No change Small increase Large Increase Small decrease Large decrease	No change Small increase Large Increase Small decrease Large decrease	No change Small increase Large Increase Small decrease Large decrease	No change Small increase Large Increase Small decrease Large decrease
Change in Temp	No change Small increase Large Increase Small decrease Large decrease	No change Small increase Large Increase Small decrease Large decrease	No change Small increase Large Increase Small decrease Large decrease	No change Small increase Large Increase Small decrease Large decrease
Change in Heat	No change Small increase Large Increase Small decrease Large decrease	No change Small increase Large Increase Small decrease Large decrease	No change Small increase Large Increase Small decrease Large decrease	No change Small increase Large Increase Small decrease Large decrease
Change in State	No change Condensation Evaporation	No change Condensation Evaporation	No change Condensation Evaporation	No change Condensation Evaporation
Change in Volume	No change Small increase Large Increase Small decrease Large decrease	No change Small increase Large Increase Small decrease Large decrease	No change Small increase Large Increase Small decrease Large decrease	No change Small increase Large Increase Small decrease Large decrease

LAB 14.4 PACKAGED UNIT REFRIGERATION SYSTEM CHARACTERISTICS

LABORATORY OBJECTIVE
The purpose of this lab is to demonstrate your ability to discuss the function of the four main refrigeration system components and describe the refrigerant characteristics throughout the system on a packaged unit.

LABORATORY NOTES
You will be assigned a packaged unit air conditioner. You should be prepared to show the instructor each of the four main refrigeration components in order and discuss their function in the system. You will also discuss the characteristics of the refrigerant as it travels through the system using pressure and temperature measurements to help you determine the condition of the refrigerant throughout the unit.

FUNDAMENTALS OF HVACR TEXT REFERENCE
Unit 14 The Refrigeration Cycle

REQUIRED MATERIALS PROVIDED BY STUDENT
Safety glasses

REQUIRED MATERIALS PROVIDED BY SCHOOL
Packaged air conditioner.
Fluke 16
Temperature clamp

PROCEDURE
Run the packaged unit and observe its operation. Note the pressure gauges that tell you what the refrigerant pressure is at every point in the system. You can measure the temperature at each location using a Fluke meter. You may also touch components to get a feel for their temperature.

CAUTION: Some components are HOT! Proceed carefully!

If you hold your hand just above a warm component you can tell if it is extremely hot or not. If you can feel radiant heat coming from it, it may be too hot to touch safely. Use the charts below to organize your data. However, the job is not complete until you TELL the instructor how the system works and respond to questions.

Completing the Refrigerant Characteristics Data Table
Record the pressure and temperature at each system location indicated on the chart. Compare the pressure and temperature using a pressure-temperature chart to determine the condition: saturated, superheated, or subcooled. Next, write in the state: gas, liquid, or mixed. Use the pressure, temperature, condition, and state to determine if the refrigerant is at a high heat content or a low heat content.

Completing the Refrigerant Changes Data Table
Look at the data to determine how the refrigerant changed as it went through each component and circle the description that best describes the changes in the conditions through each component.

Packaged Unit Refrigeration System Characteristics

Use the following table to record system operating data for the trainer.

Location	Temperature	Pressure	Condition	State	Heat Content
Compressor in					
Compressor out					
Discharge line					
Condenser in					
Condenser center					
Condenser out					
Liquid line					
Meter device in					
Meter device out					
Evaporator in					
Evaporator center					
Evaporator out					
Suction line					

Packaged Unit Refrigerant Changes Through the Components

Circle the description that best describes the changes that occur through each refrigeration system component.

No change to describe very slight or no significant change in a condition.

Small to describe a measurable change in a condition, such as 40°F to 50°F.

Large to describe a very significant change,, such as 120°F to 40°F.

Type of Change	Compressor	Condenser	Meter Device	Evaporator
Change in Pressure	No change Small increase Large Increase Small decrease Large decrease	No change Small increase Large Increase Small decrease Large decrease	No change Small increase Large Increase Small decrease Large decrease	No change Small increase Large Increase Small decrease Large decrease
Change in Temp	No change Small increase Large Increase Small decrease Large decrease	No change Small increase Large Increase Small decrease Large decrease	No change Small increase Large Increase Small decrease Large decrease	No change Small increase Large Increase Small decrease Large decrease
Change in Heat	No change Small increase Large Increase Small decrease Large decrease	No change Small increase Large Increase Small decrease Large decrease	No change Small increase Large Increase Small decrease Large decrease	No change Small increase Large Increase Small decrease Large decrease
Change in State	No change Condensation Evaporation	No change Condensation Evaporation	No change Condensation Evaporation	No change Condensation Evaporation
Change in Volume	No change Small increase Large Increase Small decrease Large decrease	No change Small increase Large Increase Small decrease Large decrease	No change Small increase Large Increase Small decrease Large decrease	No change Small increase Large Increase Small decrease Large decrease

LAB 14.5 SPLIT SYSTEM REFRIGERATION CHARACTERISTICS

LABORATORY OBJECTIVE

The purpose of this lab is to demonstrate your ability to discuss the function of the four main refrigeration system components and describe the refrigerant characteristics throughout the system on a split system air conditioner.

LABORATORY NOTES

You will be assigned a packaged unit air conditioner. You should be prepared to show the instructor each of the four main refrigeration components in order and discuss their function in the system. You will also discuss the characteristics of the refrigerant as it travels through the system using pressure and temperature measurements to help you determine the condition of the refrigerant throughout the unit.

FUNDAMENTALS OF HVACR TEXT REFERENCE

Unit 14 The Refrigeration Cycle

LABORATORY NOTES

You will be assigned a packaged unit air conditioner. You should be prepared to show the instructor each of the four main refrigeration components in order and discuss their function in the system. You will also discuss the characteristics of the refrigerant as it travels through the system using pressure and temperature measurements to help you determine the condition of the refrigerant throughout the unit.

REQUIRED MATERIALS PROVIDED BY STUDENT

Safety glasses

REQUIRED MATERIALS PROVIDED BY SCHOOL

Split-system air conditioner
Fluke 16
Temperature clamp

PROCEDURE

Run the split system and observe its operation. Note the pressure gauges that tell you what the refrigerant pressure is at every point in the system. You can measure the temperature at each location using a Fluke meter. You may also touch components to get a feel for their temperature.

CAUTION: Some components are HOT! Proceed carefully!

If you hold your hand just above a warm component you can tell if it is extremely hot or not. If you can feel radiant heat coming from it, it may be too hot to touch safely. Use the charts below to organize your data. However, the job is not complete until you TELL the instructor how the system works and respond to questions.

Completing the Refrigerant Characteristics Data Table

Record the pressure and temperature at each system location indicated on the chart. Compare the pressure and temperature using a pressure-temperature chart to determine the condition: saturated, superheated, or subcooled. Next, write in the state: gas, liquid, or mixed. Use the pressure, temperature, condition, and state to determine if the refrigerant is at a high heat content or a low heat content.

Completing the Refrigerant Changes Data Table

Look at the data to determine how the refrigerant changed as it went through each component and circle the description that best describes the changes in the conditions through each component.

Split System Refrigeration Characteristics

Use the following table to record system operating data for the trainer.

Location	Temperature	Pressure	Condition	State	Heat Content
Compressor in					
Compressor out					
Discharge line					
Condenser in					
Condenser center					
Condenser out					
Liquid line					
Meter device in					
Meter device out					
Evaporator in					
Evaporator center					
Evaporator out					
Suction line					

Split System Refrigerant Changes Through the Components

Circle the description that best describes the changes that occur through each refrigeration system component.

No change to describe very slight or no significant change in a condition.

Small to describe a measurable change in a condition such as 40°F to 50°F.

Large to describe a very significant change, such as 120°F to 40°F.

Type of Change	Compressor	Condenser	Meter Device	Evaporator
Change in Pressure	No change Small increase Large Increase Small decrease Large decrease	No change Small increase Large Increase Small decrease Large decrease	No change Small increase Large Increase Small decrease Large decrease	No change Small increase Large Increase Small decrease Large decrease
Change in Temp	No change Small increase Large Increase Small decrease Large decrease	No change Small increase Large Increase Small decrease Large decrease	No change Small increase Large Increase Small decrease Large decrease	No change Small increase Large Increase Small decrease Large decrease
Change in Heat	No change Small increase Large Increase Small decrease Large decrease	No change Small increase Large Increase Small decrease Large decrease	No change Small increase Large Increase Small decrease Large decrease	No change Small increase Large Increase Small decrease Large decrease
Change in State	No change Condensation Evaporation	No change Condensation Evaporation	No change Condensation Evaporation	No change Condensation Evaporation
Change in Volume	No change Small increase Large Increase Small decrease Large decrease	No change Small increase Large Increase Small decrease Large decrease	No change Small increase Large Increase Small decrease Large decrease	No change Small increase Large Increase Small decrease Large decrease

LAB 14.6 BENCH UNIT REFRIGERATION SYSTEM CHARACTERISTICS

LABORATORY OBJECTIVE
The purpose of this lab is to demonstrate your ability to discuss the function of the four main refrigeration system components and describe the refrigerant characteristics throughout the system on a bench unit.

LABORATORY NOTES
You will be assigned a bench unit refrigeration system. You should be prepared to show the instructor each of the four main refrigeration components in order and discuss their function in the system. You will also discuss the characteristics of the refrigerant as it travels through the system using pressure and temperature measurements to help you determine the condition of the refrigerant throughout the unit.

FUNDAMENTALS OF HVACR TEXT REFERENCE
Unit 14 The Refrigeration Cycle

REQUIRED MATERIALS PROVIDED BY STUDENT
Safety Glasses

REQUIRED MATERIALS PROVIDED BY SCHOOL
Bench-unit refrigeration system
Fluke 16
Temperature clamp

PROCEDURE
Run the bench unit refrigeration system and observe its operation. Note the pressure gauges that tell you what the refrigerant pressure is at every point in the system. You can measure the temperature at each location using a Fluke meter. You may also touch components to get a feel for their temperature.

CAUTION: Some components are HOT! Proceed carefully!

If you hold your hand just above a warm component you can tell if it is extremely hot or not. If you can feel radiant heat coming from it, it may be too hot to touch safely. Use the charts below to organize your data. However, the job is not complete until you TELL the instructor how the system works and respond to questions.

Completing the Refrigerant Characteristics Data Table
Record the pressure and temperature at each system location indicated on the chart. Compare the pressure and temperature using a pressure-temperature chart to determine the condition: saturated, superheated, or subcooled. Next, write in the state: gas, liquid, or mixed. Use the pressure, temperature, condition, and state to determine if the refrigerant is at a high heat content or a low heat content.

Completing the Refrigerant Changes Data Table
Look at the data to determine how the refrigerant changed as it went through each component and circle the description that best describes the changes in the conditions through each component.

Bench Unit Refrigeration System Characteristics

Use the following table to record system operating data for the trainer.

Location	Temperature	Pressure	Condition	State	Heat Content
Compressor in					
Compressor out					
Discharge line					
Condenser in					
Condenser center					
Condenser out					
Liquid line					
Meter device in					
Meter device out					
Evaporator in					
Evaporator center					
Evaporator out					
Suction line					

Bench Unit Refrigerant Changes Through the Components

Circle the description that best describes the changes that occur through each refrigeration system component.

No change to describe very slight or no significant change in a condition.

Small to describe a measurable change in a condition, such as 40°F to 50°F.

Large to describe a very significant change, such as 120°F to 40°F.

Type of Change	Compressor	Condenser	Meter Device	Evaporator
Change in Pressure	No change Small increase Large Increase Small decrease Large decrease	No change Small increase Large Increase Small decrease Large decrease	No change Small increase Large Increase Small decrease Large decrease	No change Small increase Large Increase Small decrease Large decrease
Change in Temp	No change Small increase Large Increase Small decrease Large decrease	No change Small increase Large Increase Small decrease Large decrease	No change Small increase Large Increase Small decrease Large decrease	No change Small increase Large Increase Small decrease Large decrease
Change in Heat	No change Small increase Large Increase Small decrease Large decrease	No change Small increase Large Increase Small decrease Large decrease	No change Small increase Large Increase Small decrease Large decrease	No change Small increase Large Increase Small decrease Large decrease
Change in State	No change Condensation Evaporation	No change Condensation Evaporation	No change Condensation Evaporation	No change Condensation Evaporation
Change in Volume	No change Small increase Large Increase Small decrease Large decrease	No change Small increase Large Increase Small decrease Large decrease	No change Small increase Large Increase Small decrease Large decrease	No change Small increase Large Increase Small decrease Large decrease

LAB 14.7 WATER COOLED SYSTEM CHARACTERISTICS

LABORATORY OBJECTIVE

The purpose of this lab is to demonstrate your ability to discuss the function of the four main refrigeration system components and describe the refrigerant characteristics throughout the system on a water cooled unit.

LABORATORY NOTES

You will be assigned a water cooled unit. You should be prepared to show the instructor each of the four main refrigeration components in order and discuss their function in the system. You will also discuss the characteristics of the refrigerant as it travels through the system using pressure and temperature measurements to help you determine the condition of the refrigerant throughout the unit.

FUNDAMENTALS OF HVACR **TEXT REFERENCE**

Unit 14 The Refrigeration Cycle

REQUIRED MATERIALS PROVIDED BY STUDENT

Safety glasses

REQUIRED MATERIALS PROVIDED BY SCHOOL

Split-system air conditioner
Fluke 16
Temperature clamp

PROCEDURE

Run the system and observe its operation. Note the pressure gauges that tell you what the refrigerant pressure is at every point in the system. You can measure the temperature at each location using a Fluke meter. You may also touch components to get a feel for their temperature.

CAUTION: Some components are HOT! Proceed carefully!

If you hold your hand just above a warm component you can tell if it is extremely hot or not. If you can feel radiant heat coming from it, it may be too hot to touch safely. Use the charts below to organize your data. However, the job is not complete until you TELL the instructor how the system works and respond to questions.

Completing the Refrigerant Characteristics Data Table

Record the pressure and temperature at each system location indicated on the chart. Compare the pressure and temperature using a pressure-temperature chart to determine the condition: saturated, superheated, or subcooled. Next, write in the state: gas, liquid, or mixed. Use the pressure, temperature, condition, and state to determine if the refrigerant is at a high heat content or a low heat content.

Completing the Refrigerant Changes Data Table

Look at the data to determine how the refrigerant changed as it went through each component and circle the description that best describes the changes in the conditions through each component.

Water Cooled Refrigeration System Characteristics

Use the following table to record system operating data for the trainer.

Location	Temperature	Pressure	Condition	State	Heat Content
Compressor in					
Compressor out					
Discharge line					
Condenser in					
Condenser center					
Condenser out					
Liquid line					
Meter device in					
Meter device out					
Evaporator in					
Evaporator center					
Evaporator out					
Suction line					

Water Cooled System Refrigerant Changes Through the Components

Circle the description that best describes the changes that occur through each refrigeration system component.

No change to describe very slight or no significant change in a condition.

Small to describe a measurable change in a condition, such as 40°F to 50°F.

Large to describe a very significant change, such as 120°F to 40°F.

Type of Change	Compressor	Condenser	Meter Device	Evaporator
Change in Pressure	No change Small increase Large Increase Small decrease Large decrease	No change Small increase Large Increase Small decrease Large decrease	No change Small increase Large Increase Small decrease Large decrease	No change Small increase Large Increase Small decrease Large decrease
Change in Temp	No change Small increase Large Increase Small decrease Large decrease	No change Small increase Large Increase Small decrease Large decrease	No change Small increase Large Increase Small decrease Large decrease	No change Small increase Large Increase Small decrease Large decrease
Change in Heat	No change Small increase Large Increase Small decrease Large decrease	No change Small increase Large Increase Small decrease Large decrease	No change Small increase Large Increase Small decrease Large decrease	No change Small increase Large Increase Small decrease Large decrease
Change in State	No change Condensation Evaporation	No change Condensation Evaporation	No change Condensation Evaporation	No change Condensation Evaporation
Change in Volume	No change Small increase Large Increase Small decrease Large decrease	No change Small increase Large Increase Small decrease Large decrease	No change Small increase Large Increase Small decrease Large decrease	No change Small increase Large Increase Small decrease Large decrease

LAB 14.8 REFRIGERANT IDENTIFICATION

LABORATORY OBJECTIVE
Identify the type and amount of refrigerant in four refrigeration systems. You will identify the refrigerant by:
- Name
- Pressure (Very High, High, or Low)
- Ozone Depletion Potential
- Global Warming Potential
- Toxicity
- Flammability
- Chemical Composition
- Formulation

LABORATORY NOTES
You will find the type of refrigerant, the amount of refrigerant, and the system test pressures on the unit data plate of each system. Then you will research the reference material to find the refrigerant characteristics.

FUNDAMENTALS OF HVACR TEXT REFERENCE
Unit 14 The Refrigeration Cycle

REQUIRED MATERIALS PROVIDED BY STUDENT
Safety glasses

REQUIRED MATERIALS PROVIDED BY SCHOOL
Four refrigeration systems (They should contain different refrigerants.)
Pencil and paper

PROCEDURE FOR EACH UNIT ASSIGNED
1. Locate the nameplate on the unit. The nameplate will be in different places on the units, you simply have to look around.
2. Find and record the unit model number in the chart below.
3. Find and record the refrigerant type in the chart below.
4. Find and record the high side test pressure on the chart below.
5. Find and record the low side test pressure on the chart below.
6. Find a cylinder of the same refrigerant as in the system and read the instructions on the cylinder.
7. Use this information to research this unit's refrigerant characteristics and record them in the chart below.

Refrigerant Characteristics

	Unit 1	Unit 2	Unit 3	Unit 4
Model Number				
Refrigerant Name				
Refrigerant Quantity				
High Side Test Pressure				
Low Side Test Pressure				
Pressure (Very High, High, Low)				
Ozone Depletion Potential				
Global Warming Potential				
Safety Rating				
Chemical Composition				
Formulation (Compound, zeotrope or azeotrope)				

Summary

1. Were there any refrigerants that did not have either ozone depletion potential or global warming potential?
2. What are the high and low side test pressures used for?
3. Which refrigerants may safely leave the cylinder as either a gas or a liquid?
4. Which refrigerants must leave the cylinder as a liquid only?
5. Which refrigerant had the highest system test pressures
6. Which refrigerant had the lowest test pressures?
7. What hazards do you need to be aware of when working with these refrigerants?
8. Which of these refrigerants will NOT be used in equipment of the future?
9. Which of these refrigerants WILL be in equipment of the future?

LAB 14.9 REFRIGERANT OIL IDENTIFICATION

LABORATORY OBJECTIVE
Identify different types of refrigeration lubricant and recommend the proper lubricant for four refrigeration systems.

LABORATORY NOTES
You will examine three different types of refrigeration lubricant and recommend where they can be used. You will recommend the correct type of lubricant for four refrigeration systems.

***FUNDAMENTALS OF HVACR* TEXT REFERENCE**
Unit 14 The Refrigeration Cycle

REQUIRED MATERIALS PROVIDED BY STUDENT
Safety glasses

REQUIRED MATERIALS PROVIDED BY SCHOOL
Three containers of refrigeration lubricant representing three different types of refrigeration lubricant. Four refrigeration systems. They should contain different refrigerants.
Pencil and paper

PROCEDURE
Examining Refrigeration Lubricant
1. Identify the type of lubricant in each container.
2. Record the type of lubricant, its viscosity, its recommended evaporator temperature range, and the refrigerants that it is compatible with in the chart below.

	Lubricant 1	Lubricant 2	Lubricant 3
Type of Lubricant			
Viscosity			
Refrigerant Compatibility			
Evaporator Temperature			

For each unit assigned:

1. Locate the nameplate on the unit. The nameplate will be in different places on the units, you simply have to look around.
2. Find and record the unit model number in the chart below.
3. Find and record the refrigerant type in the chart below.
4. Recommend a lubricant that is compatible with this unit

	Unit 1	Unit 2	Unit 3
Type of Lubricant			
Viscosity			
Refrigerant Compatibility			
Evaporator Temperature			

SUMMARY

1. Which lubricant can be used with the widest range of systems?

2. What is the difference in viscosity between lubricants designed for higher evaporator temperatures and lubricants that are designed for use in lower temperature systems?

3. What is the most common viscosity for lubricants used in air conditioning systems?

LAB 15.1 IDENTIFYING COMPRESSORS

LABORATORY OBJECTIVE
You will identify six different compressors and classify them according to their mechanical type, body, cooling method, size, and electrical characteristics.

LABORATORY NOTES
The instructor will assign six units. You will examine the compressor in each unit and record it's characteristics in the data table.

FUNDAMENTALS OF HVACR TEXT REFERENCE
Unit 15 Compressors

REQUIRED MATERIALS PROVIDED BY STUDENT
Safety glasses
Gloves
Flat blade screwdriver
Phillips screwdriver
¼" nut driver
5/16" nut driver
OR
6-in-1 with all the above
Adjustable jaw wrench

REQUIRED MATERIALS PROVIDED BY SCHOOL
Air conditioning units with compressors

PROCEDURE
Examine the compressors on the units assigned by the instructor to determine the compressor

operational type (reciprocating, rotary, etc.)

body style (hermetic, open, semi-hermetic)

motor cooling method (air cooled or refrigerant cooled)

application (low temp, medium temp, high temp)

size (horsepower and/or tonnage)

electrical data (phase and voltage)

Use the pictures in the book to identify the operational type and body style, and motor cooling method. Hermetic compressors are in welded steel cans that sit up vertically. Semi hermetic compressors are bolted together and lay horizontally. Open compressors are bolted and have a shaft sticking out. Most hermetic reciprocating compressors are oval shaped. Rotary compressors and scroll compressors are round. The rotaries are smaller with the large suction line entering near the bottom. Scrolls are larger with the large suction line entering midway, closer to the top.

The compressor electrical data comes from the compressor data plate. The compressor size is determined by the model number. Common compressor manufacturer's model number explanations are given in the following pages:

Tecumseh Hermetic Compressors						
AE	A	4	4	40	Y	XA
Family	Generation	Application	Number of Digits in BTU/Hr Capacity	First two digits	Refrigerant	Voltage

Application
1 Low Temp -10°F, Normal Start Torque
2 Low Temp -10°F, High Start Torque
3 High Temp 45°F, Normal Start Torque
4 High Temp 45°F, High Start Torque
5 Air Cond   45°F, Normal Start Torque
6 Medium Temp 20°F, Normal Start Torque
7 Medium Temp 20°F, High start torque
8 Air Cond   49°F, Normal Start Torque
9 Commercial 20°F, Normal Start Torque
A Medium/Low 20°F, Normal Start Torque

Refrigerant
A R12
B R410A
C R407C
E R22
J R502
Y R134a
Z R404A/R507

Voltage XA 115 volt 60 hz single phase
XB 230 volt 60 hz single phase
XD 208-230 volt 60 hz single phase
XF 208-230 volt 60 hz three phase
XG 460 volt 60 hz three phase
XT 200-230 volt 60 hz three phase
AB 115 volt 60 hz single phase

Size
The number after the application tells how many digits are in the BTU/hr rating and the next two numbers are the first two digits. In the example above, they compressor rating is 4000 BTU/hr – four digits with the first two being 40.

Bristol Hermetic Compressors								
H	8	2	J	193	A	B	C	A
Application	Refrigerant	Generation	Family	Capacity	Motor	Protect	Electric	Feet

Application   L Low Temp
                 H High Temp
                 M Medium Temp 7 Medium Temp 20°F, High start torque
                 R Multiple Refrigerants
                 T Two Capacity High Temp
                 V Variable Speed

Refrigerant   1 R12
                 2 R22
                 3 R134a
                 5 R502 or R402B
                 6 R404A
                 7 R407C
                 8 R410A
                 9 R407C, R22, R404A

Motor   A PSC   G 3 Phase 2 Speed
                 B CSR   J Single phase 2 speed
                 C RSCR   K 3 phase dual voltage
                 D 3 PHASE   L 3 phase wye-delta
                 E CSIR   M 3 phase variable speed
                 F 3 Phase Part Winding Start

Protection   B Internal line break overload
                 P Pilot Duty Solid State
                 R Pilot Duty Solid State 2nd generation
                 T Pilot Duty Internal Thermostat

Electrical   A 115 volt 60 hz single phase
                 B 230 volt 60 hz single phase
                 C 208-230 volt 60 hz single phase
                 D 208-230 volt 60 hz three phase
                 E 460 volt 60 hz three phase
                 L 200-230 volt 60 hz three phase
                 T 208 volts 60 hz
                 Y 208-230 volt 60 hz single phase

Size   The three numbers in the middle of the model number give the nominal compressor capacity in BTU/hr. The first two numbers are the first two digits in the capacity. The third number is the number of zeros after the first two digits. In the example above, they compressor rating is 19000 BTU/hr.

Copeland Semi-Hermetic Compressors								
4	R	R	2	3000	T	S	K	800
Family	Type	Displace	Variant	Horsepower	Motor	Protect	Electric	BOM

Type
- A Air Cooled
- D Discuss
- R Refrigerant Cooled
- T Two Stage
- W Water Cooled

Motor
- C Capacitor Run Capacitor Start
- I   Induction Run Capacitor Start
- E Three phase
- F 3 Phase Part Winding Start
- T 3 phase Specialized

Protection
- A External line break overload
- F   Internal line break overload
- H Pilot Duty Internal Thermostat and External Overload
- L Pilot Duty Internal Thermostat and 3 External Overloads
- S Internal electronic thermal sensor

Electrical
- A 115 volt 60 hz single phase
- B 230 volt 60 hz single phase
- C 208-230 volt 60 hz single phase
- D 460 volt 60 hz three phase
- H 208 volts 60 hz single phase
- V 208-230 volt 60 hz single phase

Horsepower
The four numbers in the middle of the model number give the nominal compressor capacity in horsepower. The decimal goes in the middle of the four numbers. In the example above, the compressor rating is 30 horsepower.

Copeland Hermetic Compressors									
Z	R	90	K	3	E	T	W	D	551
Family	Application	First 2 digits in BTU/hr capacity	Capacity Multiplier	Variant	Oil	Motor	Protect	Voltage	Misc

Family   C   Reciprocating
                Z   Scroll

Application   B   High/Medium Temp
                BD   High/Medium digital
                BH   High/Medium horizontal
                F   Low Temp with Injection
                FH   Low Temp Horizontal
                H   Heat Pump
                P   Air Condition 410A
                R   Air Condition
                RT   Air Condition Even Tandem
                RU   Air Condition Uneven Tandem
                S   Medium Temperature

Multiplier   K   1000
                M   10000

Oil   E   Polyol Ester (POE)
          No Letter = Mineral Oil

Motor   T   Three Phase
                P   Single Phase

Protection   F   Internal line break overload
                W External electronic

Voltage   C 208-230 volt 60 hz
                D 460 volt 60 hz
                5 200-220 volt 60 hz

Capacity   The two numbers are the first two digits of the compressor capacity in BTU/hr. They are multiplied by the capacity multiplier. In this example, the capacity is 90,000 BTU/hr 90 x 1000

LAB 15.1 COMPRESSOR IDENTIFICATION DATA SHEET						
	Unit 1	Unit 2	Unit 3	Unit 4	Unit 5	Unit 6
Type Reciprocating Rotary Scroll						
Body Open Semi-hermetic Hermetic						
Cooling Air Refrigerant						
Application AC High Medium Low						
Size Btuh or Horsepower						
Electrical Volts Phase Amps						

LAB 15.2 SEMI-HERMETIC COMPRESSOR INSPECTION

LABORATORY OBJECTIVE
You will disassemble a semi-hermetic compressor, identify its parts, discuss its operation, and reassemble it.

LABORATORY NOTES
The instructor will assign a semi-hermetic compressor for you to disassemble. You will examine the parts and discuss their function with the instructor. You will then reassemble the compressor.

FUNDAMENTALS OF HVACR TEXT REFERENCE
Unit 15 Compressors

REQUIRED MATERIALS PROVIDED BY STUDENT
Safety glasses
Gloves
6-in-1
Adjustable jaw wrench

REQUIRED MATERIALS PROVIDED BY SCHOOL
Semi-hermetic compressor
Socket set
Oil drain pan

PROCEDURE
1.   Obtain semi-hermetic compressor from instructor.

2.   Release pressure from crankcase through suction service valve.

3.   Drain oil from crankcase of compressor.

4.   Remove cylinder head and valve plate.

5.   Identify the suction and discharge valves.

6.   Mark both end bells and carefully examine their orientation on the compressor.

7.   Remove bolts from end bells.

8.   Remove end bells.

9.   Remove bolts from the bottom plate.

10.   Remove bottom plate from the crankcase.

11.   Have instructor check compressor after teardown.

12.   Examine the parts and discus their function with the instructor.

13.   Complete data sheet.

14.   Assemble compressor in reverse order of disassembly.

LAB 15.2 SEMI-HERMETIC COMPRESSOR DISASSEMBLY

Manufacturer	
Model Number	
Number of Pistons	
Crankshaft Type **(crank throw or eccentric)**	
Valve Type **(reed, ring, discuss)**	
Cooling **(Air or Refrigerant)**	
Lubrication **(Splash or Pressure)**	
Application **(AC, High, Medium, Low)**	

LAB 15.3 HERMETIC COMPRESSOR INSPECTION

LABORATORY OBJECTIVE
You will disassemble a hermetic compressor, identify its parts, discuss its operation, and reassemble it.

LABORATORY NOTES
The instructor will assign a hermetic compressor for you to disassemble. You will examine the parts and discuss their function with the instructor. You will then reassemble the compressor.

FUNDAMENTALS OF HVACR TEXT REFERENCE
Unit 15 Compressors

REQUIRED MATERIALS PROVIDED BY STUDENT
Safety glasses
Gloves
6-in-1
Adjustable jaw wrench

REQUIRED MATERIALS PROVIDED BY SCHOOL
Hermetic compressor
Socket set
Oil drain pan

PROCEDURE

Procedure for cutting open compressor shell

1. Release pressure from shell through process tube.
2. Drain oil from crankcase of compressor by inverting compressor and pouring the oil out of the process tube.
3. Purge the shell with nitrogen to prevent combustion of the oil residue while cutting open the shell.
4. Cut the shell in two at the weld seam. A reciprocating saw with a metal blade makes the cleanest and safest cut, but does leave metal filings in the shell. A grinder works quickly, but leaves a sharp edge on the metal casing. Do NOT use a cutting torch. The oxygen can mix with the oil in the compressor to create an explosion.
5. Separate the shell and cut the discharge line at a convenient point.

Procedure for disassembly

1. Remove the compressor-motor assembly from the shell.
2. Remove the compressor head and valve plate.
3. Have an instructor check your disassembly.
4. Examine the parts and discus their function with the instructor.
5. Answer the questions on the following page.
6. Reassemble valve plate and head.
7. Place compressor back in shell.

QUESTIONS

1. What type of lubrication system does this compressor use?

2. Is this compressor motor air-cooled or refrigerant cooled?

3. What is in the shell: low pressure suction gas or high pressure discharge gas?

4. What type of duty was this unit designed for? (high, medium, or low temp)

5. What size compressor is this? (horsepower and/or BTU)

6. What type of valves doe this compressor have?

7. How was the compressor mounted inside the shell?

8. Why is the discharge line shaped the way it is?

9. How does the suction gas reach the cylinders in the compressor?

LAB 15.4 ROTARY COMPRESSOR INSPECTION

LABORATORY OBJECTIVE
You will disassemble a rotary compressor, identify its parts, discuss its operation, and reassemble it.

LABORATORY NOTES
The instructor will assign a rotary compressor for you to disassemble. You will examine the parts and discuss their function with the instructor. You will then reassemble the compressor.

***FUNDAMENTALS OF HVACR* TEXT REFERENCE**
Unit 15 Compressors

REQUIRED MATERIALS PROVIDED BY STUDENT
Safety glasses
Gloves
6-in-1
Adjustable jaw wrench

REQUIRED MATERIALS PROVIDED BY SCHOOL
Rotary Compressor
Socket Set
Oil Drain Pan

PROCEDURE
1. Obtain hermetic rotary compressor from instructor.
2. Remove the hermetic casing from the rotary compressor body.
3. Remove the top plate from the compressor body.
4. Examine the blades of the compressor.
5. Turn rotor and examine action of the blades when compressor is rotated.
6. Replace top plate.
7. Replace compressor hermetic casing.
8. Answer the questions below:

      a. Is this stationary blade or rotating blade compressor?

      b. What is in the shell: low-pressure suction gas or high pressure discharge gas?

      c. How does the suction gas get into the compressor?

      d. How does the discharge gas get from the compressor to the discharge line?

LAB 15.5 SCROLL COMPRESSOR INSPECTION

LABORATORY OBJECTIVE
You will disassemble a scroll compressor, identify its parts, discuss its operation, and reassemble it.

LABORATORY NOTES
The instructor will assign a scroll compressor for you to disassemble. You will examine the parts and discuss their function with the instructor. You will then reassemble the compressor.

***FUNDAMENTALS OF HVACR* TEXT REFERENCE**
Unit 15 Compressors

REQUIRED MATERIALS PROVIDED BY STUDENT
Safety glasses
Gloves
6-in-1
Adjustable jaw wrench

REQUIRED MATERIALS PROVIDED BY SCHOOL
Scroll compressor
Socket set
Oil drain pan

PROCEDURE
1. Obtain hermetic scroll compressor from instructor.
2. Remove the hermetic casing from the scroll compressor body.
3. Remove the top plate from the compressor body.
4. Examine the scrolls of the compressor.
5. Turn rotor and examine action of the scrolls when compressor is rotated.
6. Replace top plate.
7. Replace compressor hermetic casing.
8. Answer the questions below:

   a. What is in the shell: low-pressure suction gas or high pressure discharge gas?

   b. How does the suction gas get into the compressor?

   c. How does the discharge gas get from the compressor to the discharge line?

   d. How many moving parts does this compressor have?

LAB 15.6 DETERMINING COMPRESSION RATIO

LABORATORY OBJECTIVE

You will check the suction and discharge pressures on an operating compressor and determine its compression ratio.

LABORATORY NOTES

You will install your gauges on an operating system and use the operating pressures to determine the compressor ratio. You will then block the condenser airflow, retest, and recalculate the compression ratio. Finally, you will block the evaporator airflow, retest, and recalculate the compression ratio.

***FUNDAMENTALS OF HVACR* TEXT REFERENCE**

Unit 15 Compressors

REQUIRED MATERIALS PROVIDED BY STUDENT

Safety glasses
Gloves
Two adjustable jaw wrenches
6-in-1
Refrigeration gauges
Refrigeration valve wrench
Calculator

REQUIRED MATERIALS PROVIDED BY SCHOOL

Operating refrigeration system

PROCEDURE

Operate the system assigned by the instructor until the pressures have stabilized. Measure the suction and discharge pressure and calculate the compression ratio. The compression ratio is found by dividing the absolute discharge pressure by the absolute suction pressure. This is shown in the formula:

(Discharge pressure + 15) ÷ (Suction Pressure + 15)

Temporarily restrict the airflow across the condenser.

CAUTION: Monitor the high side pressure and remove the restriction if the condenser saturation temperature reaches 140°F.

The high side pressure should increase.

Recheck the system pressures.

Record the new pressures.

Remove the restriction and allow the pressures to return to normal.

Recalculate the compression ratio using the new pressures.

What effect did blocking the condenser airflow have on the compression ratio?

Temporarily restrict the airflow across the evaporator.

The evaporator pressure should drop.

Recheck the system pressures.

Record the new pressures.

● Remove the restriction and allow the pressures to return to normal.

Recalculate the compression ratio using these new pressures.

What effect did blocking the evaporator airflow have on the compression ratio?

LAB 15.7 COMPRESSOR AUTOPSY

LABORATORY OBJECTIVE
You will disassemble a failed compressor and identify the cause of the compressor failure.

LABORATORY NOTES
You will disassemble a failed compressor, examine the parts of the compressor, and determine the type of failure. You will then recommend ways to prevent a repeat failure of the replacement compressor.

***FUNDAMENTALS OF HVACR* TEXT REFERENCE**
Unit 15 Compressors

REQUIRED MATERIALS PROVIDED BY STUDENT
Safety glasses
Gloves
Two adjustable jaw wrenches
6-in-1

REQUIRED MATERIALS PROVIDED BY SCHOOL
Failed compressor
Socket set

PROCEDURE
Release pressure from crankcase through suction service valve.

Drain oil from crankcase of compressor.

Remove cylinder head and inspect the valve plate.

Are there signs of high temperature operation? (dark varnished looking parts)

Are the discharge valves intact?

Remove the valve plate and inspect the suction valves.

Are the suction valves intact?

Mark both end bells and carefully examine their orientation on the compressor.

Remove bolts from end bells.

Remove end bells.

Remove bolts from the bottom plate.

Remove bottom plate from the crankcase.

Inspect the crank and rods.

Are all the parts intact?

Turn the crank and observe the operation of the rods and pistons.

Is there any slop in the fit between the crank, rods, and pistons?

Examine the parts and discus their function with the instructor

Assemble compressor in reverse order of disassembly.

Determine the failure type.

- Liquid flooding – broken parts, especially valves, pistons, and rods
- Overheating – dark, discolored parts, especially valve plates
- Oil Failure – scoring on bearing surfaces or seizure

Recommend ways of preventing a similar repeat failure for the replacement compressor.

LAB 16.1 AIR COOLED CONDENSER PERFORMANCE

LABORATORY OBJECTIVE
You will measure the heat rejection of an air cooled condenser.

LABORATORY NOTES
You will measure the airflow, entering air temperature, and leaving air temperature through an air-cooled condenser. Those data will be used to calculate the total condenser heat rejection.

FUNDAMENTALS OF HVACR TEXT REFERENCE
Unit 16 Condensers

REQUIRED MATERIALS PROVIDED BY STUDENT
Safety glasses
Gloves
6-in-1
Thermometer
Calculator

REQUIRED MATERIALS PROVIDED BY SCHOOL
Operating unit with air cooled condenser
Rotating vane anemometer
Condenser airflow capture device

PROCEDURE
1. Start the unit assigned by the instructor.
2. Allow the unit to operate for at least 10 minutes.
3. Place the condenser airflow capture device over the condenser discharge.
4. Measure the temperature of the air entering and leaving the condenser.
5. Calculate the temperature difference ΔT = Leaving Air Temp – Entering air Temp
6. Measure the velocity of the air leaving the condenser airflow capture device.
7. Calculate the air volume CFM = FPM (air velocity) x ft^2 area of airflow capture device.
8. Calculate the heat rejection. BTUs/hr = ΔT x CFM x 1.08
9. Complete the chart below.

Air Cooled Condenser Performance	
Entering air temperature	
Leaving air temperature	
Temperature rise	
Air Velocity in Feet per Minute FPM	
Air flow in Cubic Feet per Minute CFM	
Total Heat Rejection in BTUs/hr	

LAB 16.2 WATER COOLED CONDENSER PERFORMANCE

LABORATORY OBJECTIVE
You will measure the heat rejection of a water-cooled condenser.

LABORATORY NOTES
You will measure the water flow, entering water temperature, and leaving water temperature through an water cooled condenser. Those data will be used to calculate the total condenser heat rejection.

FUNDAMENTALS OF HVACR TEXT REFERENCE
Unit 16 Condensers

REQUIRED MATERIALS PROVIDED BY STUDENT
Safety glasses
Gloves
6-in-1
Thermometer
Calculator

REQUIRED MATERIALS PROVIDED BY SCHOOL
Operating unit with water-cooled condenser
Water flow indicator

PROCEDURE
1. Start the unit assigned by the instructor.
2. Allow the unit to operate for at least 10 minutes.
3. Measure the temperature of the water entering and leaving the condenser.
4. Calculate the temperature difference ΔT = Leaving Water Temp – Entering Water Temp
5. Observe the water flow indicator to determine the volume of water moving across the condenser in gallons per minute (GPM).
6. Calculate the total heat rejection BTUs/hr = ΔT x GPM x 8.35 lbs/gallon x 60 min/hr
7. Complete the chart below and calculate the heat rejection.

Water Cooled Condenser Performance	
Entering water temperature	
Leaving water temperature	
Temperature rise	
Water flow in GPM	
Total Heat Rejection	

LAB 17.1 INTERNALLY EQUALIZED THERMOSTATIC EXPANSION VALVES

LABORATORY OBJECTIVE
The student will disassemble an internally equalized thermostatic expansion valve and be able to describe how it operates.

LABORATORY NOTES:
For this lab exercise there should be one or more internally equalized thermostatic expansion valves available that may be disassembled.

FUNDAMENTALS OF HVACR TEXT REFERENCE
Unit 17 Metering Devices

REQUIRED TOOLS AND EQUIPMENT
Internally equalized thermostatic expansion valve

SAFETY REQUIREMENTS
None.

PROCEDURE
Step 1
Locate an internally equalized thermostatic expansion valve and examine it carefully so that you may complete the following exercise:

    A. Record all of the data that can be found on the valve such as refrigerant type, capacity, line size, refrigerant bulb charge, etc.

Thermal Bulb

Figure 17-1-1

Thermostatic Expansion Valve Data:

B. Disassemble the valve being careful not to lose components while identifying the manner in which the valve came apart so that it may then be put back together again. An internally equalized valve has only an inlet and an outlet. There is no equalizing connection.

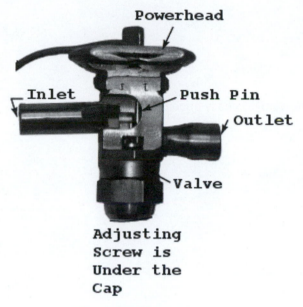

Figure 17-1-2

C. Sketch a cross sectional view of the disassembled valve in the space provided below.

D. Explain how the valve operates using the sketch you provided above.

Step 2

Using the disassembled valve, your sketch, and your description, answer the following questions:

    A. Draw a one-line representation of the diaphragm and the forces acting upon it (bulb pressure, evaporator pressure, and spring pressure).

    B. How is the diaphragm movement transmitted to the valve disk?

    C. Does the spring help to open or to close the valve?

    D. How does the evaporator pressure travel to the underside of the valve diaphragm?

    E. Does the evaporator pressure help to open or to close the valve?

F. If the bulb lost its charge would the valve fail open or closed?

G. What type of refrigerant can the valve be used for?

QUESTIONS

(Circle the letter that indicates the correct answer.)

1. A thermostatic expansion valve can be adjusted for:
   A. refrigerant flow through the evaporator.
   B. evaporator refrigerant outlet superheat.
   C. evaporator capacity.
   D. All of the above are correct.

2. An internally equalized thermostatic expansion valve senses pressure.
   A. at the evaporator outlet.
   B. at the compressor inlet.
   C. at the condenser outlet.
   D. at the evaporator inlet.

3. To increase the superheat setting of thermostatic expansion valve you would turn the adjusting screw:
   A. clockwise.
   B. counter clockwise.
   C. It can not be adjusted.
   D. back and forth.

4. If the thermal bulb on a thermostatic expansion valve lost its charge:
   A. the valve would open wide.
   B. the valve would hunt.
   C. the valve would close.
   D. the valve would frost up.

5. The thermal bulb pressure on the top of a thermostatic expansion valve diaphragm is transmitted to the valve by:
   A. the spring.
   B. the evaporator inlet pressure.
   C. the evaporator outlet pressure.
   D. push pins.

6. The thermal bulb force acting on the topside of the diaphragm in a thermostatic expansion valve is balanced by:
   A. spring force only.
   B. evaporator pressure only.
   C. spring force and evaporator pressure.
   D. an orifice plate.

7. Internally equalized thermostatic expansion valves can be used:
   A. on evaporator coils that have a minimal pressure drop.
   B. on evaporator coils that have a large pressure drop.
   C. on beverage coolers only.
   D. A & C are both correct.

8. If an internally equalized thermostatic expansion valve is used on an evaporator coil that has a large pressure drop:
   A. the evaporator coil will flood with refrigerant.
   B. the evaporator coil will freeze up.
   C. the evaporator coil will be starved.
   D. liquid will slug back to the compressor.

LAB 17.2 EXTERNALLY EQUALIZED THERMOSTATIC EXPANSION VALVES

LABORATORY OBJECTIVE
The student will disassemble an externally equalized thermostatic expansion valve and be able to describe how it operates.

LABORATORY NOTES
For this lab exercise there should be one or more externally equalized thermostatic expansion valves available that may be disassembled.

***FUNDAMENTALS OF HVACR* TEXT REFERENCE**
Unit 17 Metering Devices

REQUIRED TOOLS AND EQUIPMENT:
Externally equalized thermostatic expansion valve

SAFETY REQUIREMENTS
None.

PROCEDURE
Step 1
Locate an externally equalized thermostatic expansion valve and examine it carefully so that you may complete the following exercise:

    A. Record all of the data that can be found on the valve such as refrigerant type, capacity, line size, refrigerant bulb charge, etc.

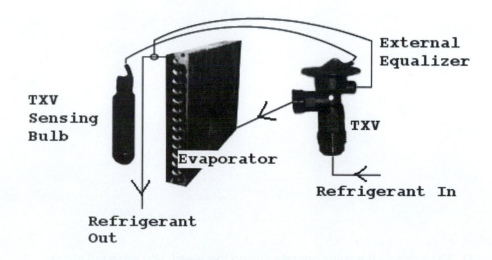

Figure 17-2-1

119

B. Disassemble the valve being careful not to lose components while identifying the manner in which the valve came apart so that it may then be put back together again. An externally equalized valve has an inlet, an outlet, and an external equalizing connection.

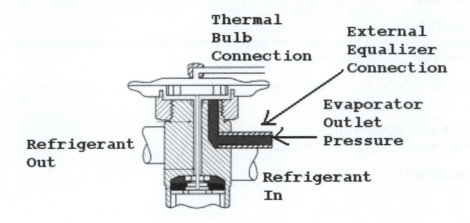

Figure 17-2-2

C. Sketch a cross sectional view of the valve in the space provided below.

D. Explain how the valve operates using the sketch you provided above.

Step 2

Using the disassembled valve, your sketch and your description, answer the following questions:

A. Draw a one-line representation of the diaphragm and the forces acting upon it (bulb pressure, evaporator pressure, and spring pressure).

B. How is the diaphragm movement transmitted to the valve disk?

C. Does the spring help to open or to close the valve?

D. How does the evaporator pressure travel to the underside of the valve diaphragm?

121

E. Does the evaporator pressure help to open or to close the valve?

F. If the bulb lost its charge would the valve fail open or closed?

G. What type of refrigerant can the valve be used for?

QUESTIONS

(Circle the letter that indicates the correct answer.)

1. A thermostatic expansion valve can be adjusted for:
    A. refrigerant flow through the evaporator.
    B. evaporator refrigerant inlet superheat.
    C. evaporator pressure.
    D. All of the above are correct.

2. An externally equalized thermostatic expansion valve senses pressure.
    A. at the evaporator outlet.
    B. at the compressor inlet.
    C. at the condenser outlet.
    D. at the evaporator inlet.

3. To decrease the superheat setting of thermostatic expansion valve you would turn the adjusting screw:
    A. clockwise.
    B. counter clockwise.
    C. It can not be adjusted.
    D. back and forth.

4. If the thermal bulb on a thermostatic expansion has a liquid charge:
    A. control will always be from the bulb.
    B. control will never be from the bulb.
    C. the valve will have a limited opening.
    D. the valve will open quicker.

5. If the thermal bulb on a thermostatic expansion has a limited liquid (gas) charge:
    A. control will always be from the bulb.
    B. control will never be from the bulb.
    C. the valve will have a limited opening.
    D. the valve will open quicker.

6. If the thermal bulb on a thermostatic expansion has a cross charge:
   A. the fluid in the bulb is the same as the refrigerant in the system.
   B. the fluid in the bulb is different than the refrigerant in the system.
   C. the suction pressure crosses the discharge pressure.
   D. the valve will open quicker.

7. Externally equalized thermostatic expansion valves can be used:
   A. on evaporator coils that have a minimal pressure drop.
   B. on evaporator coils that have a large pressure drop.
   C. on beverage coolers only.
   D. A & C are both correct.

8. If the external equalizing connection on a externally equalized thermostatic expansion valve is capped closed:
   A. the evaporator coil will flood with refrigerant.
   B. the evaporator coil will freeze up.
   C. the evaporator coil will be starved.
   D. liquid will slug back to the compressor.

9. The fluid in a straight charged thermostatic expansion valve bulb:
   A. is the same as the refrigerant in the system.
   B. is different than the refrigerant in the system.
   C. is always liquid.
   D. is always vapor.

LAB 17.3 SETTING A THERMOSTATIC EXPANSION VALVE

LABORATORY OBJECTIVE
The student will demonstrate how to properly set a thermostatic expansion valve to maintain the proper superheat at the evaporator outlet.

LABORATORY NOTES
For this lab exercise there needs to be an operating refrigeration system that has a thermostatic expansion valve. If there are no thermometers or pressure gauges currently on the unit then you must instrument the refrigeration system using a gauge manifold to measure the evaporator pressure and a thermometer to read the evaporator refrigerant temperature.

If you are unable to place probes directly into the refrigerant stream flow, you can use evaporator coil surface temperatures, however you must adjust for the heat transfer difference through the coil.

FUNDAMENTALS OF HVACR TEXT REFERENCE
Unit 17 Metering Devices

REQUIRED TOOLS AND EQUIPMENT
Operating refrigeration unit with thermostatic expansion valve
Gauge manifold & temperature sensor

SAFETY REQUIREMENTS
A. Always read the equipment manual to become familiarized with the refrigeration system and its accessory components prior to start up.
B. Wear safety goggles and gloves when working with refrigerants. Liquid refrigerant can cause frostbite when in contact with eyes and skin.
C. Use low loss hose fittings, or wrap cloth around hose fittings before removing the fittings from a pressurized system or cylinder. Inspect all fittings before attaching hoses.

PROCEDURE
Step 1
Familiarize yourself with the major components in the refrigeration system including the condenser, compressor, evaporator, and metering device.

> **A.** The thermostatic expansion (TEV) supplies the evaporator with enough refrigerant for any and all load conditions. It is <u>NOT</u> a temperature, suction pressure, humidity, or operating control.

> **B.** The thermostatic expansion valve is adjusted to control the refrigerant superheat at the outlet of the evaporator.

> **C.** Many thermostatic expansion valves come pre-set for 10°F of superheat.

> **D.** Thermostatic expansion valves should not be adjusted unless the evaporator conditions (temperature & pressure) can be measured.

Step 2
Determine the proper setting for the TEV as follows:

    A. The flow of the refrigerant through the TEV is controlled by three pressures.

        a) The evaporator pressure

        b) The spring pressure acting on the bottom of the diaphragm

        c) The bulb pressure opposing these two pressures and acting on the top of the diaphragm

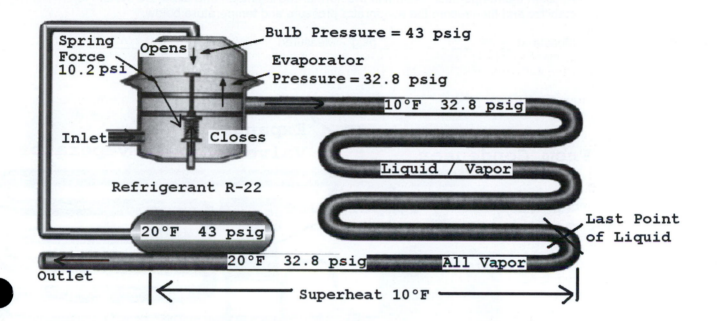

Figure 17-3-1

B. In Figure 17-3-1, these three pressures are illustrated. When the three forces are balanced and the valve is in equilibrium as shown, then there should be ten degrees of superheat at the evaporator outlet.

C. The saturation temperature for R-22 at 32.8 psig is 10°F. If the refrigerant pressure is 32.8 psig and the temperature is 20°F, then there would be what is considered as 10°F of superheat.

D. If the superheat temperature decreases, there will be a corresponding decrease in bulb pressure as the TEV would be allowing too much refrigerant to flow. The balance of the three pressures would be disrupted and the valve would begin to close until the pressures arrived at a new equilibrium point.

E. To determine the superheat value you will need to measure both the pressure and the temperature at the evaporator outlet.

Assuming we are using R-22 and the evaporator pressure is 32.8 psig as shown in Unit 10, Figure 10-12 of the *Fundamentals of HVACR* text (or Table 47-1-1 in Lab 47.1) we can look up the saturated refrigerant temperature.

From the chart at 32.8 psig, for R-22 the saturated temperature would be 10°F. You would then subtract this saturated temperature from the actual measured temperature from the thermometer to determine the total degrees of superheat.

Step 3

    A. If there are no pressure gauges currently on the unit then you must connect a gauge manifold to read the evaporator pressure. To help guide you, refer to the procedures in Lab 47.1 *Basic Refrigeration System Startup*, Steps 4, 5, & 6.

    B. If you are unable to place temperature probes directly into the refrigerant stream flow, you can use evaporator coil surface temperatures however you must adjust for the heat transfer difference through the coil.

Step 4

    A. Start the refrigeration system in the normal cooling mode and allow the system to stabilize and then record the evaporator pressure and temperature below.

Measured_____psig   Measured_____°F

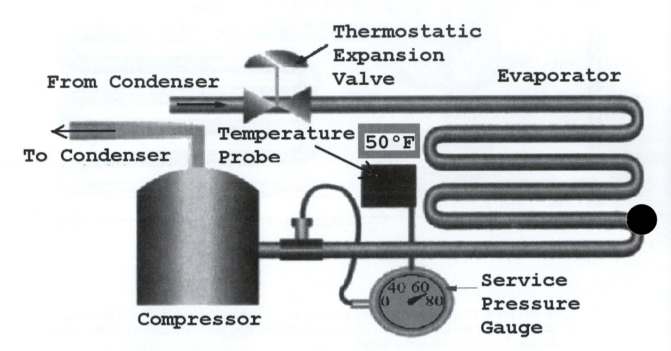

To Measure Superheat:
1. Find Suction Pressure
2. Find Matching Saturation Temperature
3. Read Temperature Leaving Evaporator
4. Superheat = Temp. Leaving - Saturation Temp.

Figure 17-3-2

    B. From the saturated pressure-temperature chart determine the saturated temperature at the measured pressure reading.

    C. Calculate the superheat.

Superheat = Measured Temp. – Saturation Temp.

126

D. Determine what adjustments need to be made, if any, and turn the adjusting screw in small increments only. This will change the spring tension that in turn will change the superheat setting.

E. On many valves the adjustment is clockwise to increase the superheat and counterclockwise to decrease the superheat. Valve instructions should be checked to be sure of correct adjustments.

F. Allow the system to run and then recheck the superheat once again before making any further changes. It may take some time before the system stabilizes and the final adjustment is complete.

Step 5

A. Continue making any necessary adjustments until the thermostatic expansion valve superheat is correctly set.

B. Allow the system to run and stabilize prior to shutting down and then carefully disconnect the gauge manifold and any instrumentation that you attached to the unit.

QUESTIONS

(Circle the letter that indicates the correct answer.)

1. Too high of a TEV superheat setting would lead to:
   A. a flooded evaporator coil.
   B. liquid slugging the compressor.
   C. a starved evaporator coil.
   D. A & B are both correct.

2. Too low of a TEV superheat setting would lead to:
   A. a flooded evaporator coil.
   B. liquid slugging the compressor.
   C. a starved evaporator coil.
   D. A & B are both correct.

3. Ice forming on the outside of a thermostatic expansion valve:
   A. indicates blockage.
   B. indicates normal operation.
   C. can crack the diaphragm.
   D. should be chipped off with an ice pick.

4. Raising the temperature of a fluid above its saturation temperature is:
   A. superheating.
   B. subcooling.
   C. sublimation.
   D. saturation absorption.

5. If the sensing bulb for a thermostatic expansion valve becomes unattached from the evaporator coil:
   A. the valve will begin to close and starve the evaporator.
   B. the valve will begin to open and starve the evaporator.
   C. the valve will begin to close and flood the evaporator.
   D. the valve will begin to open and flood the evaporator.

6. If the diaphragm on a thermostatic expansion valve fails the valve.
   A. the valve will begin to close and starve the evaporator.
   B. the valve will begin to open and starve the evaporator.
   C. the valve will begin to close and flood the evaporator.
   D. the valve will begin to open and flood the evaporator.

7. If the evaporator coil is cold at the inlet but is warm about half way down then this could indicate:
   A. a high superheat.
   B. a low superheat.
   C. the proper amount of superheat.
   D. floodback.

LAB 17.4 TYPES OF METERING DEVICES

LABORATORY OBJECTIVE
You will identify different metering devices and discuss their application.

LABORATORY NOTES
You will examine units with different metering devices, including capillary tube, orifice, and thermostatic expansion valve. You will identify the type of metering device in each unit and discuss the application in that unit.

FUNDAMENTALS OF HVACR TEXT REFERENCE
Unit 17 Metering Devices

REQUIRED MATERIALS PROVIDED BY STUDENT
Safety glasses
Gloves
6-in-1
Adjustable jaw wrench

REQUIRED MATERIALS PROVIDED BY SCHOOL
Refrigeration systems with different metering devices

PROCEDURE
1. Remove panels necessary to locate the metering device in each of the units assigned by the instructor.
2. Identify the type of metering device (cap tube, orifice, TEV).
3. Determine the number of refrigerant circuits.
4. Identify the distribution device for coils with more than one circuit.
5. Locate where the refrigerant from the metering device feeds the evaporator.

LAB 17.5 ADJUSTING AUTOMATIC EXPANSION VALVES

LABORATORY OBJECTIVE

You will adjust an automatic expansion valve to maintain a specified evaporator pressure.

LABORATORY NOTES

You will adjust the automatic expansion valve on the air dryer to the pressures specified in the lab.

***FUNDAMENTALS OF HVACR* TEXT REFERENCE**

Unit 17 Metering Devices

REQUIRED MATERIALS PROVIDED BY STUDENT

Safety glasses
Gloves
6-in-1
Adjustable jaw wrench

REQUIRED MATERIALS PROVIDED BY SCHOOL

Refrigerated Air Drier with AEV

PROCEDURE

1. Remove panels necessary to locate the automatic expansion valve.

2. Operate the system and record the suction pressure.

3. Turn the valve adjustment stem in clockwise until the suction pressure is 5 psig higher than the original reading.

4. Let the system operate for 5 minutes.

5. Turn the valve-adjusting stem out counterclockwise to set the pressure back to the original setting.

6. Explain how the adjustment on an automatic expansion valve works.

LAB 17.6 DISASSEMBLING TEVs

LABORATORY OBJECTIVE
You will examine the parts and construction of thermostatic expansion valves and discuss the difference between internally equalized valve and externally equalized valves.

LABORATORY NOTES
You will disassemble both an internally equalized valve and an externally equalized valve. You will identify the parts and discuss the difference between the two types of valves.

FUNDAMENTALS OF HVACR TEXT REFERENCE
Unit 17 Metering Devices

REQUIRED MATERIALS PROVIDED BY STUDENT
Safety glasses
Gloves

REQUIRED MATERIALS PROVIDED BY SCHOOL
Internally Equalized TEV
Externally Equalized TEV

PROCEDURE

1. Disassemble the internally equalized expansion valve.

2. Be sure to pay attention to the orientation and location of the component parts.

3. Be prepared to describe the valve operation to the instructor.

4. Pay particular attention to the path that the evaporator pressure takes to reach the underside of the diaphragm.

5. Reassemble the valve after conferring with the instructor.

6. Disassemble the externally equalized valve.

7. Be sure to pay attention to the orientation and location of the component parts.

8. Be prepared to describe the valve operation to the instructor.

9. Pay particular attention to the path that the evaporator pressure takes to reach the underside of the diaphragm.

10. Reassemble the valve after conferring with the instructor.

LAB 17.7 TEV INSTALLATION

LABORATORY OBJECTIVE
You will properly install a thermostatic expansion valve and measure the superheat.

LABORATORY NOTES
You will pump down the refrigerant into the high side of the system and remove the existing metering device. You will then install a TEV and evacuate the refrigerant lines and coil. After the lines and coil are evacuated, you will open the king valve, allowing the refrigerant to flow back into the low side. Finally, you will operate the system and measure the superheat at the expansion valve-sensing bulb.

FUNDAMENTALS OF HVACR TEXT REFERENCE
Unit 17 Metering Devices

REQUIRED MATERIALS PROVIDED BY STUDENT
Safety glasses
Gloves
Two adjustable jaw wrenches
6-in-1
Refrigeration gauges
Refrigeration valve wrench
Temperature tester

REQUIRED MATERIALS PROVIDED BY SCHOOL
Refrigeration system
TEV
Vacuum pump
Vacuum gauge
Refrigerant

PROCEDURE
1. Install refrigeration gauges on the system.
2. The compound gauge should be connected to the suction service valve.
3. The high side gauge should be connected to the king valve.
4. Purge the gauges.
5. Front seat liquid service valve.
6. Operate the system to pump down refrigerant into high side.
7. The compound gauge should pull down to 0 psig or into a vacuum.
8. The high side gauge may pull down to 0 psig or remain high, depending on the type of king valve the system has.
9. Clean the metering device connection by wiping with a clean cloth.
10. Loosen the flare connections to the existing metering device.
11. Be sure to use two properly fitting wrenches.
12. Do NOT try to loosen the flare connection with a single wrench. The tubing will twist, damaging the coil.
13. Remove the existing metering device.
14. Install thermostatic expansion valve.
15. Tighten using two wrenches, being careful not to twist the tubing.
16. Secure the TEV bulb to the suction line and insulate it.
17. Pressurize the lines and coil with nitrogen and check for leaks.
18. The king valve should REMAIN FRONTSEATED. Do NOT move it.
19. Evacuate low side of system (high side still has refrigerant in it).
20. Backseat crack the king valve to let the refrigerant back into the system.

21. Start system and adjust charge as necessary.
22. Measure and record system pressures, suction line temperature, liquid line temperature, superheat, and subcooling.
23. Have instructor check unit.

LAB 17.7 TEV INSTALLATION	
Evaporator Pressure	
Evaporator Saturation Temp	
Bulb Temperature	
Superheat (Bulb Temp –Evap Saturation)	
Condenser Pressure	
Condenser Saturation	
Liquid Line Temperature	
Subcooling Cond saturation-liquid line	

LAB 17.8 ADJUSTING VALVE SUPERHEAT

LABORATORY OBJECTIVE
You will measure the superheat of thermostatic expansion valve system and adjust it to meet manufacturer's specifications.

LABORATORY NOTES
You will use the suction pressure, suction line temperature, and PT chart to determine the operating superheat of an expansion valve system. You will then adjust the valve counterclockwise and then clockwise to observe the effects of valve adjustment. Finally, you will adjust the valve to manufacturer specification.

FUNDAMENTALS OF HVACR TEXT REFERENCE
Unit 17 Metering Devices

REQUIRED MATERIALS PROVIDED BY STUDENT
Safety glasses
Gloves
Two adjustable jaw wrenches
6-in-1
Refrigeration gauges
Refrigeration valve wrench
Temperature tester

REQUIRED MATERIALS PROVIDED BY SCHOOL
Operating refrigeration system with TEV
Refrigerant

PROCEDURE

1. Operate the system assigned by the Instructor for 15 minutes.

2. Measure the superheat of the expansion valve and record all pertinent data on the chart below.

3. Turn the adjusting stem 1 full turn counter-clockwise. Let system operate or 15 minutes.

4. Measure the superheat and record all pertinent data.

5. Turn adjusting stem 2 full turns clockwise. Let system operate for 15 minutes.

6. Measure the superheat and record all pertinent data.

7. Adjust T.E.V. for normal operating superheat. Let operate for 15 minutes.

8. Measure superheat and record all pertinent data.

9. Explain how superheat adjustment works.

10. Summarize the effect of bulb temperature on system operation.

LAB 17.8 ADJUSTING VALVE SUPERHEAT					
	Condenser Pressure	Evaporator Pressure	Saturation Temp	Bulb Temp	Superheat (Bulb –Saturation)
Starting Setting					
1 Turn CCW					
2 Turns CW					
Normal Operation					

LAB 17.9 CAPILLARY TUBE ASSEMBLY & OPERATION

LABORATORY OBJECTIVE
You will assemble, leak test, evacuate, and charge a capillary tube refrigeration system.

LABORATORY NOTES
You will pump down the system, remove the existing metering device, and install a capillary tube. You will then evacuate the system, open the liquid service valve, and operate the system. You will record all operating data including system pressures, system saturation temperatures, superheat, and subcooling.

***FUNDAMENTALS OF HVACR* TEXT REFERENCE**
Unit 17 Metering Devices

REQUIRED MATERIALS PROVIDED BY STUDENT
Safety glasses
Gloves
Two adjustable jaw wrenches
6-in-1
Refrigeration gauges
Refrigeration valve wrench
Temperature tester

REQUIRED MATERIALS PROVIDED BY SCHOOL
Operating refrigeration system
Refrigerant
Capillary tube
Vacuum pump
Torch
Brazing material

PROCEDURE

1. Install your gauges with the high side on the liquid service valve.

2. Make certain to purge your gauges.

3. Front seat liquid service valve and pump down refrigerant into high side.

4. Clean the connection.

5. Remove existing metering device.

6. Install capillary tube.

7. Pressure test system for leaks using nitrogen.

8. Evacuate low side of system (high side still has refrigerant in it).

9. Back seat crack the liquid service valve to let the refrigerant into the system.

10. Start and test system.

11. Adjust charge as necessary.

12. Record system pressures, suction line temperature, liquid line temperature, superheat, and subcooling.

13. Have instructor check unit.

LAB 17.9 DATA	
Evaporator Pressure	
Evaporator Saturation Temperature	
Suction Line Temperature	
Superheat **Suction temp – Evaporator Saturation Temp**	
Condenser Pressure	
Condenser Saturation Temperature	
Liquid Line Temperature	
Subcooling **Condenser Saturation temp – Liquid Temp**	

LAB 18.1 INSTALLING CONDENSATE DRAIN

LABORATORY OBJECTIVE
You will install an evaporator condensate drain and demonstrate that it drains properly.

LABORATORY NOTES
You will install a condensate drain on the evaporator assigned by the instructor. After installing the drain, you will pour water into the drain pan to demonstrate that the drain works properly.

***FUNDAMENTALS OF HVACR* TEXT REFERENCE**
Unit 18 Evaporators

REQUIRED MATERIALS PROVIDED BY STUDENT
Safety glasses
Gloves
6-in-1

REQUIRED MATERIALS PROVIDED BY SCHOOL
Air conditioning evaporator
PVC
PVC fittings
PVC primer and solvent
PVC shear

PROCEDURE

1. Install the condensate drain for the unit assigned by the instructor.

2. Tubing should be cut with a tubing shear.

3. If a hacksaw is used, the chips must be cleaned out before assembling the drain.

4. Joints should be primed before applying the solvent.

5. The drain line must not run up hill.

6. The drain line should not interfere with panel removal or filter replacement.

7. The drain should have a trap.

8. Verify that the drain works by pouring water into the evaporator drain pan.

LAB 19.1 IDENTIFYING REFRIGERANTS AND REFRIGERANT CHARACTERISTICS

LABORATORY OBJECTIVE
The purpose of this lab is to demonstrate your ability to identify the type of refrigerant contained in each of four different refrigeration systems.

LABORATORY NOTES
You will find the type of refrigerant, the amount of refrigerant, and the system test pressures on the unit data plate of each system. Then you will research the reference material to find the refrigerant characteristics.

You will identify the refrigerant by:

* Name
* Pressure (Very High, High, or Low)
* Ozone Depletion Potential
* Global Warming Potential
* Toxicity
* Flammability
* Chemical Composition
* Formulation

FUNDAMENTALS OF HVACR TEXT REFERENCE
Unit 19 Refrigerants and Their Properties

REQUIRED TOOLS AND EQUIPMENT
Four refrigeration systems containing different refrigerants
Refrigerant cylinders
Refrigerant MSDS

SAFETY REQUIREMENTS
Caution!!! Do not open the valves on the refrigeration cylinders to avoid possible injury due to skin contact with refrigerant. Also it is illegal to intentionally vent refrigerants comprised of CFC's and HCFC's.

PROCEDURE
Step 1
Locate the nameplate on the first unit. The nameplate will be **in different locations for each unit you inspect.**

> A. Find and record the unit model number on the provided Refrigerant Characteristics Chart for the first refrigeration system.

> B. Find and record the refrigerant type on the provided Refrigerant Characteristics Chart for the first refrigeration system.

> C. Find and record the high side test pressure the provided Refrigerant Characteristics Chart for the first refrigeration system.

D. Find and record the low side test pressure on the provided Refrigerant Characteristics Chart for the first refrigeration system.

E. Locate a cylinder of the same refrigerant type and read the instructions on the cylinder.

F. Locate the MSDS for this refrigerant type and read the information regarding the refrigerant properties.

G. Use the information from the refrigerant cylinder, MSDS and Fundamentals of HVACR text Unit 19 to complete filling out the provided Refrigerant Characteristic Chart for the first refrigeration system.

Step 2

Repeat the process from Step 1 for the other three refrigeration systems completing the provided **Refrigerant Characteristic** Chart for each one. When finished have your Instructor check the information that you recorded and then answer the questions at the end of this laboratory exercise.

<u>Refrigerant Characteristics Chart</u>

	SYSTEM #1	*SYSTEM #2*	*SYSTEM #3*	*SYSTEM #4*
Model Number				
Refrigerant Name				
Refrigerant Quantity				
High Side Test Pressure				
Low Side Test Pressure				
Pressure (Very High, High, Low)				
Ozone Depletion Potential				

Global Warming Potential				
Safety Rating				
Chemical Composition				
Formulation				

QUESTIONS

13. Were there any refrigerants that did not have either an ozone depletion potential or a global warming potential?

14. What are the high and low side test pressures used for?

15. Which refrigerants may safely leave the cylinder as either a gas or a liquid?

16. Which refrigerants must leave the cylinder as a liquid only?

17. Which refrigerant had the highest system test pressures?

18. Which refrigerant had the lowest system test pressures?

19. What hazards do you need to be aware of when working with these refrigerants?

20. Which of these refrigerants will NOT be used in equipment of the future?

21. Which of these refrigerants WILL be used in equipment of the future?

LAB 19.2 IDENTIFYING REFRIGERANT LUBRICANT CHARACTERISTICS

LABORATORY OBJECTIVE
The purpose of this lab is to demonstrate your ability to identify three different types of refrigerant lubricant and to recommend the proper lubricant for four refrigeration systems.

LABORATORY NOTES
You will examine three different types of refrigeration lubricant and recommend where they can be used. You will recommend the correct type of lubricant for four refrigeration systems.

FUNDAMENTALS OF HVACR TEXT REFERENCE
Unit 19 Refrigerants and Their Properties

REQUIRED TOOLS AND EQUIPMENT
Four refrigeration systems containing different refrigerants
Three containers of different refrigeration lubricant

SAFETY REQUIREMENTS
Caution!!! Do not open the lubricant containers and spill lubricant or allow it to contact your skin.

PROCEDURE
Step 1
Locate the first container of refrigerant lubricant.

    A. Identify the type of lubricant in the first container.
    B. Record in the Lubricant Type Chart provided:
       • Type of lubricant
       • Lubricant viscosity
       • Recommended evaporator temperature range
       • Type of refrigerants that the lubricant is compatible
    C. Repeat this process for the other lubricant types.

Lubricant Type Chart

	BRAND NAME	TYPE (MINERAL, PAG, POE, AB)	VISCOSITY	EVAPORATOR TEMPERATURE	REFRIGERANT COMPATABILITY
Container #1					
Container #2					
Container #3					

Step 2

Locate the nameplate on the first refrigeration unit. The nameplate will be **in different locations for each unit you inspect.**

      A. Find and record the unit model number on the provided System Refrigerant Lubricant Chart for the first refrigeration system.
      B. Find and record the refrigerant type on the provided System Refrigerant Lubricant Chart for the first refrigeration system.
      C. Recommend a lubricant that is compatible with this unit.

Step 3

Repeat the process from Step 2 for the other three refrigeration systems completing the provided System **Refrigerant Lubricant Characteristic** Chart for each one. When finished have your Instructor check the information that you recorded and then answer the questions at the end of this laboratory exercise.

System Refrigerant Lubricant Characteristics Chart

	SYSTEM #1	SYSTEM #2	SYSTEM #3	SYSTEM #4
Model Number				
Refrigerant Name				
Recommended Lubricant				

QUESTIONS

22. Which lubricant can be used with the widest range of systems?

23. What is the difference in viscosity between lubricants designed for higher evaporator temperatures and lubricants that are designed for use in lower temperature systems?

24. What is the most common viscosity for lubricants used in air conditioning systems?

LAB 20.1 SOLENOID VALVES

LABORATORY OBJECTIVE
The student will disassemble a solenoid valve and be able describe how it operates.

LABORATORY NOTES
For this lab exercise there should be one or more solenoid valves available that may be disassembled.

Solenoid valves can be used for many applications in a refrigeration system. They can be used to regulate flow to the evaporator and as king valves. They can also be used for the defrost cycle as well as other refrigeration applications.

FUNDAMENTALS OF HVACR TEXT REFERENCE
Unit 20 Special Refrigeration Components

REQUIRED TOOLS AND EQUIPMENT
Solenoid valves
Multimeter

SAFETY REQUIREMENTS
None.

PROCEDURE
Step 1
Locate a solenoid valve and examine it carefully so that you may complete the following exercise:

    A. Record all of the data that can be found on the valve such as volts, cycles, wattage, manufacturer, etc.

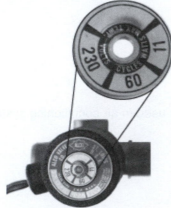

Figure 20-1-1

Solenoid Valve Data:

    B. Disassemble the valve being careful not to lose components while identifying the manner in which the valve came apart so that it may then be put back together again.

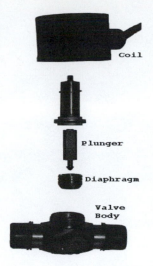

Figure 20-1-2

C. Sketch a cross sectional view of the disassembled valve in the space provided below.

D. Explain how the valve operates using the sketch you provided above.

Step 2
Measure the coil resistance as follows:

A. The coil removed from the solenoid valve will have an identification number for spare parts purposes. Notice the part identification number on the coil shown in the illustration below.

147

Figure 20-1-3

B. Using the coil part number and the manufacturer's data, you should be able to determine the normal resistance of the coil.

C. Familiarize yourself with a multimeter and zero it in

(Refer to Fundamentals of HVACR text Unit 35 *Electrical Measuring & Test Instruments*).

D. Connect the one of leads of the multimeter to each wire attached to the coil.

E. Start with the resistance reading on the highest scale.

F. Record the resistance reading _____Ohms.

G. Compare this reading to the manufacturer's data to see if the coil is satisfactory.

Step 3
Re-assemble the solenoid valve, making sure that <u>none of the parts are lost and each is in its proper order.</u>

QUESTIONS

1. A common failure with solenoid valves is that the coil eventually burns out. Regarding the valve you disassembled, if the coil burned out, would the valve fail open or closed?

2. If the measured resistance of the coil is infinite, what would this indicate?

3. If the measured resistance of the coil is zero, what would this indicate?

(Circle the letter that indicates the correct answer.)

4. If the coil on the solenoid supplying refrigerant to an evaporator burned out:
   A. The refrigerated space would warm up.
   B. The refrigerated space would cool down.
   C. The compressor would shut down on high pressure.
   D. The compressor motor would overload.

5. If a solenoid valve is energized.
   A. it will always be closed.
   B. then the oil must be burned out.
   C. it may feel warm to the touch.
   D. then the circuit is faulty.

6. If the measured resistance of a solenoid coil is infinite:
   A. then the coil is satisfactory.
   B. the coil may have a short.
   C. the coil may have an open.
   D. the coil will run hot.

7. If the measured resistance of a solenoid coil is infinite:
   A. then the coil is satisfactory.
   B. the coil may have a short.
   C. the coil may have an open.
   D. the coil will run hot.

8. Many refrigerant solenoid valves often fail in the:
   A. neutral position.
   B. reverse flow position.
   C. open position.
   D. closed position.

9. When the coil of a solenoid valve is energized:
   A. it will spin.
   B. it will act as a magnet.
   C. its polarity will be reversed.
   D. it will slowly rotate.

10. An energized solenoid valve:
   A. will act as a magnet.
   B. may make a slight humming sound.
   C. may feel warm to the touch.
   D. All of the above are correct.

11. Many solenoid valves have a pilot orifice to assist in valve lift:
   A. to speed up the action of the valve.
   B. to reverse the flow through the valve.
   C. for manual operation if the coil burns out.
   D. to allow for smaller size coils to be used.

12. Before removing a coil from a solenoid valve:
   A. make sure that you have a spare.
   B. twist the ends of the wires together.
   C. tap it gently with a wrench.
   D. de-energize the circuit.

LAB 20.2 SETTING AN EVAPORATOR PRESSURE REGULATOR

LABORATORY OBJECTIVE
The student will demonstrate how to properly set an evaporator pressure regulator that is controlling the temperature of an evaporator coil for a refrigerated space.

LABORATORY NOTES
For this lab exercise there needs to be an operating refrigeration system that has an evaporator pressure regulator on the outlet of an evaporator for the space being cooled. If there are no thermometers or pressure gauges currently on the unit then you must instrument the refrigeration system using a gauge manifold to measure the evaporator pressure and a thermometer to read the evaporator refrigerant temperature.

If you are unable to place probes directly into the refrigerant stream flow, you can use evaporator coil surface temperatures, however you must adjust for the heat transfer difference through the coil.

Multiple evaporator systems are commonly found in supermarkets. A single compressor may be used to control a number of different case or fixture temperatures.

Without an evaporator pressure regulator (EPR) all of these spaces would have a common evaporator pressure and temperature. The refrigerant would need to be cold enough for the low temperature boxes but this would make it undesirable for the warmer boxes.

The refrigerant in the evaporator coil should be approximately 15°F lower than the space being cooled. If the refrigerant temperature is excessively low, then this will tend to rob the food of its moisture and dry it out. This is particularly applicable to fruits and vegetables. An EPR will elevate the evaporator pressure and thus the refrigerant temperature to bring it on more in line with the box temperature.

***FUNDAMENTALS OF HVACR* TEXT REFERENCE**
Unit 20 Special Refrigeration Components

REQUIRED TOOLS AND EQUIPMENT
Operating refrigeration unit with evaporator pressure regulator
Gauge manifold and temperature sensor

SAFETY REQUIREMENTS
A. Always read the equipment manual to become familiarized with the refrigeration system and its accessory components prior to start up.
B. Wear safety goggles and gloves when working with refrigerants. Liquid refrigerant can cause frostbite when in contact with eyes and skin.
C. Use low loss hose fittings, or wrap cloth around hose fittings before removing the fittings from a pressurized system or cylinder. Inspect all fittings before attaching hoses.

PROCEDURE
Step 1
Familiarize yourself with the major components in the refrigeration system including the condenser, compressor, evaporator, and metering device.

> **A.** Evaporator pressure regulators (EPRs) are placed at the outlets of the suction lines for the warmer temperature evaporators. These are adjusted to maintain the desired evaporator pressure and thereby controlling evaporator temperature.

B. A check valve is installed at the outlet of the suction line for the coldest evaporator coil. This prevents migration of the refrigerant from the higher temperature coils to the low temperature coil.

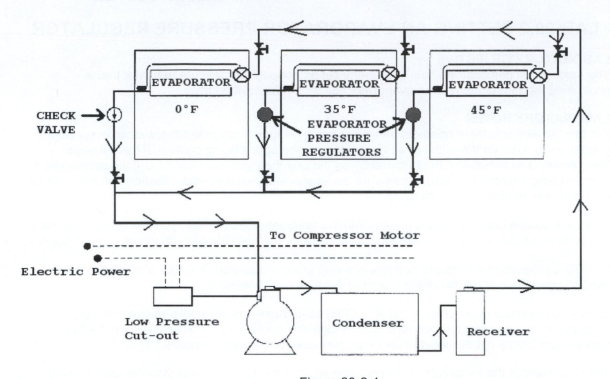

Figure 20-2-1

Step 2
Determine the proper setting for the EPR as follows:

> **A.** You should be able to determine the proper evaporator pressure based upon the desired space temperature. As an Example, use the 45°F controlled space temperature in Figure 20-2-1. The approximate refrigerant temperature should be 45°F - 15°F = 30°F.
>
> **B.** You should also consult any operating manual parameters in regard to the expected temperature difference between the space and the evaporator as recommended by the Manufacturer. We will be using a 15°F temperature differential for this Example which is most common, however different types of systems may require other settings.
>
> **C.** Assuming we are using R-22 then from the P-T chart, Table 47-1-1, depicted in Lab 47.1 *Basic Refrigeration System Startup* and from Figure 10-12 in Unit 10 of the *Fundamentals of HVACR* text, we can look up the saturated refrigerant temperature.
>
> Inside the evaporator coil the condition of the refrigerant will always be saturated as long as there is liquid and vapor present together.
>
> From the chart at 30°F, for R-22 the EPR set pressure would be approximately 55 psig. For another type of refrigerant such as R-134a the expected evaporator pressure would be somewhat lower at approximately 26 psig.

Step 3

> **A.** If there are no pressure gauges currently on the unit then you must connect a gauge manifold to read the evaporator pressure.

B. If you are unable to place temperature probes directly into the refrigerant stream flow, you can use evaporator coil surface temperatures, however you must adjust for the heat transfer difference through the coil.

Step 4

A. Start the refrigeration system in the normal cooling mode and allow the system to stabilize and then record the evaporator pressure and temperature below.

_____ psig _____ °F

B. From your readings, determine what adjustments need to be made if any and set the evaporator pressure regulator by turning the adjusting screw with a screwdriver or Allen wrench dependent on EPR type.

C. Allow the system to run and stabilize prior to shutting down and then carefully disconnect the gauge manifold and any instrumentation that you attached to the unit.

QUESTIONS

(Circle the letter that indicates the correct answer.)

1. The temperature in an evaporator coil is directly related to:
   A. the temperature of the space to be cooled.
   B. the temperature of the condenser.
   C. the pressure in the evaporator.
   D. the compressor pressure.

2. If an evaporator pressure regulator fails shut.
   A. the space will freeze up.
   B. the system will still operate as normal.
   C. the compressor will run continuously.
   D. the space will warm up.

3. An evaporator pressure regulator:
   A. should be used on the warmer evaporators.
   B. should be used on the colder evaporators.
   C. is a safety device.
   D. unloads the compressor at high space temperatures.

4. On a multiple box system, a check valve is usually installed at the outlet of the:
   A. coldest evaporator.
   B. warmest evaporator.
   C. compressor.
   D. condenser.

LAB 20.3 IDENTIFYING ACCESSORIES

LABORATORY OBJECTIVE
You will correctly identify refrigeration accessories on refrigeration systems and explain their function.

LABORATORY NOTES
You will locate the refrigeration accessories on systems assigned by the instructor. You will then point them out to the instructor and describe their function.

FUNDAMENTALS OF HVACR TEXT REFERENCE
Unit 20 Refrigeration Accessories

REQUIRED MATERIALS PROVIDED BY STUDENT
Safety glasses
Gloves

REQUIRED MATERIALS PROVIDED BY SCHOOL
Systems with refrigeration accessories

PROCEDURE

Locate the following accessories on systems in the shop.

LAB 20.3 IDENTIFYING ACCESSORIES	
Refrigeration Accessory	
Liquid Line Filter Drier	
Suction Line Filter Drier	
Muffler	
Sight Glass	
Moisture Indicator	
Receiver	
Accumulator	
Crankcase Heater	
Vibration Eliminator	
Solenoid Valve	
Heat Exchanger	
Crankcase Pressure Regulator	

LAB 20.4 INSTALLING ACCESSORIES

LABORATORY OBJECTIVE
You will demonstrate your ability to install a filter drier on a refrigeration system.

LABORATORY NOTES
You will front-seat the king valve or liquid line service valve, operate the system, and pump down the refrigerant into the high side of the system. You will then install a liquid line filter-drier in the liquid line. After installing the drier, yow will leak test the joints with nitrogen, evacuate the lines and coil, and open the king valve to let the system refrigerant into the rest of the system.

FUNDAMENTALS OF HVACR **TEXT REFERENCE**
Unit 20 Refrigeration Accessories

REQUIRED MATERIALS PROVIDED BY STUDENT
Safety glasses
Gloves
Two adjustable jaw wrenches
6-in-1
Refrigeration gauges
Refrigeration valve wrench

REQUIRED MATERIALS PROVIDED BY SCHOOL
Refrigeration system with king valve or liquid line service valve
Liquid line filter drier
Vacuum pump

PROCEDURE
1. Follow the procedures given in Lab 28.9 to pump down the system.

2. Use two wrenches to loosen the flare nuts on the existing filter-drier.

3. One wrench holds the filter to keep it from turning while the other turns the flare nut.

4. Remove the existing filter.

5. Remove the protective caps from the new filter.

Note: Do NOT remove the caps until just before installing the filter. If the caps are removed ahead of time and the filter is allowed to sit open in the air, it will absorb moisture from the air, reducing its ability to absorb moisture from the system.

6. Install the new filter with the arrow pointing towards the metering device.

7. Tighten the flare nuts using two wrenches.

8. Leak test with nitrogen and soap bubbles.

9. Release nitrogen and evacuate the lines and coil.

10. Open the king valve or liquid line valve to let refrigerant into the system.

11. Operate the system and check system pressures

LAB 21.1 DESCRIBE THE REFRIGERATION SYSTEM CHARACTERISTICS USING A REFRIGERATION TRAINER

LABORATORY OBJECTIVE: The purpose of this lab is to demonstrate your ability to discuss the function of the four main refrigeration system components and describe the refrigerant characteristics throughout the system on the refrigeration trainer.

LABORATORY NOTES: You will be assigned a refrigeration trainer. You should be prepared to show the instructor each of the four main refrigeration components in order and discuss their function in the system. You will also discuss the characteristics of the refrigerant as it travels through the system using pressure and temperature measurements to help you determine the condition of the refrigerant throughout the trainer.

FUNDAMENTALS OF HVACR TEXT REFERENCE
Unit 21 Plotting the Refrigeration Cycle

REQUIRED TOOLS AND EQUIPMENT
Instrumented refrigerant trainer
Pressure temperature chart
Temperature sensor with temperature clamp and thermocouple clamp

SAFETY REQUIREMENTS
None

PROCEDURE
Step 1
Identify the components on the refrigeration trainer.

Locate the compressor, condenser, metering device, and evaporator.

Draw a sketch of the trainer.

Label the components.

Draw arrows to indicate the direction of refrigerant flow.

Identify high pressure and low pressure circuits.

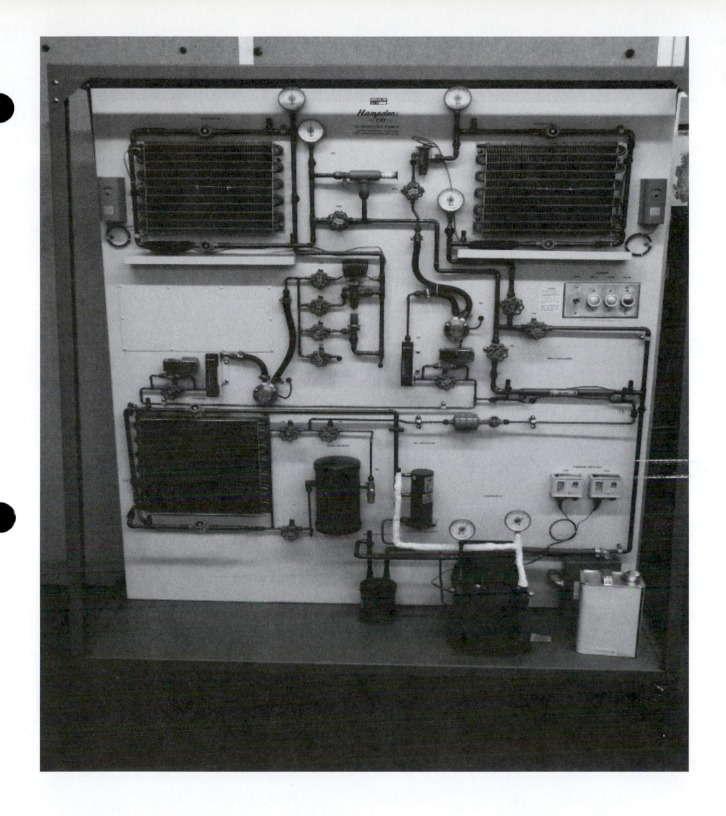

Figure 21-1-1

DRAW SKETCH OF REFRIGERATION TRAINER IN THE SPACE PROVIDED

INSTRUCTOR CHECK BEFORE PROCEEDING TO NEXT STEP _____

DRAW SKETCH OF REFRIGERATION TRAINER IN THE SPACE PROVIDED

Step 2

Line up the components and then start the refrigeration trainer using the trainer's instruction manual if necessary. If you are not sure how to start the trainer, then check with your Instructor.

Step 3

Run the trainer and observe its operation. Note the pressure gauges that tell you what the refrigerant pressure is at every point in the system.

A. If there are no installed temperature sensors then you can measure the temperature at each location using a portable temperature sensor.

B. You may also touch components to get a feel for their temperature.

CAUTION: Some components are HOT! Proceed carefully!

C. If you hold your hand just above a warm component you can tell if it is extremely hot or not. If you can feel radiant heat coming from it, it may be too hot to touch safely.

Step 4

Use the provided chart to record and organize your data. However, the job is not complete until you TELL the instructor how the system works and respond to questions.

A. Record the pressure and temperature at each system location indicated on the chart.

B. Compare the pressure and temperature using a pressure-temperature chart as found in Appendix A of the HVACR text, to determine the condition. These are saturated, superheated, or subcooled.

C. Write in the state: gas, liquid, or mixed.

D. Use the pressure, temperature, condition, and state to look up the heat content on an enthalpy chart such as found in Unit 21 of the Fundamentals of HVACR text.

LOCATION	TEMPERATURE	PRESSURE	CONDITION	STATE	HEAT CONTENT
Compressor in					
Compressor out					
Discharge line					
Condenser in					
Condenser center					

Condenser out					●
Liquid line					
Metering device in					
Metering device out					
Evaporator in					
Evaporator center					●
Evaporator OUT					
Suction line					

Step 5

Use the following terms to describe the changes in the refrigerant through each of the four major components and fill in the chart provided.

Stable – to describe no significant change in a condition

Small increase – to describe an increase of 30% or less

Large increase – to describe an increase of 200% or more

Small decrease – to describe a decrease of 30% or less

Large decrease – to describe a decrease of 200% or more

160

TYPE OF CHANGE	COMPRESSOR	CONDENSER	METERING DEVICE	EVAPORATOR
Change in pressure				
Change in temperature				
Change in heat				
Change in State				
Change in volume				

161

LAB 22.1 DISPOSABLE REFRIGERANT CYLINDERS

LABORATORY OBJECTIVE
The student will demonstrate an understanding of disposable refrigerant cylinder types and handling procedures.

LABORATORY NOTES
For this lab exercise there should be a number of different disposable refrigerant cylinders for the students to inspect. Empty cylinders are acceptable as the student will not be removing any refrigerant from the cylinder. It would be preferable to also have cylinders of different sizes.

***FUNDAMENTALS OF HVACR* TEXT REFERENCE**
Unit 22 Refrigeration Safety

REQUIRED TOOLS AND EQUIPMENT
Disposable refrigerant cylinders of different sizes and types

SAFETY REQUIREMENTS
Caution!!! Do not open the valves on the refrigeration cylinders to avoid possible injury due to skin contact with refrigerant. Also it is illegal to intentionally vent refrigerants comprised of CFC's and HCFCs.

PROCEDURE
Step 1
In 1990 the Congress of the United States passed a series of amendments to the Clean Air Act that greatly affected the refrigeration and air conditioning industry. The Act establishes a set of standards and requirements for the use and disposal of certain common refrigerants containing chlorine.

Review the Information on the Clean Air Act from the *Fundamentals of HVACR* text Unit 28 and answer the following:

    A. Can you knowingly release CFC's, HCFCs, or HFC's while repairing appliances?

    B. When did it become illegal to vent CFC's and HCFCs?

    C. Do you need to be certified to service, maintain, or dispose of appliances containing refrigerants? If yes, when did this become mandatory?

    D. How much can you or your company be fined per day for violating Section 608 of the Federal Clean Air Act?

    E. How much is the bounty for turning someone in?

Step 2

Locate a <u>disposable</u> refrigerant cylinder and examine it carefully so that you may complete the following exercise.

●

A. The refrigerant type, chemical designation of the refrigerant, cylinder color code, refrigerant boiling temperature,
normal discharge (head) pressure, normal suction pressure, and latent heat value.

B. Refrigerant toxicity, flammability, corrosive tendency, and rated refrigerant weight when cylinder is full.

C. What is the maximum temperature that this cylinder can be exposed to?

D. How often must this cylinder be checked (DOT regulation)?

●

E. Can you remove both liquid and vapor refrigerant from this cylinder? If yes, then describe how you would remove vapor and how you would remove liquid refrigerant.

F. Can you re-use this cylinder in an emergency?

G. What type of protection against bursting due to excessive pressure does the cylinder have?

H. Draw a rough sketch of the protection device in the space provided below.

●

QUESTIONS

(Circle the letter that indicates the correct answer.)

1. A disposable cylinder:
- A. can be refilled <u>only</u> with the proper type of refrigerant.
- B. can never be refilled.
- C. is always color coded yellow and grey.
- D. Both B & C are correct.

2. Cylinders over 4.5 inches in diameter and over 12 inches long:
- A. will always be disposable cylinder.
- B. will always be recovery cylinders.
- C. can not be used for refrigerant purposes.
- D. must have some type of pressure relief device.

3. When transporting a refrigerant cylinder:
- A. it should be properly secured in an upright position.
- B. it should be properly secured laying down.
- C. it should be properly secured in an inverted position.
- D. All of the above are correct.

4. Refrigerant cylinder temperatures should not exceed:
- A. 92.5 °F.
- B. 99.8 °F.
- C. minus 25 °F.
- D. 125 °F.

5. The cardboard boxes that contain new refrigerant cylinders:
- A. can be used to prop the bottle upright when charging.
- B. may contain important safety information.
- C. may need to be kept on the job site.
- D. B & C are both correct.

6. To charge a liquid using a disposable cylinder:
- A. it must be heated.
- B. it must be cooled.
- C. it must be shaken.
- D. it must be inverted.

7. The burst disk on a disposable refrigerant cylinder.
- A. will always reseat.
- B. will completely drain the cylinder.
- C. will only drain the liquid refrigerant.
- D. acts like a relief valve.

LAB 22.2 REFILLABLE REFRIGERANT RECOVERY CYLINDERS

LABORATORY OBJECTIVE
The student will demonstrate an understanding of refillable refrigerant recovery cylinder types and handling procedures.

LABORATORY NOTES
For this lab exercise there should be a refillable refrigerant recovery cylinder for the students to inspect. Empty cylinders are acceptable as the student will not be removing any refrigerant from the cylinder.

FUNDAMENTALS OF HVACR TEXT REFERENCE
Unit 22 Refrigeration Safety

REQUIRED TOOLS AND EQUIPMENT:
Refillable refrigerant recovery cylinder

SAFETY REQUIREMENTS:
Caution!!! Do not open the valves on the refrigeration cylinders to avoid possible injury due to skin contact with refrigerant. Also it is illegal to intentionally vent refrigerants comprised of CFC's and HCFCs.

PROCEDURE
Step 1
Locate a refillable refrigerant recovery cylinder and examine it carefully so that you may complete the following exercise.

 A. Sketch a cross sectional view of the cylinder including fittings and valves in the space provided below.

B. Explain how can you remove either liquid or vapor from this cylinder.

C. What is the cylinder weight empty (T.W. - Tare Weight)?

D. What is the maximum refrigerant weight that the cylinder can hold?

E. What is the weight of a full cylinder including the metal?

F. What type of protection against bursting due to excessive pressure does the cylinder have?

G. How often must this cylinder be inspected?

H. What type of refrigerant can be stored in this cylinder?

QUESTIONS

(Circle the letter that indicates the correct answer.)

1. A refillable refrigerant recovery cylinder:
   A. can be refilled <u>only</u> with the proper type of refrigerant.
   B. can never be refilled.
   C. is always color coded yellow and grey.
   D. Both A & C are correct.

2. A refillable refrigerant recovery cylinder must meet DOT approval:
   A. True.
   B. False.

LAB 22.3 REFRIGERANT CYLINDERS

LABORATORY OBJECTIVE

You will examine two non-refillable refrigerant cylinders and a refrigerant recovery cylinder and identify the important information found on the cylinders.

LABORATORY NOTES

You will examine two non-refillable refrigerant cylinders and identify the type of refrigerant, the amount of refrigerant the cylinder holds when new, the position of the cylinder when removing refrigerant from it, and the potential fine for refilling it. You will then examine a ref refrigerant recovery cylinder and identify important data found on the cylinder; including, tare weight, water capacity, service pressure, the type of refrigerant in the cylinder, the amount of refrigerant currently in the cylinder, and the maximum amount of refrigerant the cylinder can hold.

***FUNDAMENTALS OF HVACR* TEXT REFERENCE**

Unit 22 Refrigeration Safety
Unit 28 Refrigerant Management

REQUIRED MATERIALS PROVIDED BY STUDENT

Safety Glasses

REQUIRED MATERIALS PROVIDED BY SCHOOL

One R-22 non-refillable refrigerant cylinder
One R-410A non-refillable refrigerant Cylinder
Refrigerant Recovery Cylinder
Digital Scale

PROCEDURE

Non-refillable Cylinders

Examine the two non-refillable cylinders and complete the data sheet

Non-Refillable Refrigerant Cylinders		
	Cylinder 1	Cylinder 2
Cylinder Color		
Type of refrigerant in Cylinder		
Amount of Refrigerant in new cylinder		
Position of cylinder during charging		
Fine for refilling and transporting		

Refrigerant Recovery Cylinder

Examine a recovery cylinder and complete the data sheet.

Refrigerant Recovery Cylinder	
Cylinder Color	
Cylinder Tare weight (TW)	
Cylinder Water capacity (WC)	
Cylinder maximum service pressure	
At what pressure does the pressure relief valve open?	
Date when this cylinder must be re-certified	
Type of refrigerant in Cylinder	
What color is the handle on the vapor valve?	
What color is the handle on the liquid valve?	
Type of Refrigerant in Cylinder	
Maximum weight of refrigerant cylinder can safely hold	
Maximum safe gross weight (tare weight + maximum safe fill)	
Weigh the cylinder to determine the current Gross cylinder weight	
Weight of refrigerant in cylinder (Gross weight – tare weight)	

Show safe fill level calculations below using formula

Safe Fill = Water Capacity x 0.8 x Refrigerant Specific Gravity

LAB 23.1 REFRIGERANT GAUGE MANIFOLDS

LABORATORY OBJECTIVE
The student will demonstrate how to properly use a refrigerant gauge manifold.

LABORATORY NOTES
For this lab exercise there should be a typical gauge manifold available for students to inspect.

FUNDAMENTALS OF HVACR TEXT REFERENCE
Unit 23 Refrigerant System Servicing and Testing Equipment

REQUIRED TOOLS AND EQUIPMENT
Gauge Manifold

SAFETY REQUIREMENTS:
None

PROCEDURE
Step 1
Locate a refrigerant gauge manifold and examine it carefully so that you may complete the following exercise.

    A. Identify the two gauges that are on the manifold including the minimum and maximum readings.

    B. What is meant by compound gauge?

    C. What is the range of the pressure scale on the compound gauge?

    D. List three operations that can be performed with a gauge manifold.

    E. What do the colored scales in the center of the gauge indicate?

F. Sketch a cross sectional view of a gauge manifold, labeling all parts and connections in the space provided below.

QUESTIONS

(Circle the letter that indicates the correct answer.)

1. When not in use:
   A. gauge manifolds should be stored in a plastic wrap cloth.
   B. gauge manifolds should be press charged with an inert gas.
   C. the ports and charging lines should be capped.
   D. Both A & C are correct.

2. Gauge manifolds:
   A. have a small adjustment crew that allows the gauge to be calibrated.
   B. can never be calibrated.
   C. are oiled every week.
   D. Both A & C are correct.

3. Gauge manifolds should only be used with:
   A. refrigerant.
   B. clean oil.
   C. Both A & B are correct.
   D. None of the above is correct.

4. A compound gauge measure both pressure and vacuum:
   A. True.
   B. False.

LAB 23.2 REFRIGERATION EQUIPMENT FAMILIARIZATION

LABORATORY OBJECTIVE
You will demonstrate how to use a vacuum pump, vacuum gauge, digital scale, and recovery unit.

LABORATORY NOTES
You will use a vacuum pump to pull a vacuum on your gauges and check the vacuum with a vacuum gauge.
You will weigh a set of screws with a digital scale, zero the scale, and then weigh the screws removed.
You will use a recovery unit to transfer refrigerant from one recovery cylinder to another.

FUNDAMENTALS OF HVACR TEXT REFERENCE
Unit 23 Refrigerant System Servicing and Testing Equipment / Unit 28 Refrigerant Management / Unit 30 Evacuation / Unit 31 Charging

REQUIRED MATERIALS PROVIDED BY STUDENT
Safety Glasses
Refrigeration Gauges

REQUIRED MATERIALS PROVIDED BY SCHOOL
Vacuum Pump
Vacuum Gauge
Refrigerant Recovery Unit
2 Refrigerant Recovery Cylinders with the same refrigerant.
Digital Scale
Assortment of small weights (can be a bucket of screws)
Extra refrigeration hoses
Infrared thermometer

PROCEDURE
1. Tighten the low side hose (blue) and the high side hose (red) on the hanger so they will not leak.
2. Connect the middle hose to the vacuum gauge.
3. Connect another hose from the vacuum gauge to the vacuum pump.
4. *(Depending upon the vacuum gauge used, a separate tee and hose may be required)*
5. Close the manifold gauge handwheels by turning both of them clockwise.
6. Open the ballast valve and isolation valves on the vacuum pump.
7. Turn on the vacuum gauge.
8. Start the vacuum pump. It will gurgle loudly at first, then start to get quieter.
9. Close the ballast valve on the vacuum pump.
10. The vacuum gauge reading should drop to less than 100 microns.
11. Open the manifold gauge handwheels.
12. The pump will get louder again for a minute or so and then quiet down.
13. The vacuum gauge reading will go up and then fall back down.
14. Close the vacuum pump isolation valve and turn off the vacuum pump and vacuum gauge.
(Always close the isolation valve BEFORE turning off the vacuum pump)
15. Turn on the scale and zero it by pushing the "zero" button.

16. Place the bucket of screws on the scale.
17. Push the "Units" button to see the weight displayed in pounds, pounds and ounces, and kilograms.
18. Zero the scale with the screws still on it.
19. Remove some of the screws – the weight of the screws you removed should be displayed on the scale.
20. Connect the vapor valve of one of the recovery cylinders to the inlet of the recovery unit.
21. Connect the outlet of the recovery unit to the vapor valve of the other recovery cylinder.
22. Purge the hoses and the recovery unit.
   (If the recovery unit was last used on the same type of refrigerant it should not need purging.)
23. Place the recovery cylinder connected to the recovery unit outlet valve on the scale.
24. Zero the scale.
25. Open the valves on the recovery cylinders.
26. Set the recovery unit valves to recover vapor.
27. Turn on the recovery unit.
28. Check the cylinder temperatures periodically as the unit operates.
29. The cylinder connected to the recovery unit inlet should drop in temperature while the cylinder connected to the recovery unit outlet should increase in temperature.
30. The scale should show that the cylinder on the scale is increasing in weight.
31. Turn off the recovery unit and close the recovery unit valves.
32. Turn the cylinder connected to the recovery unit inlet valve upside down.
33. Position the recovery unit valves to equalize the pressures in the recovery unit.
34. *(This can be done on many units by setting the valves to the purge position)*
35. Set the valves on the recovery unit to recover liquid.
36. Open the valves on the recovery unit.
37. Start the recovery unit.
38. The scale should indicate that the cylinder weight is increasing faster.
39. Close the valve on the cylinder connected to the recovery unit inlet valve.
40. The recovery unit suction gauge should drop into a vacuum and the recovery unit should shut off.
   (Not all recovery units have auto shutoff.)
41. Turn off the recovery unit.
42. Set the valves on the recovery unit to purge.
43. Start the recovery unit an operate until it has cleared itself.
44. Close the valves on the recovery unit, and turn it off.
45. Make sure the valves are closed on both cylinders before disconnecting the hoses.

LAB 24.1 IDENTIFYING FITTINGS

LABORATORY OBJECTIVE

You will identify the following types of fittings and their sizes:

- pipe fittings
- sweat fittings
- flare fittings
- PVC fittings

LABORATORY NOTES

You will receive an assortment of fittings. You will identify each fitting and list its name and size.

***FUNDAMENTALS OF HVACR* TEXT REFERENCE**

Unit 24 Piping and Tubing

REQUIRED MATERIALS PROVIDED BY STUDENT

Safety Glasses
Ruler

REQUIRED MATERIALS PROVIDED BY SCHOOL

Assorted fittings

PROCEDURE

Give the name and size of each fitting in the space provided. Line up the fittings and keep them out so that you may show ALL fittings with proper name and size to the instructor.

#	FITTING NAME	SIZE	#	FITTING NAME	SIZE
1			21		
2			22		
3			23		
4			24		
5			25		
6			26		
7			27		
8			28		
9			29		
10			30		
11			31		
12			32		
13			33		
14			34		
15			35		
16			36		
17			37		
18			38		
19			39		
20			40		

LAB 24.2 FLARING

LABORATORY OBJECTIVE

You will construct a leak-proof 2-foot long assembly using $\frac{1}{4}$ inch, $\frac{3}{8}$ inch, $\frac{1}{2}$ inch, and $\frac{5}{8}$ inch copper tubing with flare connections.

LABORATORY NOTES

You will build a flared assembly that contains four sizes of tubing connected by flare fittings. With the instructor's help, you will then leak test the assembly using nitrogen and soap bubbles.

FUNDAMENTALS OF HVACR TEXT REFERENCE

Unit 24 Piping and Tubing

REQUIRED MATERIALS PROVIDED BY STUDENT

Safety Glasses
Ruler
Tubing Cutter
Flaring Tool
Large adjustable wrench

REQUIRED MATERIALS PROVIDED BY SCHOOL

Copper Tubing
Flare Fittings
Fixed wrenches

PROCEDURE

1. Obtain proper tools and materials from the tool room needed to make a 2-foot long assembly using $\frac{1}{4}$ inch, $\frac{3}{8}$ inch, $\frac{1}{2}$ inch, and $\frac{5}{8}$ inch copper tubing with flare connections. The $\frac{5}{8}$ inch end will be sealed with a plug and the $\frac{1}{4}$ inch end will be attached to a refrigerant drum for leak testing.
2. Fabricate the assembly.
3. Seal the $\frac{5}{8}$ inch end with a plug and connect the $\frac{1}{4}$ inch end to the regulator on a nitrogen cylinder.
4. Instructor will help you test for leaks in the assembly using a soap solution.

SAFETY NOTE
When pressure testing the assembly, make sure it is not pointed in another person's direction or towards you. Crack the valve slowly to avoid a sudden pressure surge on your assembly wear safety glasses since your assembly will be under pressure.

LAB 24.3 SWAGING

LABORATORY OBJECTIVE

You will make a total of 60 swage joints, 15 of each of $\frac{1}{4}$ inch, $\frac{3}{8}$ inch, $\frac{1}{2}$ inch, and $\frac{5}{8}$ inch copper tubing.

LABORATORY NOTES

You will make swage joints to be used for the soldering and brazing labs.

FUNDAMENTALS OF HVACR TEXT REFERENCE

Unit 24 Piping and Tubing

REQUIRED MATERIALS PROVIDED BY STUDENT

Safety Glasses
Ruler
Tubing Cutter
Flaring Tool
Swaging Tool

REQUIRED MATERIALS PROVIDED BY SCHOOL

Copper Tubing
Hammer

PROCEDURE

Cut 15 4" pieces of each size copper tubing: $\frac{1}{4}$ inch, $\frac{3}{8}$ inch, $\frac{1}{2}$ inch, and $\frac{5}{8}$ inch. Hold the loose

end of the roll on the bench and roll out the tubing on the bench to form a long, straight section. It is far easier to straighten the tubing in one long section. If you cut the tubing while it is still rolled you will not be able to straighten the short 4" sections. The swage joint should be as long as the diameter of the tubing. The tubing should fit snugly into the swage joint. Loose joints will make soldering a painful experience. Time spent making good joints will save time in the long run. Save these joints because you will use them in the soldering section coming up in Lab 25.4 and the brazing in Lab 25.2.

LAB 25.1 SOLDERING COPPER PIPE

LABORATORY OBJECTIVE
The student will demonstrate how to properly solder a copper pipe connection.

LABORATORY NOTES
For this lab exercise a fitting will be soldered to copper piping. The fitting will then be tested to determine if it is satisfactory.

Soldering takes place below 840 °F and is used for water or condensate drains. Solder is not approved for refrigerant line joints.

Soldering to join copper pipes uses an air acetylene torch, air Mapp torch, or air propane torch. Oxyacetylene is not recommended for soldering due to high flame temperatures.

TABLE 25-1-1 Common Soldering and Brazing Metal and Fluxes Showing Base Metals That Can Be Joined

	Alloy	Flux Type	Base Metal
S O L D E R	95-5 Tin antimony solder[1]	C-Flux	Copper pipe, brass, steel
	95-5 Tin antimony solder	Rosin	Copper pipe, copper wiring, brass
	95-5 Tin antimony solder	Acid	Copper pipe, brass, steel, galvanized sheet metal
	98-2 Tin silver solder	Mineral based flux	Copper pipe, brass, steel
	40-60 Cadmium zinc solder	Specific flux from solder manufacturer	Aluminum
B R A Z I N G	Copper phosphorus silver brazing BCuP	1% to 15% Silver no flux required	Copper pipe
	Copper phosphorus silver brazing BCuP	1% to 15% Silver mineral based flux	Copper pipe to brass, brass, steel
	Copper silver brazing BCuP	45% Silver mineral flux	Copper pipe to steel, brass, steel

[1]The percentages of the materials in the flux are given in the numbers, for example 95% tin, 5% antimony.

FUNDAMENTALS OF HVACR TEXT REFERENCE
Unit 25 Soldering and Brazing

REQUIRED TOOLS AND EQUIPMENT
Air Acetylene or Air Propane/Mapp Torch
Copper Piping, Fitting, Solder
Hacksaw, Pliers and Flat Screwdriver

SAFETY REQUIREMENTS
A. One hundred percent cotton or leather clothing is the best material to wear while brazing, soldering, or welding.

B. Shirts should have long sleeves and work gloves must be worn (cloth, leather palm, or all leather).

C. Lead solder should be avoided.

PROCEDURE
Step 1
Prepare the copper piping and fitting for use.

A. Use an abrasive sanding cloth or wire brush to remove all contaminants from the surface of the pipe.

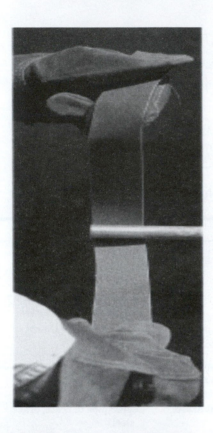

Figure 25-1-1

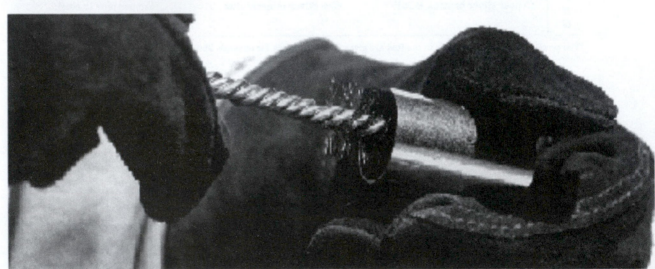

Figure 25-1-2

B. Do not touch the cleaned surfaces with you hands as oil from your fingers can prevent solder from flowing completely into a joint.

C. Use a brush to apply the flux to the end of the pipe.

D. <u>Do not apply flux to the very end of the pipe</u>.

E. Apply the flux to approximately 1/16 to 1/8 from the end of the pipe to avoid flux contamination into the system.

Step 2
Once the joint is prepared then continue as follows:

A. Three quarters of an inch of solder is adequate to make soldered joints in 1/2 inch through 1 inch diameter copper pipe. Bend the solder approximately three quarters of an inch from the end.

B. Use an air fuel torch and begin by heating the pipe.

C. As the pipe begins to be heated, move the torch onto the fitting and pipe.

D. Periodically test the back side of the joint with the tip of the solder.

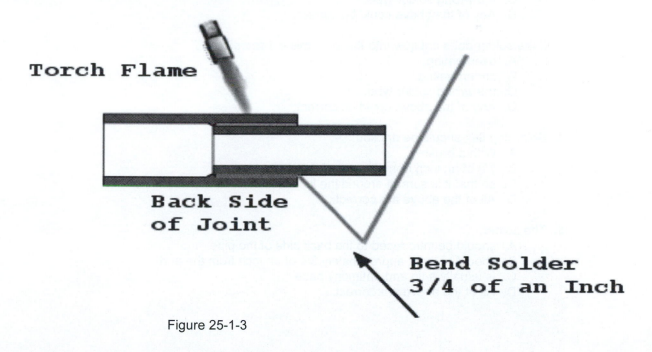

Figure 25-1-3

E. Once the solder begins to melt, remove the torch.

F. Continue adding the solder until the joint is filled.

G. Ideally a small fillet of solder should be left at the joint surface.

H. Pushing more solder into the joint simply results in solder bb's being formed inside the piping.

Step 3
Destructively test the soldered joint as follows:

A. Once cooled, use a hacksaw to slice a 45 degree angle through the fitting but not the pipe.

B. Place a flat bladed screwdriver into the slot and twist to release the pipe fitting from the copper pipe.

C. Using a pair of pliers, peel back the copper fitting to expose the soldered surface.

D. If the surface is smooth and has no large voids, the soldering was successful.

QUESTIONS

(Circle the letter that indicates the correct answer.)

1. Some of the common problems with soldering include:
   A. overheating.
   B. underheating.
   C. Both A & B are correct.
   D. None of the above is correct.

2. Tiny bubbles in the solder indicate:
   A. overheating.
   B. underheating.
   C. the wrong solder type.
   D. Any of the above could be correct.

3. If the solder does not flow into the joint, this indicates:
   A. overheating.
   B. underheating.
   C. the wrong solder type.
   D. Any of the above could be correct.

4. Soldering flux should be applied:
   A. With a brush.
   B. 1/8 of an inch from the end of the pipe.
   C. so that it is spread around the joint.
   D. All of the above are correct.

5. The solder:
   A. should be introduced to the back side of the pipe.
   B. should be bent approximately 3/4 of an inch from the end.
   C. is usually a tin and antimony base.
   D. All of the above are correct.

180

LAB 25.2 BRAZING

LABORATORY OBJECTIVE
The student will demonstrate how to properly braze a copper pipe connection.

LABORATORY NOTES
For this lab exercise a fitting will be brazed to copper piping. The fitting will then be tested to determine if it is satisfactory.

Brazing takes place at temperatures above 840 °F and Codes require that refrigerant joints be made using silver brazing alloys.

The most commonly used torch for this type of brazing is the air acetylene torch. An air acetylene flame burns at approximately 4220 °F. An oxyacetylene torch is generally used for making joints in copper pipes 1/2 inch or larger in diameter.

TABLE 25-2-1 Common Soldering and Brazing Metal and Fluxes Showing Base Metals That Can Be Joined

	Alloy	Flux Type	Base Metal
S O L D E R	95-5 Tin antimony solder[1]	C-Flux	Copper pipe, brass, steel
	95-5 Tin antimony solder	Rosin	Copper pipe, copper wiring, brass
	95-5 Tin antimony solder	Acid	Copper pipe, brass, steel, galvanized sheet metal
	98-2 Tin silver solder	Mineral based flux	Copper pipe, brass, steel
	40-60 Cadmium zinc solder	Specific flux from solder manufacturer	Aluminum
B R A Z I N G	Copper phosphorus silver brazing BCuP	1% to 15% Silver no flux required	Copper pipe
	Copper phosphorus silver brazing BCuP	1% to 15% Silver mineral based flux	Copper pipe to brass, brass, steel
	Copper silver brazing BCuP	45% Silver mineral flux	Copper pipe to steel, brass, steel

[1]The percentages of the materials in the flux are given in the numbers, for example 95% tin, 5% antimony.

FUNDAMENTALS OF HVACR TEXT REFERENCE
Unit 25 Soldering and Brazing

REQUIRED TOOLS AND EQUIPMENT
Air Acetylene Torch
Copper Piping, Fitting, Brazing Rod
Hacksaw

SAFETY REQUIREMENTS
A. One hundred percent cotton or leather clothing is the best material to wear while brazing, soldering, or welding.
B. Shirts should have long sleeves and work gloves must be worn (cloth, leather palm, or all leather).
C. When using Acetylene, torch pressure should be approximately 5 psig and the cylinder valve should be open no more than one and 1 1/2 turns.

PROCEDURE
Step 1
Prepare the copper piping and fitting for use.

A. Use an abrasive sanding cloth or wire brush to remove all contaminants from the surface of the pipe.

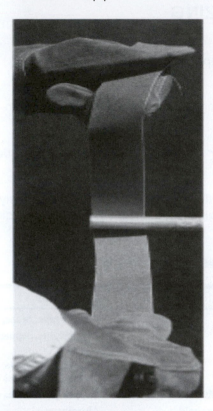

Figure 25-2-1

Figure 25-2-2

B. Do not touch the cleaned surfaces with you hands as oil from your fingers can prevent solder from flowing completely into a joint.

C. Use a brush to apply the flux to the end of the pipe.

D. <u>Do not apply flux to the very end of the pipe.</u>

E. Apply the flux to approximately 1/16 to 1/8 from the end of the pipe to avoid flux contamination into the system.

Step 2
Prepare to light the torch as follows:

A. Once the air acetylene torch and regulator have been attached to the acetylene cylinder, the acetylene cylinder valve is opened one quarter turn.

B. OSHA requires that a non-adjustable wrench be used on any cylinder valve stems that are not equipped with a hand wheel.

C. Hold the cylinder steady with one hand while opening the valve with the other hand.

D. Never open the acetylene cylinder valve more than 1 1/2 turns.

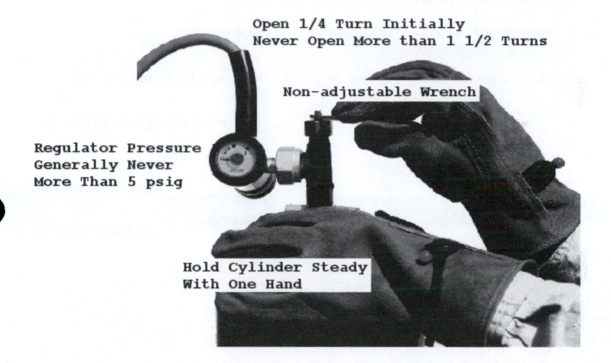

Figure 25-2-3

E. With the pressure adjusted properly according to the manufacturer's recommendations for that torch tip, turn the acetylene valve slightly to light the flame.

F. The only safe devices to use to light any torch are those specifically designed for that purpose, such as spark lighters or flint lighters.

G. When using a flint lighter hold it slightly off to one side of the torch tip. Holding the lighter over the end of the tip may cause a pop when the torch is lit.

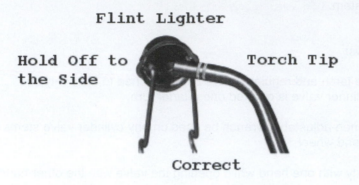

Flint Lighter

Hold Off to
the Side

Torch Tip

Correct

Do Not Hold
Directly
Over Tip

Incorrect

Figure 25-2-4

H. Open the hand wheel of the torch body completely.

I. Failure to provide the torch tip with the proper gas flow will result in the tip being overheated.

J. With low gas flows rates, the flame gets closer and closer to the tip and it will become hot.

Correct Flame

Flame Too Close
to Tip

Figure 25-2-5

Step 3
Begin soldering the joint as follows:

A. Heat the pipe first and then the fitting.

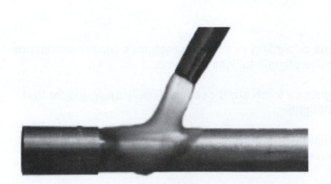

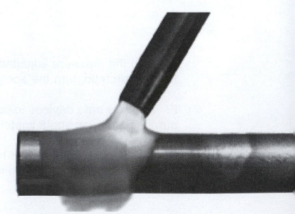

First Heat the Pipe

Then Heat the Fitting

Figure 25-2-6

B. Once the pipe has become hot, approaching a dull red color, move the torch flame onto the fitting so that it envelops both the fitting and the pipe.

C. Occasionally touch the pipe surface with the tip of the brazing rod as a test of temperature readiness.

D. Continue heating the pipe until the brazing metal begins to flow evenly over the surface.

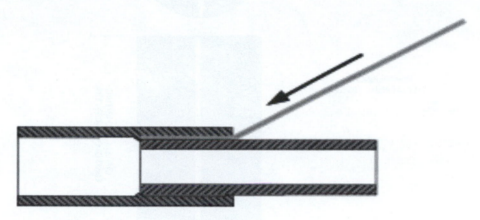

Figure 25-2-7

E. If you watch carefully, you can see a slight change in the fitting's color as the filler metal flows into the joint space.

F. Bringing the torch down on the fitting will help draw the filler metal completely into the joint.

G. Once the joint gap has been filled, continue adding small amounts of filler braze metal until a fillet of metal surrounds the joint.

Step 4
Destructively test the joint as follows:

A. Use a hacksaw to saw off the copper fitting at a point just beyond the depth that the pipe was inserted into the joint.

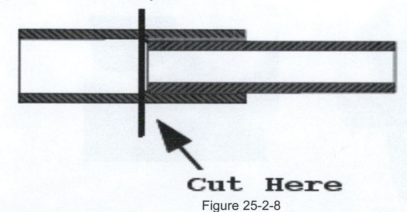

Cut Here

Figure 25-2-8

B. Once the fitting has been cut completely apart, clamp the pipe in a vise and cut straight down through the entire joint into the pipe.

C. Rotate the pipe ninety degrees and repeat the process so that there are four cuts.

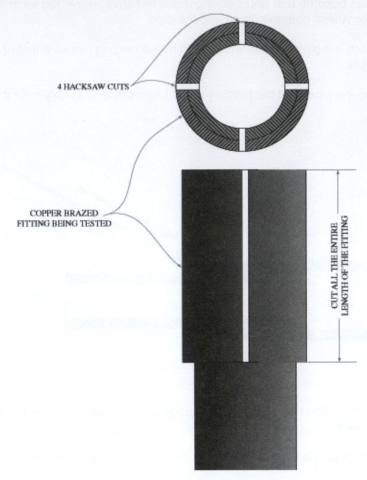

4 HACKSAW CUTS

COPPER BRAZED
FITTING BEING TESTED

CUT ALL THE ENTIRE
LENGTH OF THE FITTING

Figure 25-2-9

D. Bend each quarter out with a wrench.

E. Using a hammer and an anvil, flatten each of the four corners.

Figure 25-2-10

F. As the joint pieces are flattened, it is easy to see areas where 100% joint penetration did not occur.

QUESTIONS

(Circle the letter that indicates the correct answer.)

1. Acetylene gas must be stabilized in the cylinder with:
    A. propane.
    B. oxygen.
    C. acetone.
    D. nitrogen.

2. It is illegal to operate a torch with an acetylene pressure:
    A. greater than 15 psi.
    B. less than 15 psi.
    C. greater than 5 psi.
    D. greater than 3 psi.

3. Turning the acetylene pressure up beyond the recommended pressure value for a specific torch will:
    A. make the torch operate better.
    B. cause it to "pop".
    C. not make it work better.
    D. cause it to "crack".

4. It is against OSHA regulations to open any acetylene cylinder more than:
    A. 1 1/2 turns.
    B. 1 turn.
    C. 2 turns.
    D. 2 1/2 turns.

5. When transporting an acetylene cylinder:
    A. lay it on its side.
    B. leave the cylinder "cracked" open.
    C. always remove the regulator.
    D. leave the cylinder cap off.

LAB 25.3 LIGHTING THE TORCH

LABORATORY OBJECTIVE
You will demonstrate your ability to safely adjust, light, and shut down an air acetylene torch.

LABORATORY NOTES
The instructor will demonstrate how to safely adjust, light, and shut off an air acetylene torch.
You will then demonstrate your ability to safely adjust, light, and shut off an air acetylene torch.

FUNDAMENTALS OF HVACR TEXT REFERENCE
Unit 25 Soldering and Brazing

REQUIRED MATERIALS PROVIDED BY STUDENT
Safety Glasses

REQUIRED MATERIALS PROVIDED BY SCHOOL
Air Acetylene Torch
Torch Key
Sparker

PROCEDURE
Adjusting
1. Put on your safety glasses!
2. The instructor will demonstrate the proper way to handle and light an air acetylene torch.
3. You will demonstrate your ability for the instructor using the following procedure.
4. The regulator should be adjusted all the out counter-clockwise.
5. The torch handle valve should be closed.
6. Place the torch key on the valve stem and open ½ turn counter clockwise.
7. The tank pressure gauge should show the cylinder fill level.
8. Adjust the regulator by turning it in clockwise approximately 2 ½ turns.

Lighting
1. Before lighting the torch, practice sparking the striker.
2. Open the torch handle, hold the striker in front of the torch, and spark it.

LAB 25.4 SOLDERING COPPER TUBING

LABORATORY OBJECTIVE
You will make 9 solder joints each of 1/4", 3/8", 1/2", and 5/8" copper using the swage joints you made in Lab 24.3. You will cut open the joints to verify the solder penetration and coverage.

LABORATORY NOTES
The instructor will demonstrate how to soft solder. You will then soft solder 3 joints in the vertical, horizontal, and upside down positions for ¼", 3/8", ½", and 5/8" copper using the swages you made earlier.

FUNDAMENTALS OF HVACR TEXT REFERENCE
Unit 25 Soldering and Brazing

REQUIRED MATERIALS PROVIDED BY STUDENT
Safety Glasses
Gloves

REQUIRED MATERIALS PROVIDED BY SCHOOL
Air Acetylene Torch
Torch Key
Sparker
Sand tape
Soft solder flux
Soft Solder
Vise

PROCEDURE
The instructor will demonstrate the correct way to solder a joint and how to cut it open for inspection.

You are to solder 3 joints in a horizontal position, 3 joints in a vertical position, and 3 joints in an upside down position for 1/4", 3/8", 1/2", and 5/8" copper. Use the swages that you made in the earlier section. Joints will be graded on the following basis:

1. There should be no solder past the end of the joint inside the tubing.
2. There must be a complete seal around the entire circumference of the joint.
3. There should be few pockets, or bubbles, where voids exist in the solder.
4. The joint should be neat in appearance.

To judge the penetration of the solder, you will be required to cut each joint longitudinally and peel back the outer layer of copper base metal. This will reveal all "cannon balls" and bubbles. Be sure to save ALL joints.

If you use up all the swages that you prepared earlier, you may make some more. The joints in each position should be presented to the instructor for grading.

LAB 25.5 OXY-ACETYLENE TORCH SAFETY

LABORATORY OBJECTIVE

You will demonstrate your ability to safely adjust, light, and shut down an oxy- acetylene torch.

LABORATORY NOTES

The instructor will demonstrate how to safely adjust, light, and shut off an oxy-acetylene torch. You will then demonstrate your ability to safely adjust, light, and shut off an oxy-acetylene torch.

***FUNDAMENTALS OF HVACR* TEXT REFERENCE**

Unit 25 Soldering and Brazing

REQUIRED MATERIALS PROVIDED BY STUDENT

Safety Glasses
Gloves

REQUIRED MATERIALS PROVIDED BY SCHOOL

Oxy Acetylene Torch
Torch Key
Sparker

PROCEDURE

The following is a short discussion of oxy-acetylene torch safety.

READ AND ABSORB THESE RULES CAREFULLY. FAILURE TO DO SO IS PUNISHABLE BY DEATH!!!

BLOW OUT CYLINDER VALVE BEFORE ATTACHING REGULATORS.

> Why do we have this rule? It is simply to get the dust or combustible dirt out of the cylinder valve. Is dust a combustible? Have you ever read of a grain elevator explosion? Almost every case has been pinned down to dust. Combustion requires a fuel, an oxygen source, and a temperature source to cause ignition. The better the combustible material is mixed with oxygen, the better the chance of combustion. In the pure oxygen environment of the oxygen regulator, dust is downright explosive!

RELEASE THE ADJUSTING SCREW ON REGULATOR BEFORE OPENING CYLINDER VALVE.

> Why? Oxygen in the cylinder is compressed to pressures in excess of 2000 psig. When the adjusting screw is released, the seat of the regulator is in contact with the nozzle with sufficient pressure to hold 2000 psig. The oxygen travels only a short distance through the regulator this way. The regulator is not subjected to a sudden rush of pressure and less likely to rupture. If the pressure is allowed to blast into the regulator nozzle, any dust or oil can be easily ignited. This will burn out the diaphragm of the regulator and cause a potentially hazardous situation.

STAND TO THE SIDE OF THE REGULATOR WHEN OPENING THE VALVE.

> Where could a man go who stands in front of his regulator when opening the valve? Possibly to the hospital or the morgue. The weakest point of any regulator is the front and back. If one is going to blow, the force will travel forwards and/or backwards. Your chance of surviving a regulator blowout is much greater if you are not standing directly in front of or behind the regulator.

OPEN THE CYLINDER VALVE SLOWLY.

There are two reasons for this one. The first is that if we open the cylinder valve slowly, the heat made from the gas travel is very small. This reduces the possibility of an explosion. The second reason is to reduce shock on the regulator. Slow opening of the valve will increase regulator life.

PURGE OXYGEN AND FUEL LINES INDIVIDUALLY BEFORE LIGHTING TORCH.

Are acetylene and oxygen an explosive mixture? They sure are, especially when they are confined in an area as they are in a torch. Have you ever seen a blown up torch? What causes these blown up torches? Analyzing the cause of the explosion, you know that fuel, oxygen, and ignition temperatures must be present at the point of the explosion. A mixture of acetylene and oxygen can be a potentially fatal brew. By purging the lines, you prevent mixture of the two gasses other than at the tip.

LIGHT THE ACETYLENE FIRST.

Here we go to a short study of the burning rate of acetylene. Acetylene emits from the tip orifice at the correct velocity using this method: open acetylene valve only and light. Adjust the flow between these two points: 1) when the flame flares out in a fan and is luminously bright and 2) when the flame leaves the tip. The flame-burning rate should be set with the acetylene valve only. After adjusting the burn rate, open the oxygen valve and adjust to the desired flame.

DO NOT USE OXYGEN AS A SUBSTITUTE FOR AIR.

How many people have you seen use oxygen to blow dirt off of their clothes? Clothes are a fuel; oxygen greatly increases the chance of combustion. A lighted cigarette could burn you up faster than if you had poured gasoline over your body and lit it.

DO NOT USE OIL ON REGULATORS OR TORCH FITTINGS.

Why shouldn't we use oil to lubricate this equipment? An oil-oxygen mixture can explode. Every gauge made to be used with oxygen has this information printed on the side of it. Oxygen cylinders have as much as 2200 psig pressure on them. When the oxygen is released into the regulator under pressure, it travels faster than the speed of sound. This generates heat due to the friction of the rapidly moving compressed gas. If the smallest amount of combustible material is present, combustion may ensue. Even the oil from your skin is enough to cause an explosion.

DO NOT USE ACETYLENE AT PRESSURES ABOVE 15 PSIG.

Acetylene gas is composed of two atoms of hydrogen and two molecules of carbon C2H2. The two chemicals do not have a great affinity for each other at pressures above 15 psig, making acetylene very unstable at high pressures. When molecules break down into their constituent elements, the energy holding them together is released. When acetylene breaks down under pressure, it releases enough heat to kindle a chain reaction of chemical breakdown, giving off large quantities of heat. In other words, acetylene can explode at pressures in excess of 15 psig WITHOUT oxygen. This is a very dangerous situation and should be avoided.

KEEP THE FLAME AWAY FROM COMBUSTABLES.

This rule is usually taken for granted, but it is imperative to make sure that you will not ignite something else when lighting and using a torch.

The instructor will demonstrate the proper way to adjust an oxy-acetylene torch.
You will demonstrate your ability to safely adjust an oxy-acetylene torch.
The instructor will demonstrate the proper way to safely light an oxy-acetylene torch.
You will demonstrate your ability to safely light an oxyacetylene torch.

LAB 25.6 BRAZING

LABORATORY OBJECTIVE
You will demonstrate how to braze.

LABORATORY NOTES
The instructor will demonstrate how to braze. You will use copper-phos to braze 5 joints of ¼", 3/8", ½", and 5/8" copper using the swages you made earlier in Lab 24.3. You will then braze one joint of ¼", 3/8", ½", and 5/8" copper using high silver content braze material. Finally, you will braze copper to steel and copper to brass using high silver content braze material.

FUNDAMENTALS OF HVACR TEXT REFERENCE
Unit 25 Soldering and Brazing

REQUIRED MATERIALS PROVIDED BY STUDENT
Safety Glasses
Gloves
Five swage joints for 1/4", 3/8", 1/2", and 5/8" copper pipe

REQUIRED MATERIALS PROVIDED BY SCHOOL
Oxy acetylene torch
Torch key
Sparker
Oxyacetylene torch
Copper-phos brazing rod
Silver brazing alloy
Silver brazing flux
Plumber's tape
One piece of steel and one piece of brass

PROCEDURE

COPPER-PHOS BRAZING
1. Demonstrate your ability to light the oxy-acetylene torch.
2. Instructor will demonstrate copper-phos brazing technique.
3. Braze one joint using copper-phos brazing alloy.
4. Braze the remaining joints of each size pipe.
5. Show all joints to instructor.

SILVER BRAZING
1. Instructor will demonstrate silver brazing technique.
2. Braze one joint of each size tubing using silver solder and the oxy-acetylene torch.
3. Cut a short 2 inch section of 5/8" steel round stock and a short 2" section of 5/8" brass round stock
4. Braze a swaged 5/8" copper tube to the brass round stock
5. Braze a swaged 5/8" copper tube to the steel round stock

LAB 27.1 SCHRADER, PIERCING, & SERVICE VALVES

LABORATORY OBJECTIVE
The student will demonstrate how to properly replace the core on a refrigerant Schrader Valve, install a piercing valve and operate a service valve.

LABORATORY NOTES
For this lab exercise there should be a typical Schrader valve assembly connected to a low pressure air line as depicted below. There should also be a piercing valve and a service valve available for the student to use.

***FUNDAMENTALS OF HVACR* TEXT REFERENCE**
Unit 27 Accessing Sealed Refrigeration Systems

REQUIRED TOOLS AND EQUIPMENT
Refrigerant Schrader Valve and Tubing
Refrigerant Piercing Valve
Refrigerant Service Valve

PROCEDURE
Step 1
The inner core for a typical refrigerant Schrader valve will be replaced.

A. Familiarize yourself with the operation of a Schrader valve. Notice the valve opens when the valve stem is depressed as shown in Figure 27-1-1. When the stem is depressed, the system is open and at that point refrigerant could leak out of a system or if at a negative pressure air could leak in.

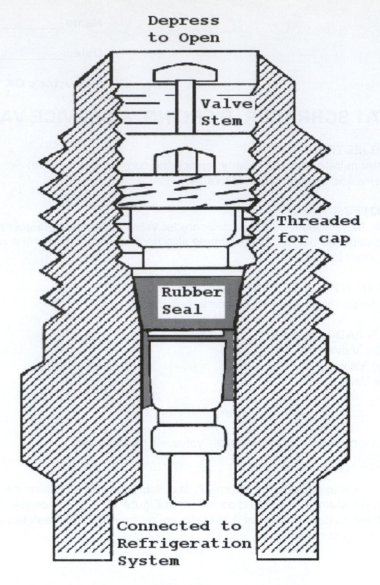

Depress
to Open

Valve
Stem

Threaded
for cap

Rubber
Seal

Connected to
Refrigeration
System

Figure 27-1-1

B. The Schrader valve should be connected to a low pressure air supply with a shutoff valve a shown in Figure 27-1-2. Close the air supply isolation valve and remove the core from the Schrader valve with the removal tool or if the valve cap is so designed, turn it around and it can be used to remove the core (unscrew counterclockwise).

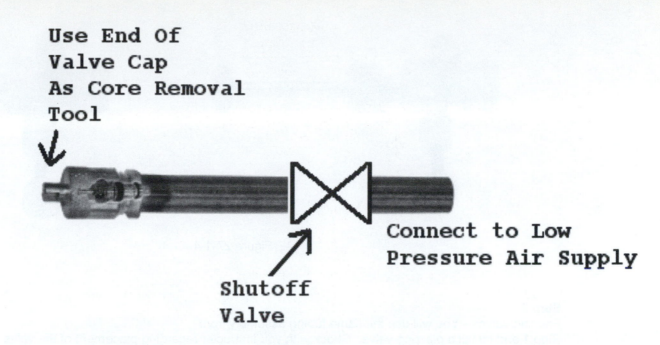

Use End Of
Valve Cap
As Core Removal
Tool

Connect to Low
Pressure Air Supply

Shutoff
Valve

Figure 27-1-2

C. Install a replacement core and then test for leakage by turning the air supply back on to the assembly and apply a soap solution to the stem area and watch for bubbles forming.

Schrader Valve
Cores

Figure 27-1-3

D. Some core removal tools as shown in Figure 27-1-4, allow you to change the valve core while the system is under pressure. If you have such a tool, then repeat Parts B & C above without shutting off the air supply.

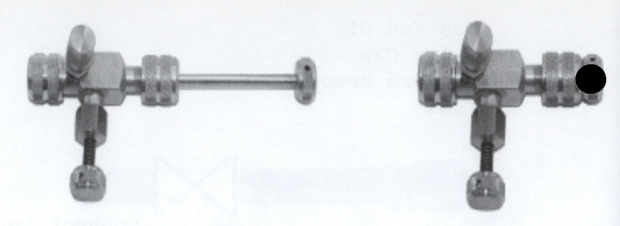

Figure 27-1-4

Step 2
Piercing valves - You will use the same tubing assembly from
Step 1 and install a piercing valve. Check with you Instructor regarding placement of the valve so
as not to waste too much tubing.

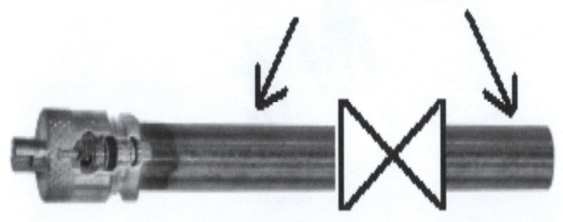

Install Piercing
Valve Wherever
Most Convenient

Figure 27-1-5

A. Piercing valves are clamped to the tubing, sealed by a bushing gasket, and then they
pierce the tube with a tapered needle. Some examples are shown in Figure 27-1-6.

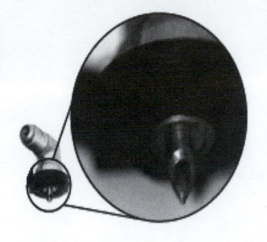  

Figure 27-1-6

B. Once the piercing valve is in place, it should be tested for leaks by applying a soap solution around the seal area and watching for bubbles.

Step 3

Service valves – Locate a service valves and examine it carefully so that you may complete the following exercise.

A. Draw a cross sectional sketch of the service valve, labeling the components and ports in the space provided below.

B. Why is it called a service valve?

C. Why does the valve have a protective cap?

D. What type of packing does the valve have?

E. What is meant by back-seating the valve and what is its purpose?

QUESTIONS

(Circle the letter that indicates the correct answer.)

1. When a service valve is backseated:
   A. the gauge manifold can be attached.
   B. normal flow through the valve will be shut off.
   C. there is less chance for leakage along the valve stem.
   D. Both A & C are correct.

2. A Schrader valve core can never be replaced while under pressure.
   A. True.
   B. False.

3. Access-piercing valves must be removed once the source of the sealed system malfunction has been located.
   A. True.
   B. False.

4. When a service valve is said to be in the mid- position:
   A. there is flow through the service port only.
   B. there is flow through the main valve only.
   C. there is flow through the service port and through the main valve.
   D. None of the above is correct.

5. The thing to always remember when conforming to EPA refrigerant management requirements is to:
   A. always release refrigerant to the atmosphere whenever possible.
   B. only vent refrigerant a little at a time.
   C. put service caps back on all access ports.
   D. None of the above is correct.

LAB 27.2 SERVICE VALVE INSPECTION

LABORATORY OBJECTIVE
You will examine several refrigerant access valves, identify the type of access valve, and discuss how each valve is used.

***FUNDAMENTALS OF HVACR* TEXT REFERENCE**
Unit 27 Accessing Sealed Refrigeration Systems

REQUIRED MATERIALS PROVIDED BY STUDENT
Safety glasses
Service valve wrench
Adjustable jaw wrench
6-in-1 tool

REQUIRED MATERIALS PROVIDED BY SCHOOL
Assortment of refrigeration service access valves
Stem type access valve (not on unit)
Refrigeration Systems with different refrigeration access valves

PROCEDURE
Use the manual stem service valve and your refrigeration valve wrench to demonstrate the following valve positions:

> front seated back seated
> front seat cracked
> back seat cracked
> mid position
> the position the valve should be in before removing gauges
> the position the valve should be in to check system pressure
> the position the valve should be in to evacuate a system

Locate the following types of valves on systems in the shop:

Suction service valve _____

Discharge service valve _____

Liquid receiver service valve _____

Liquid line service valve _____

Schrader service valve _____

Piercing valve _____

Split system shutoff valve _____

LAB 27.3 INSTALLING GAUGES ON SCHRADER VALVES

LABORATORY OBJECTIVE

You will demonstrate your ability to safely install gauges on four operating refrigeration systems with Schrader valves, obtain readings, and safely remove the gauges from the systems.

LABORATORY NOTES

You will safely install gauges, get pressure readings, and remove your gauges on four operating systems: 2 packaged air conditioners and 2 split system air conditioners.

***FUNDAMENTALS OF HVACR* TEXT REFERENCE**

Unit 27 Accessing Sealed Refrigeration Systems

REQUIRED MATERIALS PROVIDED BY STUDENT

Safety glasses
Gloves
6-in-1 tool
Refrigeration gauges
Positive shutoff hose or adapter (optional)

REQUIRED MATERIALS PROVIDED BY SCHOOL

Operating refrigeration systems with Schrader valves

PROCEDURE

You will check the operating refrigerant pressures on 2 packaged units and 2 split systems assigned by the instructor and complete the Data Sheet. The procedure is listed below.

Installing Gauges
1. Put on your safety glasses and gloves.
2. Turn the unit off.
3. Check to see that the gauge hand wheels are closed – fully clockwise.
4. Remove the low side valve cap.
5. Take the blue hose end with the Schrader core depressor (usually the crooked end) and connect to the low side Schrader valve.
6. Hold the hose with one hand (the left hand for right handed people) and quickly tighten the knurled end with your most kinesthetically developed hand (the right hand for right handed people).
7. You should see a reading on your low side gauge (the blue one).
8. Purge the air and residual refrigerant out of your low side hose and manifold by loosening the end of the middle hose on your gauge set and cracking the low side gauge hand wheel for approximately 1 second.
9. You should hear air escaping out of the end of the middle hose.
10. Tighten the middle hose back and close the low side hand wheel.
11. Remove the high side valve cap.
12. Take the red hose end with the Schrader core depressor (usually the crooked end) and connect to the high side Schrader valve.
13. You should see a reading on your high side gauge (the red one).
14. Purge the air and residual refrigerant out of your high side hose and manifold by loosening the end of the middle hose on your gauge set and cracking the low side gauge hand wheel for approximately 1 second.

15. You should hear air escaping out of the end of the middle hose.
16. Tighten the middle hose back and close the high side hand wheel.
    ### a. *Reading Pressures*
17. The gauge manifold gauge handwheels should both be closed.
18. Record the pressures on the data sheet.
19. Start the unit.
20. The low side should drop and the high side should rise.
21. Let the unit run long enough for the pressures to stabilize and quit moving, normally 10 to 15 minutes.
22. Record the pressures on the data sheet.

Removing Gauges Without Positive Shutoff Adapter
If your gauges have positive shutoff hoses skip down to next procedure

1. Remove the low side (blue) hose with the system running.
2. Turn the unit off and wait for the high side pressure to quit dropping.
3. Quickly disconnect the high side hose from the Schrader valve. **Refrigerant will spray out. Hold your hand and fingers to the side as you loosen the hose to avoid having them in the direct spray of refrigerant!**
4. Loosen the middle hose on your gauges to release the refrigerant trapped in the middle hose.
5. Replace the service valve caps.

Removing Gauges With a Positive Shutoff Adapter
Use this procedure if your gauges have positive shutoff hoses or you have a positive shutoff adapter.

1. The system should remain running.
2. Disconnect the high side hose or close the shutoff valve on the end of the high side hose. (depending on the type of positive shutoff hoses you have.
3. Crack both hand wheels on your gauges.
4. The high side pressure should drop, the low side gauge will rise.
5. Both gauges will drop to the operating suction pressure.
6. Remove your low side gauge.
7. Turn the unit off.
    a. If this procedure is done correctly, very little refrigerant will escape.
8. Replace the service valve caps.

Schrader Service Valve Readings				
Unit	Refrigerant	Equalized Pressure	Suction Pressure	Discharge Pressure
Packaged Unit				
Packaged Unit				
Split System				
Split System				

LAB 27.4 INSTALLING MANIFOLD GAUGES ON MANUAL STEM SERVICE VALVES

LABORATORY OBJECTIVE
You will demonstrate your ability to safely install gauges on systems with manual stem service valves, read system pressures, and safely remove gauges from the system.

LABORATORY NOTES
You will first practice installing gauges on an inoperative system with manual service valves. You will then safely install gauges on four operating systems with manual service valves, obtain pressure readings, and safely remove the gauges.

FUNDAMENTALS OF HVACR TEXT REFERENCE
Unit 27 Accessing Sealed Refrigeration Systems

REQUIRED MATERIALS PROVIDED BY STUDENT
Safety glasses
Gloves
6-in-1 tool
Adjustable jaw wrench
Refrigeration wrench
Refrigeration gauges

REQUIRED MATERIALS PROVIDED BY SCHOOL
Operating refrigeration systems with manual service valves and record your readings in the data sheet.

PROCEDURE
Installing Gauges
1. Put on your safety glasses and gloves.
2. Turn the unit off.
3. Use the adjustable jaw wrench to remove the valve cover caps.
4. Use the service valve wrench and make certain the valves are backseated – turned all the way out counter clockwise.
5. Remove the valve caps.
6. Check to see that the gauge hand wheels are closed – fully clockwise.
7. Take the blue hose end with the Schrader core depressor (usually the crooked end) and connect to the low side manual service valve.
8. Place the service valve wrench on the valve stem and turn clockwise ½ turn.
9. You should see a reading on your low side gauge (the blue one).
10. Purge the air and residual refrigerant out of your low side hose and manifold by loosening the end of the middle hose on your gauge set and cracking the low side gauge hand wheel for approximately 1 second.
11. You should hear air escaping out of the end of the middle hose.
12. Tighten the middle hose back and close the low side hand wheel.
13. Take the red hose end with the Schrader core depressor (usually the crooked end) and connect to the high side manual service valve.
14. Place the service valve wrench on the valve stem and turn clockwise ½ turn.
15. You should see a reading on your high side gauge (the red one).
16. Purge the air and residual refrigerant out of your high side hose and manifold by loosening the end of the middle hose on your gauge set and cracking the low side gauge hand wheel for approximately 1 second.

17. You should hear air escaping out of the end of the middle hose.
18. Tighten the middle hose back and close the high side hand wheel.

Reading Pressures
1. The gauge manifold gauge handwheels should both be closed.
2. Record the pressures on the data sheet.
3. Start the unit.
4. The low side should drop and the high side should rise.
5. Let the unit run long enough for the pressures to stabilize and quit moving, normally 10 to 15 minutes.
6. Record the pressures on the data sheet.

Removing Gauges (quick and dirty way)
This procedure is legal, quick, and easy, but releases more refrigerant.
1. With the unit running, use the service valve wrench to backseat the suction service valve by turning it counter clockwise. Note: this should only take ½ turn.
2. Turn the unit off and wait for the high side pressure to quit dropping.
3. Use the service valve wrench to backset the discharge service valve by turning the valve counter-clockwise. Note: this should only take ½ turn.
4. Loosen each hose at the service valves to let out trapped refrigerant.
5. **Refrigerant will spray out. Hold your hand and fingers to the side as you loosen the hose to avoid having them in the direct spray of refrigerant!**
6. Remove the hoses from the service valves.
7. Loosen the middle hose on your gauges to release the refrigerant trapped in the middle hose.
8. Replace the service valve caps and covers.

Removing Gauges by Returning Refrigerant to the System
This procedure takes a little longer, but puts most of the refrigerant back into the system, releasing less refrigerant.
1. The system should remain running.
2. Use the service valve wrench to backseat the discharge service valve.
3. Crack both hand wheels on your gauges.
4. The high side pressure should drop, the low side gauge will rise.
5. Both gauges will drop to the operating suction pressure.
6. Use the service valve wrench to backseat the suction service valve.
7. Turn the unit off.
8. Remove the hoses from the service valves.
9. If this procedure is done correctly, very little refrigerant will escape.
10. Loosen the middle hose on your gauges to release the refrigerant trapped in the middle hose.
11. Replace the service valve caps and covers.

Manual Service Valve Pressure Readings				
Unit	Refrigerant	Equalized Pressure	Suction Pressure	Discharge Pressure
Unit 1				
Unit 2				
Unit 3				
Unit 4				

LAB 28.1 RECOVERY UNIT

LABORATORY OBJECTIVE
The student will demonstrate how to properly use a refrigerant recovery unit.

LABORATORY NOTES
For this lab exercise there should be a typical recovery unit for the student to inspect.

***FUNDAMENTALS OF HVACR* TEXT REFERENCE**
Unit 28 Refrigerant Management and the EPA

REQUIRED TOOLS AND EQUIPMENT
Refrigerant recovery unit

SAFETY REQUIREMENTS
A. None

PROCEDURE
Step 1
Locate a refrigerant recovery unit and examine it carefully so that you may complete the following exercise.

    A. What types of refrigerants can the recovery unit be used for?

    B. Does the unit have a signal device to indicate when recovery has been completed?

    C. Can the unit recover vapor refrigerant

D. Draw a sketch representing a vapor recovery process in the space provided below.

E. If the recovery unit has a fan switch, what is it used for?

F. If you used this recovery unit to recover refrigerants, would this be considered system dependent (passive) or self contained (active)?

G. Could you perform liquid recovery with this unit?

H. Draw a sketch representing a vapor recovery process in the space provided below.

I. What is the primary difference between using a recovery unit as compared to a vacuum pump?

QUESTIONS

(Circle the letter that indicates the correct answer.)

1. A refrigerant recovery unit will stop on:
       A. low pressure.
       B. high pressure.
       C. Both A & B are correct.
       D. None of the above is correct.

2. Refrigerant recovery units:
       A. can only draw in vapor.
       B. can only draw in liquid.
       C. can draw in both liquids and vapors.
       D. All of the above are correct.

3. A liquid recovery is also known as the pull method:
       A. True.
       B. False.

LAB 28.2 REFRIGERANT VAPOR RECOVERY SELF CONTAINED ACTIVE

LABORATORY OBJECTIVE
The student will demonstrate the correct procedure for the self contained active method of recovering vapor refrigerant from a refrigeration system.

LABORATORY NOTES
This lab is intended to help students practice the removal of refrigerant from a system using the self contained active method of vapor recovery. Refrigerant recovery must be done on any and all systems prior to any refrigerant side repairs or dismantling.

FUNDAMENTALS OF HVACR TEXT REFERENCE
Unit 28 Refrigerant Management and the EPA

REQUIRED TOOLS AND EQUIPMENT
Gloves & goggles
Recovery cylinder & hoses
Recovery unit
Refrigerant scale
Disabled refrigeration system

SAFETY REQUIREMENTS
A. Wear safety goggles and gloves when working with refrigerants. Liquid refrigerant can cause frostbite when in contact with eyes and skin.

B. Use low loss hose fittings, or wrap cloth around hose fittings before removing the fittings from a pressurized system or cylinder. Inspect all fittings before attaching hoses.

PROCEDURE
Step 1
Familiarize yourself with the refrigerant recovery unit that will be used for the lab and carefully read the instructions that came with the unit. It is important to remember that the recovery unit utilizes its own built in compressor (self contained). Compressors can only pump vapors. Therefore liquid refrigerant must <u>NEVER</u> be introduced into the recovery unit!

Step 2
Connect and tighten both ends of a refrigerant hose from the disabled refrigeration system to the recovery unit. Include a filter drier in the line if available.

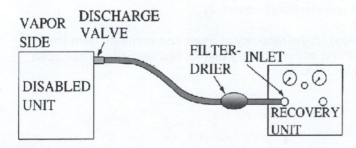

Figure 28-2-1

Step 3
Familiarize yourself with the refrigerant recovery cylinder.

       A. Recovery cylinders must be hydrostatically tested every five years and are stamped on the neck with an expiration date. Recovery cylinders have a yellow neck and grey body and new cylinders are either sealed in a vacuum or with an inert gas. <u>NEVER</u> mix refrigerants! Once the cylinder is first used, the refrigerant type should be clearly marked on the yellow neck with a permanent marker.

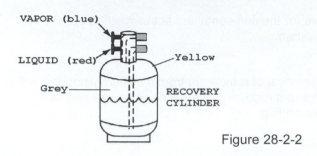

                                                  Figure 28-2-2

       B. The recovery cylinder has two valves that are the blue vapor valve and the red liquid valve. Red indicates danger as liquid refrigerant has a greater potential to cause frostbite to eyes and exposed skin.

       C. Recovery cylinders must never be filled to a level of more than 80% full. Refrigerants are always measured by weight. Stamped on the neck of the cylinder you will find the tare weight (T.W.) and the water capacity (W.C.). The maximum amount of refrigerant that the cylinder can hold is equal to 0.8 x W.C.

       D. The tare weight is the weight of an empty cylinder. If you are weighing a recovery cylinder on a refrigerant scale then you need to include this in your calculation. Therefore the total maximum weight of any recovery cylinder is equal to 0.8 x W.C. + T.W.

Step 4
Connect and purge the lines as follows:

       A. Tighten both ends of a refrigerant hose from the outlet of the recovery unit to the vapor valve (blue) on the recovery cylinder and place the cylinder on top of a refrigerant scale.

       B. Crack open the discharge of the disabled unit and purge air from the refrigerant hose by loosening the fitting at the recovery unit for two seconds. Then open the disabled unit discharge valve fully.

       C. Crack open the vapor valve (blue) on the recovery cylinder and purge air from the refrigerant hose by loosening the fitting at the recovery unit for two seconds. Then open the vapor valve on the recovery cylinder fully.

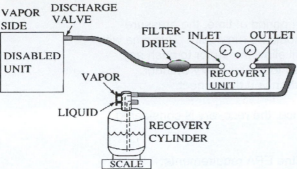

Figure 28-2-3

Step 5
Proceed to start the recovery unit according to the manufacturer's instructions.

A. The recovery unit may have a bypass switch that allows it to start without a load.

B. It may also have a built in condenser fan which may need to be switched on.

C. You will be drawing vapor into the recovery unit where it will be passed through a condenser. Therefore liquid refrigerant will discharge from the recovery unit and pass through the connected hose to the vapor valve on the recovery cylinder and drop from the vapor connection into the remaining liquid at the bottom.

D. You are recovering vapor, however you will be sending liquid to the recovery cylinder.

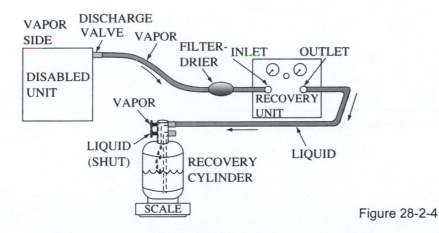

Figure 28-2-4

Step 6
The recovery unit is equipped with a low pressure cutout.

A. These controls are usually set below 29 in Hg for low pressure refrigerants and at 10 in Hg for high pressure refrigerants.

B. When the machine shuts off on the low pressure cutout, the recovery is not necessarily completed.

C. If the machine remains idle for a short period of time, the pressure may creep up. The recovery machine should be run again. This time with the condenser fan off if possible. This will allow the vapor to push any remaining liquid out of the hose connected to the recovery cylinder, thereby minimizing any refrigerant loss when you disconnect.

D. Usually when the machine cuts off twice, the recovery is considered complete.

E. Laboratory Manual Table 28-1-2 lists the EPA requirements.

Step 7
After the recovery is complete close the valve on the disabled unit and the vapor valve on the recovery cylinder and carefully disconnect the hoses.

TABLE 28-1-2 Required Levels of Evacuation for High Pressure Appliances

Recover Equipment Refrigerant and Charge	Maufacturing Date	
	Before 11/15/93 (in Hg)	After 11/15/93 (in Hg)
R-22 Appliance, <200 lb charge	0*	0
R-22 Appliance, >200 lb charge	4	10
Other high pressure appliance <200 lb charge	4	10
Other high pressure appliance >200 lb charge	4	15
Very high pressure equipment	0	0

*A zero (0) vacuum is atmospheric pressure. A perfect vacuum is 30 in Hg.

QUESTIONS

1. Given a recovery cylinder with a water capacity of 26.5 pounds and a tare weight of 14.6 pounds, calculate the maximum weight of a full recovery cylinder (remember: recovery cylinders can only be filled to 80% capacity).

2. Given a recovery cylinder with a water capacity of 32.6 pounds and a tare weight of 15.7 pounds. The cylinder already has some refrigerant in it and its total weight is 27.3 pounds. How many more pounds of refrigerant will the cylinder be able to hold?

(Circle the letter that indicates the correct answer.)

3. The active method of refrigerant recovery:
    A. uses the small appliance compressor to recover the refrigerant.
    B. uses the pressure in the system to recover the refrigerant.
    C. uses self contained recovery equipment.
    D. can be used on high pressure systems only.

4. The recovery cylinder:
    A. must have DOT approval.
    B. can never be refilled.
    C. can only be filled to 95% capacity.
    D. is color coded blue.

5. Very high pressure equipment contains refrigerant that:
    A. has a pressure above 1000 psig.
    B. has a pressure in excess of 152 psig.
    C. has a boiling point below -50 °C (-58 °F) at atmospheric pressure.
    D. can never be recovered due to excessive pressure.

6. A grey cylinder with a yellow top:
    A. is color coded for a disposable cylinder.
    B. is always used with propane.
    C. is always used with nitrogen.
    D. is color coded for a recovery cylinder.

7. Whenever a refrigeration circuit needs to be opened up for service:
    A. If there is still any refrigerant in the system, the refrigerant
        charge must be recovered before the system is opened.
    B. You can simply release all remaining refrigerant to atmosphere.
    C. notify the EPA before servicing.
    D. you can only release the refrigerant vapor to atmosphere.

8. The date stamped on the neck of a recovery cylinder:
    A. indicates when it was manufactured.
    B. indicates the last date that the cylinder can be used for recovering,
        storing, or handling refrigerants.
    C. Both A and B are correct.
    D. Neither A or B is correct.

9. In an emergency, a disposable cylinder can be used as a replacement for a recovery cylinder.
    A. True.
    B. False.

10. When brand new, recovery cylinders are 80% full of refrigerant.
    A. True.
    B. False.

11. Recovery cylinders:
    A. can be used for mixtures of refrigerants.
    B. should be used to hold one refrigerant type only.
    C. must always be stored laying on their sides.
    D. Both A & C are correct.

12. When a refrigerant is first introduced into a recovery cylinder:
    A. the cylinder should be clearly marked with the refrigerant
        designation using a permanent marker.
    B. a tag should be placed around the cylinder valve.
    C. a piece of tape should be placed on the bottom of the cylinder.
    D. nothing needs to be done.

13. When brand new, recovery cylinders:
    A. may have a press charge of an inert gas.
    B. may be under a vacuum.
    C. Both A & B are correct.
    D. None of the above is correct.

14. Recovery cylinders have two valves:
    A. and one is the inlet while the other is the outlet.
    B. and one is for refrigerant while the other is for an inert gas.
    C. and one is for liquid while the other is for vapor.
    D. None of the above is correct.

15. A recovery cylinder liquid valve:
    A. is usually red.
    B. is connected to an inner tube that extends to the bottom of the
        cylinder.
    C. allows for liquid without needing to turn the cylinder upside down.
    D. All of the above are correct.

LAB 28.3 LIQUID RECOVERY SELF CONTAINED ACTIVE

LABORATORY OBJECTIVE
The student will demonstrate the correct procedure for the self contained active method of recovering liquid refrigerant from a refrigeration system.

LABORATORY NOTES
This lab is intended to help students practice the removal of refrigerant from a system using the self contained active method of liquid recovery. The liquid is forced out of the disabled unit using the recovery machine to lower the pressure in the recovery cylinder and increase the pressure in the disabled unit. This causes rapid movement of the liquid. The advantage of the liquid recovery method of transfer is that liquid recovery is much faster than the vapor recovery method. However the final evacuation still must be done by the vapor recovery method.

FUNDAMENTALS OF HVACR TEXT REFERENCE
Unit 28 Refrigerant Management and the EPA

REQUIRED TOOLS AND EQUIPMENT
Gloves & goggles
Recovery cylinder & hoses
Recovery unit
Refrigerant scale
Disabled refrigeration system

SAFETY REQUIREMENTS
A. Wear safety goggles and gloves when working with refrigerants. Liquid refrigerant can cause frostbite when in contact with eyes and skin.
B. Use low loss hose fittings, or wrap cloth around hose fittings before removing the fittings from a pressurized system or cylinder. Inspect all fittings before attaching hoses.

PROCEDURE
Step 1
Familiarize yourself with the refrigerant recovery unit that will be used for the lab and carefully read the instructions that came with the unit. It is important to remember that the recovery unit utilizes its own built in compressor (self contained). Compressors can only pump vapors. Therefore liquid refrigerant must <u>NEVER</u> be introduced into the recovery unit!

Step 2
Connect and tighten both ends of a refrigerant hose from the vapor side of the disabled refrigeration system to the recovery unit.

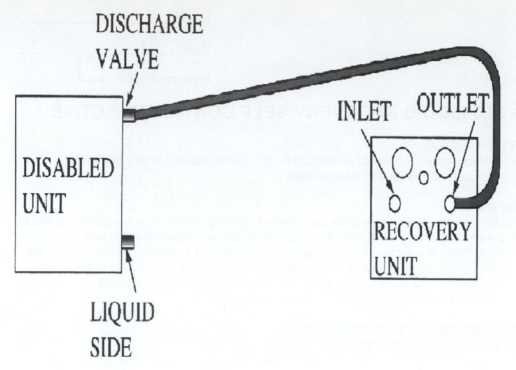

Figure 28-3-1

Step 3
Familiarize yourself with the refrigerant recovery cylinder.

Step 4
Connect and tighten both ends of a refrigerant hose from the inlet of the recovery unit to the vapor valve (blue) on the recovery cylinder and place the cylinder on top of a refrigerant scale.

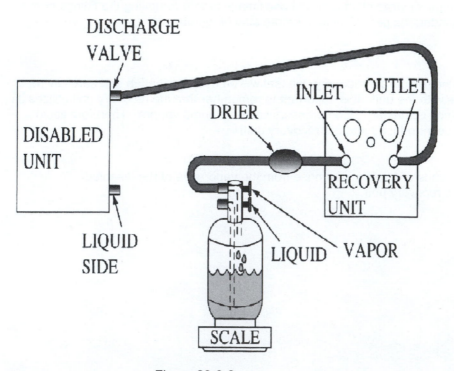

Figure 28-3-2

Step 5
Connect and tighten both ends of a refrigerant hose including a sight glass from the liquid side of the disabled unit to the liquid valve (red) on the recovery cylinder.

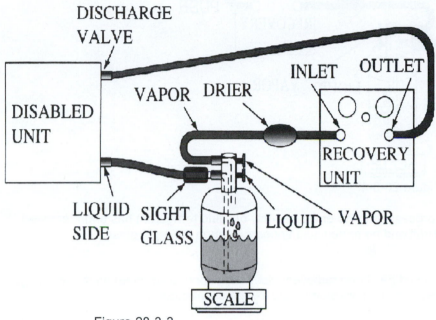

Figure 28-3-3

Step 6
Purge the lines of air as follows:

A. Crack open the discharge of the disabled unit and purge air from the refrigerant hose by loosening the fitting at the recovery unit for two seconds. Then open the disabled unit discharge valve fully.

B. Crack open the vapor valve (blue) on the recovery cylinder and purge air from the refrigerant hose by loosening the fitting at the recovery unit for two seconds. Then open the vapor valve on the recovery cylinder fully.

C. Crack open the liquid valve (red) on the recovery cylinder and purge air from the refrigerant hose by loosening the fitting at the disabled unit for two seconds. Then open the disabled unit liquid side valve fully.

Step 7
Proceed to start the recovery unit according to the manufacturer's instructions.

A. The recovery unit may have a bypass switch that allows it to start without a load.

B. It may also have a built in condenser fan which should remain OFF.

C. You will be drawing vapor into the recovery unit where its pressure will be increased and sent to the disabled unit. The vapor entering the disabled unit will push the liquid through the sight glass and into the recovery cylinder (PUSH).

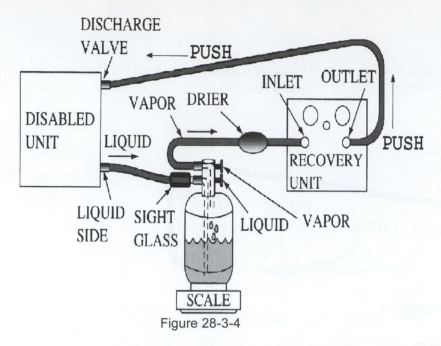

Figure 28-3-4

D. Continue this process until the entire quantity of liquid refrigerant has been removed from the disabled unit and no more liquid appears through the sight glass.

Step 8

A. After all of the liquid has been removed, close the vapor valve (blue) on the recovery cylinder and the recovery unit will stop on the low pressure cutout.

B. When the recovery unit has stopped close the discharge valve and the liquid side valve on the disabled unit and the liquid valve (red) on the recovery cylinder.

C. Carefully disconnect the hoses and then reinstall them to proceed with a vapor recovery (PULL).

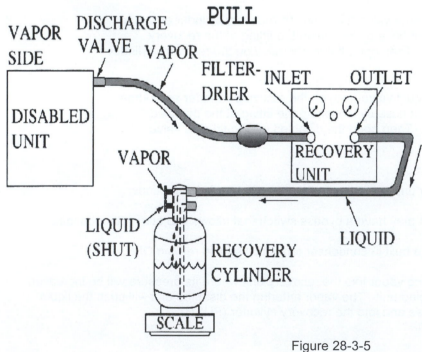

Figure 28-3-5

Step 9
When the machine shuts off on the low pressure cutout, the recovery is not necessarily completed.

    A. If the machine remains idle for a short period of time, the pressure may creep up. The recovery machine should be run again.

    B. This time with the condenser fan off if possible. This will allow the vapor to push any remaining liquid out of the hose connected to the recovery cylinder, thereby minimizing any refrigerant loss when you disconnect.

    C. After the recovery is complete close the valve on the disabled unit and the vapor valve on the recovery cylinder and carefully disconnect the hoses.

QUESTIONS

(Circle the letter that indicates the correct answer.)

1. The liquid recovery procedure will recover about:
   A. 45% of the entire charge of the refrigeration unit.
   B. 95% of the entire charge of the refrigeration unit.
   C. 75% of the entire charge of the refrigeration unit.
   D. the entire charge of the refrigeration unit.

2. A liquid recovery:
   A. is always followed by a vapor recovery.
   B. is followed by a vapor recovery about 50% of the time.
   C. directs liquid into the inlet of the recovery unit.
   D. is slower than a vapor recovery.

3. To recover the refrigerant from a centrifugal chiller:
   A. always pull all the vapor first.
   B. always push out all of the liquid first.
   C. always push out all of the vapor first.
   D. None of the above is correct.

4. Pulling vapor from a centrifugal chiller before removing the liquid:
   A. is a proper method.
   B. is extremely fast.
   C. is always recommended.
   D. can damage the chiller tubes due to the chill water freezing up.

5. When recovering refrigerant from a centrifugal chiller:
   A. the chill water pump should be running to reduce the possibility of freeze up.
   B. the chill water may be warmed slightly to reduce the possibility of freeze up.
   C. the Push – Pull method should be used.
   D. All of the above are correct.

6. When performing a liquid recovery:
   A. a sight glass should be installed in the discharge line of the recovery unit.
   B. a drier should be installed in the discharge line of the recovery unit.
   C. a sight glass should be installed in the liquid line connected to the recovery cylinder.
   D. All of the above are correct.

LAB 28.4 LIQUID RECOVERY SYSTEM DEPENDENT PASSIVE

LABORATORY OBJECTIVE
The student will demonstrate the correct procedure for the system dependent passive method of recovering liquid refrigerant from a refrigeration system.

LABORATORY NOTES
This lab is intended to help students practice the removal of refrigerant from a system using the system dependent passive method of liquid recovery. The liquid is pumped out of the refrigeration unit by using its own compressor. The only recovery equipment required is a recovery cylinder. With this method, 90% of the refrigerant must be recovered.

FUNDAMENTALS OF HVACR TEXT REFERENCE
Unit 28 Refrigerant Management and the EPA

REQUIRED TOOLS AND EQUIPMENT
Gloves & Goggles
Recovery Cylinder
Refrigerant Scale
Service Valve Wrench
Gauge Manifold
Operating Refrigeration System

SAFETY REQUIREMENTS:
A. Wear safety goggles and gloves when working with refrigerants. Liquid refrigerant can cause frostbite when in contact with eyes and skin.

B. Use low loss hose fittings, or wrap cloth around hose fittings before removing the fittings from a pressurized system or cylinder. Inspect all fittings before attaching hoses.

PROCEDURE
Step 1
Familiarize yourself with the major components in the refrigeration system including the condenser, compressor, evaporator, and metering device. Determine where the high side and low side connections for the system are located.

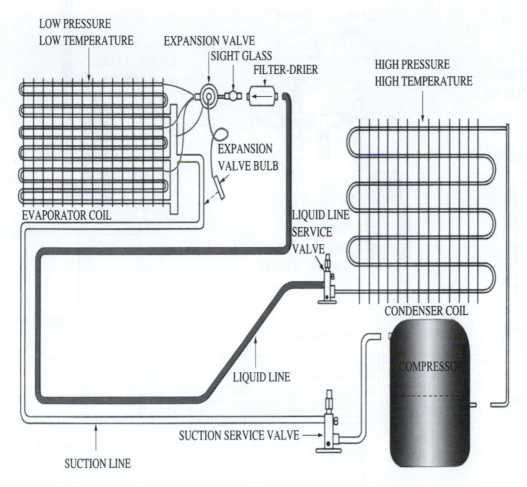

Figure 28-4-1

Step 2
Connect the gauge manifold as shown below and purge the lines of air). The recovery cylinder hose should be connected to the liquid side (red) valve on the cylinder.

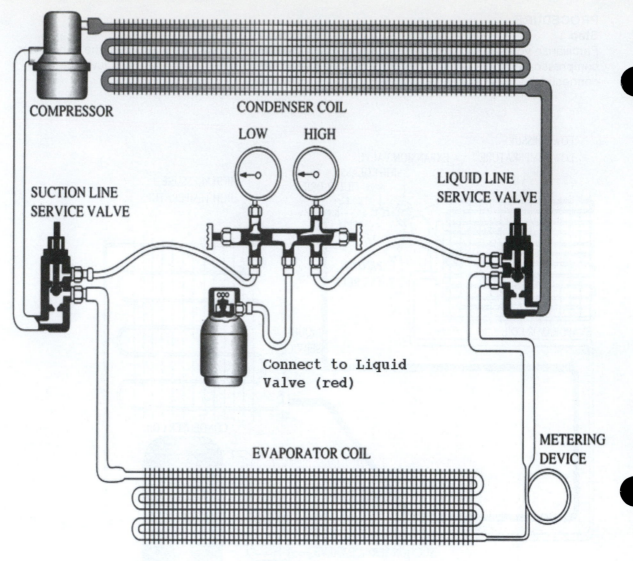

COMPRESSOR

CONDENSER COIL

LOW   HIGH

SUCTION LINE
SERVICE VALVE

LIQUID LINE
SERVICE VALVE

Connect to Liquid
Valve (red)

EVAPORATOR COIL

METERING
DEVICE

Figure 28-4-2

Step 3
After the lines have been connected and purged, place the recovery cylinder on a scale.
Familiarize yourself with the refrigerant recovery cylinder.

Step 4
To prepare for recovery, line up the system as follows:

   A. The liquid line service valve should be in the mid position to allow flow through the
   service port.

   B. The high side gauge manifold valve should be closed.

   C. The liquid valve (red) on the recovery cylinder should be wide open.

   D. The low side gauge manifold valve should be closed.

   E. The suction line service valve should be open one turn off its back
   seat.

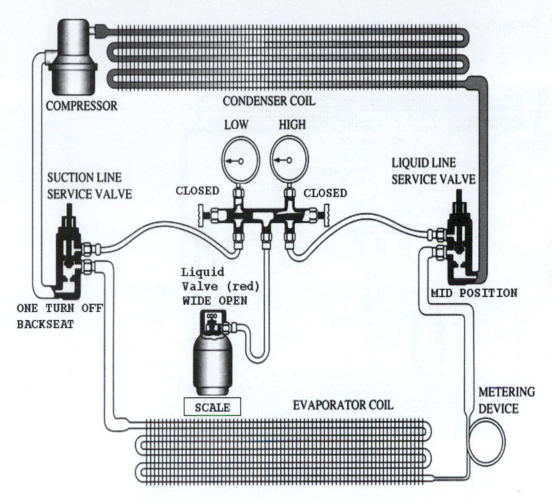

COMPRESSOR

CONDENSER COIL

LOW   HIGH

SUCTION LINE
SERVICE VALVE

CLOSED

CLOSED

LIQUID LINE
SERVICE VALVE

Liquid
Valve (red)
WIDE OPEN

MID POSITION

ONE TURN OFF
BACKSEAT

METERING
DEVICE

SCALE

EVAPORATOR COIL

Figure 28-4-3

Step 5
Begin recovery as follows:

A. Start condenser cooling fan or cooling water flow as normal.

B. Start compressor and monitor the high side and low side pressures.

C. Slowly open the gauge manifold high side valve wide open and refrigerant should begin flowing into the recovery cylinder.

D. Do not allow discharge pressure to exceed the normal recovery cylinder maximum pressure (usually no greater than 250 psig).

E. Monitor the cylinder weight (Remember no more than 80% full).

F. Cool the recovery cylinder with cold water or ice to help speed recovery if necessary.

G. When the low side pressure reaches 0 psig, the recovery is complete.

H. Close the liquid fill valve on the recovery cylinder and then shut down the refrigeration system.

I. Back seat the high side and low side service valves.

J. Carefully disconnect the hoses.

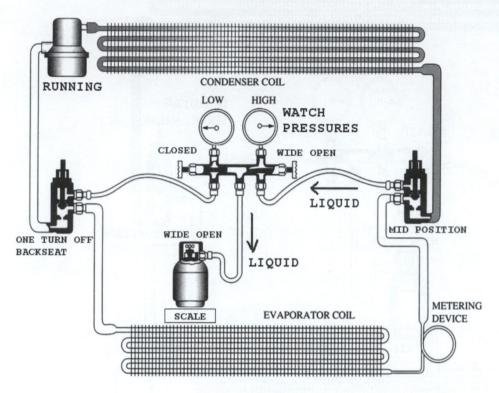

Figure 28-4-4

QUESTIONS

(Circle the letter that indicates the correct answer.)

1. The system dependent passive method of recovery with the compressor running should be capable of recovering:
    A. 80% of the entire charge of the refrigeration unit.
    B. 90% of the entire charge of the refrigeration unit.
    C. 75% of the entire charge of the refrigeration unit.
    D. the entire charge of the refrigeration unit.

2. The system dependent passive method of recovery <u>without</u> the compressor running should be capable of recovering:
    A. 80% of the entire charge of the refrigeration unit.
    B. 90% of the entire charge of the refrigeration unit.
    C. 75% of the entire charge of the refrigeration unit.
    D. the entire charge of the refrigeration unit.

3. The Push – Pull method of recovery is considered to be:
    A. system dependent active.
    B. system dependent passive.
    C. self contained passive
    D. self contained active

222

4. If you do not have a recovery unit and the system compressor does not work:
    A. you do not have to recover the refrigerant.
    B. the "grandfather' clause allows you to simply vent the charge.
    C. try to run the compressor by hand.
    D. you still must recover the refrigerant.

5. If you recover liquid refrigerant using the system dependent passive method:
    A. you will still need to use an appropriate recovery cylinder.
    B. you can use any type of refrigerant cylinder.
    C. you must use a disposable refrigerant cylinder.
    D. None of the above is correct.

6. The method of recovery used in a system dependent passive procedure:
    A. is most closely related to Push.
    B. is most closely related to Push - Pull.
    C. is most closely related to Pull.
    D. All of the above are somewhat correct.

7. If you have no recovery unit and the system compressor will not operate:
    A. just cut the refrigerant lines to release the charge.
    B. cut the lines but be very careful not to get frostbite.
    C. always recover the required amount of refrigerant prior to opening the
       system.
    D. None of the above is correct.

8. If the system compressor does not operate then the only alternative is to remove 80% of the charge using the system dependent passive method of recovery.
    A. True.
    B. False.

LAB 28.7 PACKAGED UNIT RECOVERY

LABORATORY OBJECTIVE
You will demonstrate your ability to recover refrigerant from a packaged air conditioning unit.

LABORATORY NOTES
You will operate a packaged air conditioning unit and record its operating pressures. Then you will shut off the unit and recover the refrigerant to the EPA specified level.

FUNDAMENTALS OF HVACR TEXT REFERENCE
Unit 28 Refrigerant Management and the EPA

REQUIRED MATERIALS PROVIDED BY STUDENT
Safety Glasses
Gloves
Manifold gauges
Refrigeration wrench

REQUIRED MATERIALS PROVIDED BY SCHOOL
Operating Packaged air conditioning unit
Refrigerant recovery unit
Refrigerant recovery cylinder
Scale
Extra refrigerant hose

PROCEDURE
(The following assumes the high side gauge port is on the discharge line, not the liquid line.)

Select Proper Equipment
Note the type and amount of refrigerant in the system and record on the data sheet.
Determine the EPA specified recovery level based on the refrigerant type and amount.
Record the required recovery level on the data sheet.
Choose a recovery Unit, recovery cylinder, and extra hose that are compatible with that refrigerant.
Calculate the cylinder safe fill weight. Tare + (Water Capacity x 0.8 x Refrigerant Specific Gravity)
Weigh the cylinder.
Determine the amount of space in the cylinder. Safe fill weight – current cylinder weight.
Make sure the amount of space in the cylinder exceeds the total unit charge.

Connecting Equipment
Connect your refrigeration gauges to the unit.
Operate the unit long enough for the pressures to stabilize and the compressor to warm up.
Read the operating pressures and record on the data sheet.
Connect the middle hose of your gauges to the inlet of the recovery unit.
Connect the outlet of the recovery unit to the recovery cylinder.
Purge the connections to the recovery unit and the recovery unit.
(This does NOT involve operating the recovery unit or air conditioning unit. The exact procedure to purge the recovery unit and hoses can vary by model. If the recovery unit was last used on the

same type of refrigerant, it may not need purging, but the hoses will still need purging.)

Zero the scale.

Place the recovery cylinder on the scale and record the cylinder weight.

Zero the scale with the cylinder on it.

Set the recovery unit to recover vapor. *(Exact valve positions will vary by model)*

Open the valve on the recovery cylinder.

Open both manifold hand wheels.

Recovering the Refrigerant

Start the recovery unit.

Monitor the cylinder weight to insure that it does not exceed the safe fill level.

Operate until the pressure in the air conditioning unit is slightly lower than the EPA specified level.

Close the hand wheels on your gauges and wait to see if the recovery level will hold.

Set the recovery unit to purge.

Operate the recovery unit until it shuts itself off or the inlet gauges indicate a vacuum.

Close all valves including the hand wheels on your gauges, the inlet and outlet valves of the recovery unit, and the valve on the recovery cylinder.

Turn off the recovery unit.

Check and record the amount of refrigerant recovered.

Remove Cylinder from the scale and zero the scale.

Weigh and record the final cylinder weight.

Leave gauges connected for Lab 22 Deep Evacuation

Package Unit System Refrigerant Recovery	
Refrigerant Type	
Refrigerant Quantity	
Required EPA recovery level	
Operating Suction Pressure	
Operating Discharge Pressure	
Cylinder Tare Weight (TW)	
Cylinder Gross Weight	
Water Capacity (WC)	
Safe refrigerant fill weight Tare + (WC x 0.8 x Ref SG)	
Actual Cylinder Weight Before Recovery	
Amount of Room in Cylinder (Safe Fill Weight – Actual Weight)	
Amount of Refrigerant Recovered	
Final Cylinder Weight	

LAB 28.8 SPLIT SYSTEM RECOVERY

LABORATORY OBJECTIVE
You will demonstrate your ability to recover refrigerant from a split system air conditioning unit.

LABORATORY NOTES
You will operate a split system air conditioning unit and record its operating pressures. Then you will shut off the unit and recover the refrigerant to the EPA specified level.

***FUNDAMENTALS OF HVACR* TEXT REFERENCE**
Unit 28 Refrigerant Management and the EPA

REQUIRED MATERIALS PROVIDED BY STUDENT
Safety Glasses
Gloves
Manifold gauges
Refrigeration wrench

REQUIRED MATERIALS PROVIDED BY SCHOOL
Operating split system air conditioning unit
Refrigerant recovery unit
Refrigerant recovery cylinder
Scale
Extra refrigerant hose

PROCEDURE

Select Proper Equipment
Note the type and amount of refrigerant in the system and record on the data sheet.
Determine the EPA specified recovery level based on the refrigerant type and amount.
Record the required recovery level on the data sheet.
Choose a recovery Unit, recovery cylinder, and extra hose that are compatible with that refrigerant.
Calculate the cylinder safe fill weight. Tare + (Water Capacity x 0.8 x Refrigerant Specific Gravity)
Weigh the cylinder.
Determine the amount of space in the cylinder. Space = Safe fill weight – current cylinder weight.
Make sure the amount of space in the cylinder exceeds the total unit charge.

Connecting Equipment
Connect your refrigeration gauges to the unit.
Operate the unit long enough for the pressures to stabilize and the compressor to warm up.
Read the operating pressures and record on the data sheet.
Connect the middle hose of your gauges to the inlet of the recovery unit.
Connect the outlet of the recovery unit to the recovery cylinder.
Purge the connections to the recovery unit and the recovery unit.
(This does NOT involve operating the recovery unit or air conditioning unit. The exact procedure to purge the recovery unit and hoses can vary by model. If the recovery unit was last used on the same type of refrigerant, it may not need purging, but the hoses will still need purging.)
Zero the scale.
Place the recovery cylinder on the scale and record the cylinder weight.

Zero the scale with the cylinder on it.

Recovering the Refrigerant

Set the recovery unit to recover liquid. *(Exact valve positions will vary by model)*

Open the valve on the recovery cylinder.

Open only the high side manifold hand wheel.

Start the recovery unit.

Monitor the cylinder weight to insure that it does not exceed the safe fill level.

Operate in liquid recovery mode until there is no more liquid refrigerant.

(connections at the gauge ports will no longer be cold)

Set the recovery unit to recover vapor. *(Exact valve positions will vary by model)*

Open both manifold valve hand wheels.

Operate until the pressure in the air conditioning unit is slightly lower than the EPA specified level.

Close the hand wheels on your gauges, turn off recovery unit, and wait to see if the recovery level will hold.

Set the recovery unit to purge. *(Exact valve positions will vary by model)*

Operate the recovery unit until it shuts itself off or the inlet gauges indicate a vacuum.

Close all valves including the hand wheels on your gauges, the inlet and outlet valves of the recovery unit, and the valve on the recovery cylinder.

Turn off the recovery unit.

Check and record the amount of refrigerant recovered.

Remove Cylinder from the scale and zero the scale.

Weigh and record the final cylinder weight.

Split System Refrigerant Recovery	
Refrigerant Type	
Refrigerant Quantity	
Required EPA recovery level	
Operating Suction Pressure	
Operating Discharge Pressure	
Cylinder Tare Weight (TW)	
Cylinder Gross Weight	
Water Capacity (WC)	
Safe refrigerant fill weight Tare + (WC x 0.8 x Ref SG)	
Actual Cylinder Weight Before Recovery	
Amount of Room in Cylinder (Safe Fill Weight – Actual Weight)	
Amount of Refrigerant Recovered	
Final Cylinder Weight	

LAB 28.9 SYSTEM PUMPDOWN

LABORATORY OBJECTIVE
You will demonstrate your ability to recover all the system refrigerant into the high side.

FUNDAMENTALS OF HVACR TEXT REFERENCE
Unit 28 Refrigerant Management and the EPA

REQUIRED MATERIALS PROVIDED BY STUDENT
Safety glasses
Gloves
Adjustable jaw wrench
6-in-1
Refrigeration gauges
Refrigeration valve wrench

REQUIRED MATERIALS PROVIDED BY SCHOOL
Refrigeration system with king valve or liquid line service valve

LABORATORY NOTES
You will front-seat the king valve or liquid line service valve, operate the system, and pump down the refrigerant into the high side of the system.

PROCEDURE
1. Install your gauges.
2. The high pressure gauge should be installed on the king valve or the liquid line service valve.
3. Purge your gauges.
4. Front seat the king valves or liquid line service valve.
5. Operate the unit.
6. The low side should pull into a vacuum.
7. The high side will either pull down to 0 psig, or its pressure will remain, depending upon the orientation of the king valve.

LAB 29.1 ELECTRONIC LEAK DETECTOR

LABORATORY OBJECTIVE
The student will demonstrate how to properly use an electronic leak detector.

LABORATORY NOTES
For this lab exercise there should be a typical electronic leak detector for the student to inspect.

FUNDAMENTALS OF HVACR TEXT REFERENCE
Unit 29 Refrigerant Leak Testing

REQUIRED TOOLS AND EQUIPMENT
Electronic Leak Detector

SAFETY REQUIREMENTS
None

PROCEDURE
Step 1
Locate an electronic leak detector and examine it carefully so that you may complete the following exercise.

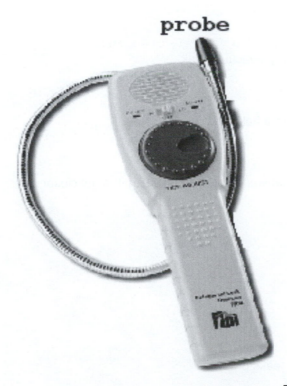

probe

Figure 29-1-1

A. What types of refrigerant can the leak detector be used for?

B. Should the probe tip be positioned above or below the suspected leak?

C. How quickly should the probe tip be moved across a suspected leak area?

D. How do you adjust the sensitivity of the meter?

E. Would the wind affect you reading outside?

QUESTIONS

(Circle the letter that indicates the correct answer.)

1. An electronic leak detector is capable of detecting a vacuum leak.
  A. True.
  B. False.

2. An electronic leak detector is more sensitive than a halide torch:
  A. True.
  B. False.

3. An electronic needs its sensitivity adjusted so that you can keep re-setting it to get closer and closer to the source of the leak.
  A. True.
  B. False.

4. An electronic leak detector has a:
  A. copper button.
  B. titanium flapper.
  C. platinum diode.
  D. All of the above are correct.

5. If a system is pressurized with an inert gas:
  A. an electronic leak detector will find any leaks present.
  B. an electronic leak detector will not work.
  C. an electronic leak detector might work.
  D. None of the above is correct.

230

6. In order to use an electronic leak detector:
  A. the refrigeration system must be under a vacuum.
  B. the refrigeration system must be completely drained.
  C. the refrigeration system must contain CFCs.
  D. the refrigeration system must be under a positive pressure.

7. An electronic leak detector can detect very small refrigerant leaks.
  A. True.
  B. False.

8. When using an electronic leak detector to find a leak in a system that has been pressurized with an inert gas:
  A. a trace amount of R-22 must be added.
  B. a trace amount of argon must be added.
  C. a slight amount of paraffin based oil must be added.
  D. the system will need to be heated.

9. When raising the pressure on an air conditioning chiller to check for leaks:
  A. Gag the pressure relief valve.
  B. warm the chill water loop.
  C. never exceed the rupture disk pressure.
  D. Both B & C are correct.

LAB 29.2 HALIDE TORCH LEAK DETECTOR

LABORATORY OBJECTIVE
The student will demonstrate how to properly use a halide torch leak detector.

LABORATORY NOTES
For this lab exercise there should be a typical halide torch leak detector for the student to inspect.

FUNDAMENTALS OF HVACR TEXT REFERENCE
Unit 29 Refrigerant Leak Testing

REQUIRED TOOLS AND EQUIPMENT
Halide Torch Leak Detector

SAFETY REQUIREMENTS:
A. None

Step 1
Locate a halide torch leak detector and examine it carefully so that you may complete the following exercise.

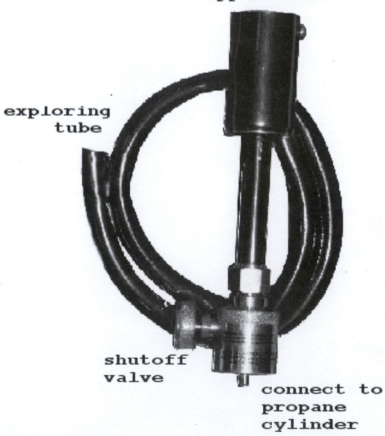

Figure 29-2-1

A. What color would you expect the flame to be when not exposed to refrigerants?

B. What type of metal is exposed to the flame area?

C. What color would you expect the flame to be when exposed to a halide refrigerant?

D. According to EPA regulations, do systems with less than fifty pounds of refrigerant have to be repaired?

E. Could the halide torch be used to detect ammonia refrigerant?

QUESTIONS

(Circle the letter that indicates the correct answer.)

1. A halide torch can be used for detecting:
   A. a small leak in a large area.
   B. a large leak in a small area.
   C. ammonia leaks.
   D. vacuum leaks.

2. A halide torch can be used for detecting:
   A. HFC refrigerants.
   B. ammonia.
   C. halocarbon refrigerants.
   D. inert gas leaks.

3. If the halide torch detects a refrigerant the flame color will change:
   A. green to blue.
   B. yellow to blue.
   C. blue to yellow.
   D. blue to green.

4. If the exploring tube of a halide torch is partially blocked:
  A. the flame will go out.
  B. the flame will turn white.
  C. the flame will smoke black.
  D. None of the above is correct.

5. The fuel for a halide torch can be:
  A. propane.
  B. butane.
  C. methyl alcohol.
  D. All of the above are correct.

6. When a halocarbon vapor passes over the hot copper element in a halide torch, the flame changes from normal color to bright green or purple.
  A. True.
  B. False.

LAB 29.3 NITROGEN PRESSURE TEST

LABORATORY OBJECTIVE
The student will demonstrate the correct procedure for a nitrogen pressure test to check for leaks in a refrigeration system.

LABORATORY NOTES
After a system has been repaired and before the final charge has been installed, the system needs to be leak tested. This is done by pressurizing the system with an inert gas such as nitrogen or carbon dioxide mixed with a trace amount of 5% or 10% R-22.

Oxygen should not be used, since it can cause an explosion.

Under current EPA regulations, the mixture of R-22 and inert gas is not considered to be refrigerant. Therefore it can be vented to atmosphere following the leak check.

However, the EPA does not look favorably on the addition of the inert gas to the refrigerant already present in a system. This is the reason why you must completely evacuate the remaining refrigerant in the system before introducing the trace gas.

This lab is intended to allow student practice on basic leak detection procedures. Since it is not the intention to recharge the system at the completion of this lab, an unrepaired leak could be set up and left for students to find. On such a system it would be impossible to pass a micron evacuation or any vacuum pressure drop test.

FUNDAMENTALS OF HVACR TEXT REFERENCE
Unit 29 Refrigerant Leak Testing

REQUIRED TOOLS AND EQUIPMENT
Gloves & Goggles
Service Valve Wrench
Gauge Manifold
Electronic leak detector / Halide Torch
Cylinder of nitrogen gas with regulator
Cylinder of Refrigerant R-22
Refrigeration System that has a leak

SAFETY REQUIREMENTS
A. Wear safety goggles and gloves when working with refrigerants. Liquid refrigerant can cause frostbite when in contact with eyes and skin.

B. Use low loss hose fittings, or wrap cloth around hose fittings before removing the fittings from a pressurized system or cylinder. Inspect all fittings before attaching hoses.

PROCEDURE
Step 1
Prior to performing this procedure, all of the refrigerant should be recovered..

Step 2
Connect the R-22 refrigerant cylinder, gauge manifold, and the gas cylinder as shown in Figure 29-3-1.

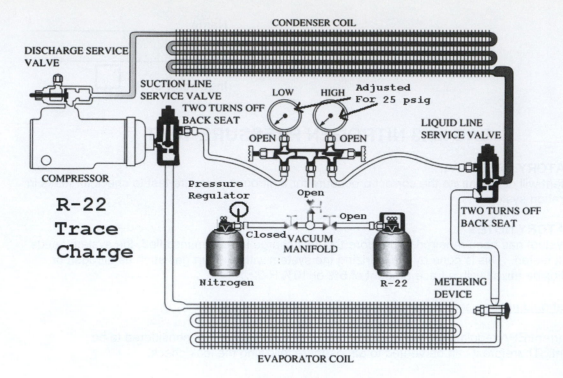

Figure 29-3-1

A. The center hose is connected to a vacuum manifold assembly. This is simply a three valve operation for attaching the refrigerant cylinder and the gas cylinder, each with a shutoff valve. If you do not have this type of an arrangement, you can connect each item separately.

B. Purge any air from the hose connection to the gauge manifold prior to introducing the refrigerant into the system.

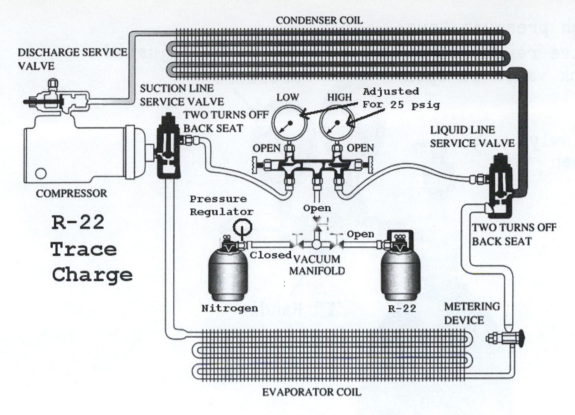

CONDENSER COIL

DISCHARGE SERVICE VALVE

SUCTION LINE SERVICE VALVE

TWO TURNS OFF BACK SEAT

LOW   HIGH   Adjusted For 25 psig

OPEN   OPEN

LIQUID LINE SERVICE VALVE

COMPRESSOR

R-22 Trace Charge

Pressure Regulator

Open

Open

Closed   VACUUM MANIFOLD

Nitrogen   R-22

TWO TURNS OFF BACK SEAT

METERING DEVICE

EVAPORATOR COIL

Figure 29-3-2

Step 3
Admit the trace charge as follows:

A. After the lines have been purged, slowly admit R-22 vapor to the system through both the high side and low side valves as a "trace charge" of 25 psig.

B. When the trace charge has been added close the cylinder the gauge manifold valves, and the vacuum manifold valve.

C. You are only adding a small amount of refrigerant so that it can be detected by the leak detector.

D. If only pure nitrogen were added, it would be difficult to find the leak. With pure nitrogen only, you would need to use the soap and bubble method or an ultrasonic leak detector.

Step 4
Open the nitrogen cylinder and adjust the pressure regulator as follows:

A. Make sure that the nitrogen cylinder pressure regulator is turned all the way out (counterclockwise).

B. Slowly open the cylinder valve fully open to backseat it. The tank pressure should register on the regulator high pressure gauge. The pressure in the tank can be in excess of 2,000 psi. <u>Do not stand in front of the regulator "T" handle</u>.

C. Slowly turn the regulator "T" handle inward (clockwise) until the regulator adjusted pressure reaches approximately 150 psig.

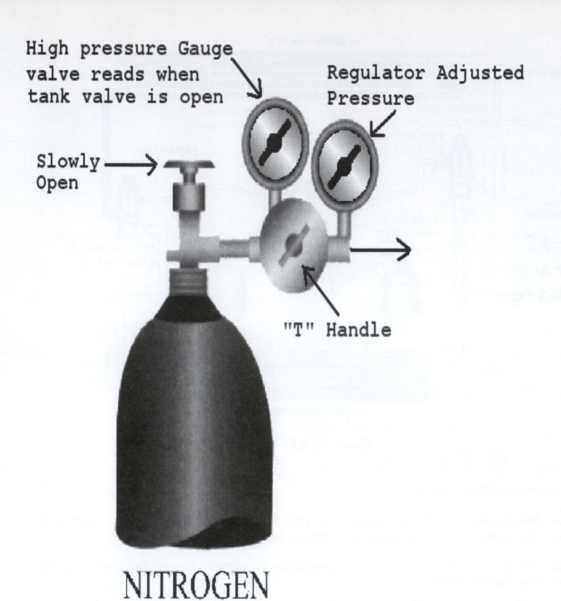

High pressure Gauge
valve reads when
tank valve is open

Regulator Adjusted
Pressure

Slowly
Open

"T" Handle

NITROGEN

Figure 29-3-3

Step 5
Begin adding nitrogen to the system as follows:

    A. Close both valves on the gauge manifold.

    B. Open the vacuum manifold isolation valve leading to the gauge manifold
and purge any air in the hose between the cylinder pressure regulator and
the gauge manifold.

    C. Open both gauge manifold valves slowly and obtain the desired nitrogen
pressure of approximately 150 psig. Then close the gauge manifold valves.

238

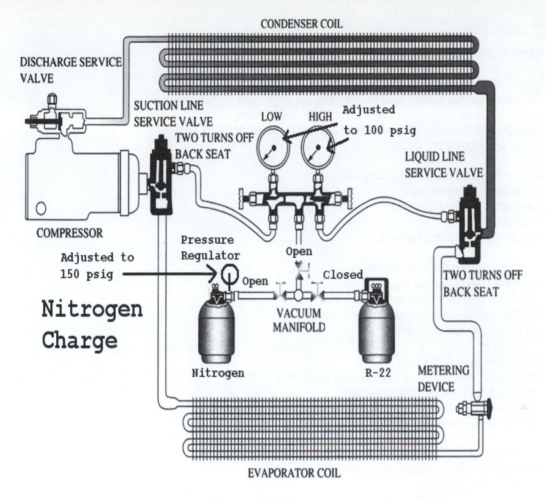

CONDENSER COIL

DISCHARGE SERVICE VALVE

SUCTION LINE SERVICE VALVE

TWO TURNS OFF BACK SEAT

LOW   HIGH

Adjusted to 100 psig

LIQUID LINE SERVICE VALVE

COMPRESSOR

Pressure Regulator

Open

Adjusted to 150 psig

Open   Closed

TWO TURNS OFF BACK SEAT

Nitrogen Charge

VACUUM MANIFOLD

Nitrogen   R-22   METERING DEVICE

EVAPORATOR COIL

Figure 29-3-4

Step 6

A. With the system pressure now at 150 psig, you may begin checking for leaks using an electronic leak detector or halide torch.

B. Record all leak locations and consult with the Lab Instructor before making any repairs.

Step 7

A. After the all of the leaks have been located, you may drain the system.

B. Close the shut off valve on the nitrogen cylinder. Carefully disconnect the hose from the R-22 Refrigerant cylinder and open the closed valve on the vacuum manifold to bleed the system pressure to 0 psig.

C. Once the pressure has been bled, you may back all the way off on the "T" handle (counterclockwise) for the pressure regulator on the Nitrogen cylinder.

QUESTIONS

(Circle the letter that indicates the correct answer.)

1. According to EPA regulations:
   - A. nitrogen can be added to a charged system for leak testing.
   - B. pure CFCs or HCFs can be released for leak testing.
   - C. nitrogen mixed with 5% or 10% R-22 can be used for leak testing.
   - D. All of the above are correct.

2. To check for refrigerant leaks you may use:
   - A. a halide leak detector.
   - B. an electronic leak detector.
   - C. a bubble test.
   - D. All of the above are correct.

3. If you use 100% nitrogen for leak testing:
   - A. a halide leak detector must be used.
   - B. a bubble test must be performed.
   - C. an electronic leak detector may be used.
   - D. All of the above are correct.

4. An ultrasonic type leak detector:
   - A. can detect any gas leaking through an orifice.
   - B. detects pressure or vacuum leaks.
   - C. detects ultrasonic noise from arcing electrical switchgear.
   - D. All of the above are correct.

5. An electronic leak detector works by:
   - A. drawing air over a platinum diode.
   - B. a magnetic impulse.
   - C. sending out an electric charge.
   - D. None of the above is correct.

LAB 29.4 DETECTING LEAKS USING NITROGEN AND SOAP BUBBLES

LABORATORY OBJECTIVE
You will demonstrate your ability to find leaks in refrigeration piping and components using nitrogen and soap bubbles.

LABORATORY NOTES
You will safely pressurize a system with nitrogen and use soap bubbles to locate leaks in the system.

FUNDAMENTALS OF HVACR TEXT REFERENCE
Unit 29 Refrigerant Leak Testing

REQUIRED MATERIALS PROVIDED BY STUDENT
Safety Glasses   Refrigeration wrench
Gloves   Soap bubbles
Manifold gauges

REQUIRED MATERIALS PROVIDED BY SCHOOL
Refrigeration system
Nitrogen cylinder
Nitrogen pressure regulator
Soap bubbles

PROCEDURE
1. Install your gauges on the high and low sides of the system you are checking.
2. If the pressure is above 0 psig you will need to recover the refrigerant from the system.
3. Find the system's testing pressures on the data plate.
4. Connect the outlet of the nitrogen regulator to the middle hose on your gauges.
5. Adjust the nitrogen regulator T-handle out counterclockwise until there is no spring pressure on it. NOTE: It will come completely out if you adjust it out too far.
6. Stand behind the nitrogen cylinder on the side opposite the regulator and crack the valve open counterclockwise. Once the valve has been cracked open you can continue to open the valve on the nitrogen cylinder the rest of the way.
7. Adjust the T-handle on the regulator in clockwise slowly until the pressure on the regulator gauge is equal to the low side testing pressure of the system.
8. Crack open gauges and allow nitrogen pressure to enter the system until the pressure in the system is at least 50 psig but not over the system's low side testing pressure.
9. Close your gauges and listen. Really large leaks will make an audible hiss.
10. If you do not hear any leaks, apply soap bubbles to all areas likely to have a leak. Be sure to check:
    a. Brazed and soldered connections
    b. Mechanical connections
    c. Gaskets
    d. Service valves
11. Really large leaks will likely just blow the soap solution away because of the velocity of the gas leaving the hole. You should be able to hear these leaks. Moderately large leaks will blow visible bubbles, and very small leaks will form a white froth or foam around the leak after several minutes.
12. Show the instructor the leaks you have located and be prepared to demonstrate how you found them.

LAB 29.5 DETECTING LEAKS USING ULTRASONIC LEAK DETECTOR

LABORATORY OBJECTIVE
The purpose of this lab is to demonstrate your ability to find leaks in refrigeration piping and components using an ultrasonic leak detector.

LABORATORY NOTES
You will pressurize a system with nitrogen and use the ultrasonic leak detector to locate leaks in the system.

FUNDAMENTALS OF HVACR TEXT REFERENCE
Unit 29 Refrigerant Leak Testing

REQUIRED MATERIALS PROVIDED BY STUDENT
Safety glasses
Gloves
Manifold gauges
Refrigeration wrench

REQUIRED MATERIALS PROVIDED BY SCHOOL
Refrigeration system
Nitrogen cylinder
Nitrogen pressure regulator
Ultrasonic leak detector

PROCEDURE
1. Install your gauges on the high and low sides of the system you are checking.
2. If the pressure is above 0 psig you will NOT need to add nitrogen. If the system is empty, proceed with steps 3-9. If the system already has pressure, proceed to step 10.
3. Find the system's testing pressures on the data plate.
4. Connect the outlet of the nitrogen regulator to the middle hose on your gauges.
5. Adjust the nitrogen regulator T-handle out counterclockwise until there is no spring pressure on it. NOTE: It will come completely out if you adjust it out too far.
6. Stand behind the nitrogen cylinder on the side opposite the regulator and crack the valve open counterclockwise. Once the valve has been cracked open you can continue to open the valve on the nitrogen cylinder the rest of the way.
7. Adjust the T-handle on the regulator in clockwise slowly until the pressure on the regulator gauge is equal to the low side testing pressure of the system.
8. Crack open you gauges an allow nitrogen pressure to enter the system until the pressure in the system is at least 50 psig but not over the system's low side testing pressure.
9. Close your gauges and listen. Really large leaks will make an audible hiss.
10. Use the ultrasonic detector to check all areas likely to have a leak. Be sure to check:
    a. Braze and solder connections
    b. Mechanical connections
    c. Gaskets
    d. Service valves
11. Really large leaks may not be picked up by the ultrasonic detector, you should be able to hear these leaks with your ears.
12. Show the instructor the leaks you have located and be prepared to demonstrate how you found them.

LAB 29.6 DETECTING LEAKS USING A HALIDE TORCH

LABORATORY OBJECTIVE
The purpose of this lab is to demonstrate your ability to find leaks in refrigeration piping and components using a halide torch.

LABORATORY NOTES
You will demonstrate your ability to use a halide torch for locating leaks in a refrigeration system.

FUNDAMENTALS OF HVACR TEXT REFERENCE
Unit 29 Refrigerant Leak Testing

REQUIRED MATERIALS PROVIDED BY STUDENT
Safety Glasses
Gloves
Manifold gauges
Refrigeration wrench

REQUIRED MATERIALS PROVIDED BY SCHOOL
Refrigeration system
Halide Torch
Matches

PROCEDURE
1. Find the system's refrigerant type and testing pressures on the data plate.
2. Verify that the type of refrigerant the system uses can be detected with a halide torch.
3. Install your gauges on the high and low sides of the system you are checking.
4. If the pressure is below 50 psig with the system OFF, you will need to add refrigerant.
5. Connect the center hose of your gauges to a refrigerant cylinder and purge your hoses and gauges. Crack open your gauges and allow refrigerant vapor to enter the system until the pressure in the system is at least 50 psig but not over the system's low side testing pressure.
6. Close your gauges and listen.
7. Light the halide torch and wait for the copper disk to turn red.
8. Remember that the torch will set things on fire. Be careful about the placement of the torch tip while you are working!
9. Move the sniffer hose slowly around all likely leak spots.
10. Be sure to check:
    a. Braze and solder connections
    b. Mechanical connections
    c. Gaskets
    d. Service valves
11. Really large leaks will create a brilliant blue flame. Moderately large leaks will create an obvious bright green flame, and very small leaks will create a faint green flame.
12. Show the instructor the leaks you have located and be prepared to demonstrate how you found them using the halide torch.

LAB 29.7 DETECTING LEAKS USING AN ELECTRONIC SNIFFER

LABORATORY OBJECTIVE
The purpose of this lab is to demonstrate your ability to find leaks in refrigeration piping and components using an electronic leak detector.

LABORATORY NOTES
You will demonstrate your ability to use an electronic leak detector for locating leaks in a refrigeration system.

FUNDAMENTALS OF HVACR TEXT REFERENCE
Unit 29 Refrigerant Leak Testing

REQUIRED MATERIALS PROVIDED BY STUDENT
Safety glasses   Manifold gauges
Gloves   Refrigeration wrench

REQUIRED MATERIALS PROVIDED BY SCHOOL
Refrigeration System
Electronic Leak Detector (corona discharge type) TIF
Electronic Leak Detector (heated diode type) DeTek

PROCEDURE
1. Find the system's refrigerant type and testing pressures on the data plate.
2. Verify that the type of refrigerant the system uses can be detected with leak detectors you are using.
3. Install your gauges on the high and low sides of the system you are checking.
4. If the pressure is below 50 psig with the system OFF, you will need to add refrigerant.
5. Connect the center hose of your gauges to a refrigerant cylinder and purge your hoses and gauges. Crack open your gauges and allow refrigerant vapor to enter the system until the pressure in the system is at least 50 psig but not over the system's low side testing pressure.
6. Close your gauges.
7. Hold the leak detector away from the system and turn it on.
8. It will squeal or tick as it adjusts itself to the surrounding environment, when the continuous squeal or tick stops it is ready to use.
9. Move the tip slowly around all likely leak spots.
10. Be sure to check:
    a. Braze and solder connections
    b. Mechanical connections
    c. Gaskets
    d. Service valves
11. When the detector finds a leak it will squeal loudly. For really small leaks it may just tick faster.
12. After locating a leak, back the detector away from the leak area and wait for the detector calm down.
13. Go back to the same spot again to verify the leak.
14. Try both types of detectors: corona discharge and heated diode.
15. Show the instructor the leaks you have located and be prepared to demonstrate how you found them using the leak detectors.

LAB 29.8 DETECTING LEAKS USING A FLUORESCENT DYE AND A BLACK LIGHT

LABORATORY OBJECTIVE
The purpose of this lab is to demonstrate your ability to find leaks in refrigeration piping and components using a fluorescent dye and a black light.

FUNDAMENTALS OF HVACR TEXT REFERENCE
Unit 29 Refrigerant Leak Testing

REQUIRED MATERIALS PROVIDED BY STUDENT
Safety Glasses
Gloves
Manifold gauges
Refrigeration wrench

REQUIRED MATERIALS PROVIDED BY SCHOOL
2 small hand valves with ¼" flare connections
Dye tube
Extra refrigeration hose
Black light

PROCEDURE
Dye Injection
1. Find the system's refrigerant type and testing pressures on the data plate.
2. Verify that the type of refrigerant AND the type of lubricant in the system is compatible with the dye you are using.
3. Connect the small hand valves on each end of the dye tube and make sure that the hand valves are closed.
4. Connect a refrigerant hose from the hand valve on the outlet side of the tube to the suction side of the system.
5. Purge the hose by loosening the end connected to the hand valve for 1 second.
6. A small amount of refrigerant should escape.
7. Tighten the hose.
8. Connect a refrigerant hose from the hand valve on the inlet side of the tube to the high side of the system.
9. Purge the hose by loosening the end connected to the hand valve for 1 second.
10. A small amount of refrigerant should escape.
11. Tighten the hose.
12. Operate the system.
13. After the system has operated a few minutes to establish suction and discharge pressures, open the cylinder valve first on the outlet side of the dye tube, and then on the inlet side of the dye tube. This limits the pressure on the dye tube.
14. Watch the dye in the tube. Let the refrigerant gas continue to flow for 15 seconds after all the dye has left the tube. This will help flush it out of the connecting hose.
15. Close the valve on the tube inlet and wait another 15 seconds to allow the tube pressure to drop to the system suction pressure.
16. Close the valve on the tube outlet, shut off the system, and remove the hoses and tube.
17. You will need to clean up the area where the hoses connect. Spray from the dye will cause a false positive leak indication.

Black Light Procedure
1. The system must operate for several hours with the dye in it before the dye will indicate a leak. A full day is recommended.
2. Large mercury vapor black lights must be plugged I and warmed up.
3. Direct black light on suspected leak areas including
4. Braze and solder connections
    a. Mechanical connections
    b. Gaskets
    c. Service valves
5. Leaks will show up as bright yellow.
6. Show the instructor the leaks you have located and be prepared to demonstrate how you found them.

LAB 30.1 VACUUM PUMP

LABORATORY OBJECTIVE
The student will demonstrate how to properly use a vacuum pump.

LABORATORY NOTES
For this lab exercise there should be a typical vacuum pump for the student to inspect.

***FUNDAMENTALS OF HVACR* TEXT REFERENCE**
Unit 30 Refrigerant System Evacuation

REQUIRED TOOLS AND EQUIPMENT
Vacuum Pump

SAFETY REQUIREMENTS
None

PROCEDURE
Step 1
Locate a vacuum pump and examine it carefully so that you may complete the following exercise.

A. What is the rating of the vacuum pump in CFM?

B. Is the pump most suitable for residential work, small appliance work, or commercial work? How can you tell?

C. Is the pump a deep vacuum pump?

D. Is it a single stage or a double stage pump?

E. Many vacuum pumps measure vacuum in microns. One inch of vacuum is how many microns?

F. What is a greater vacuum, 750 or 500 microns?

G. A deep vacuum pump should be at least how many microns?

QUESTIONS

(Circle the letter that indicates the correct answer.)

1. A vacuum pump is used to remove:
   A. non condensable gases.
   B. refrigerant.
   C. oil.
   D. All of the above are correct.

2. Many vacuum pumps:
   A. use a water seal.
   B. have oil that needs to be periodically changed.
   C. are cooled by refrigerant.
   D. All of the above are correct.

3. Before evacuating a refrigeration system:
   A. fill it completely with an inert gas.
   B. run the vacuum pump for at least one hour to warm it up.
   C. recover all of the refrigerant from the system.
   D. None of the above is correct.

4. The deeper the vacuum pulled on a system:
   A. the wetter it will become.
   B. the drier it will become.
   C. the hotter it will become.
   D. the more brittle it will become.

5. At a low pressure:
   A. water will condense.
   B. oil will condense.
   C. water will vaporize.
   D. the vacuum pump will start automatically.

6. 20,080 microns is equal to about:
   A. 20,080 psia.
   B. 20,080 kPa.
   C. 40 psia.
   D. 0.4 psia.

7. Water can boil at:
   A. 212 °F only.
   B. 100 °C only.
   C. Both A & B are correct.
   D. at a temperature of minus 60 °F at a pressure of 25 microns.

LAB 30.2 DEEP METHOD OF EVACUATION

LABORATORY OBJECTIVE
The student will demonstrate the correct procedure for the deep method of evacuation of a refrigeration system.

LABORATORY NOTES
This lab is intended to help students practice the evacuation of a refrigeration system. Proper evacuation of a system will remove non-condensable gases (air, water vapor, and inert gases) and assure a tight, dry system before charging.

FUNDAMENTALS OF HVACR TEXT REFERENCE
Unit 30 Refrigerant System Evacuation

REQUIRED TOOLS AND EQUIPMENT
Gloves & Goggles
Deep Vacuum Pump & Micron Gauge
Service Valve Wrench
Gauge Manifold
Refrigeration System

SAFETY REQUIREMENTS
Wear safety goggles and gloves when working on refrigeration systems. Oil in the system can become acidic over time and can cause acid burns to the skin and eyes.

PROCEDURE
Step 1
Familiarize yourself with the deep vacuum pump. It is somewhat like an air compressor in reverse.

> **A.** They are rated according to free air displacement in cubic feet per minute (CFM), or liters per minute (l/pm) in the SI system.

> **B.** The degree of vacuum the pump can achieve is expressed in microns. One micron is equivalent to 1/25,400 of an inch of mercury.

> **C.** Deep vacuum pumps achieve levels of 500 microns or less and a reading of 0 microns would be equal to a perfect vacuum. These pressures are too small to read on a standard gauge manifold therefore most deep vacuum pumps include a micron gauge as shown in Figure 30-2-1.

Micron
Gauge

Figure 30-2-1

D. To illustrate the measurement of microns refer to Table 30-2-1 shown below.

TABLE 30-2-1 Comparison of Three Different Pressure Measuring Systems

Boiling Point of Water		Unit of Absolute Pressure		Units of Vacuum (in Hg)
°F	°C	psia	Microns of Mercury	
212	100	14.7	—	0
79	26	0.5	25,400	28.9
72	22	0.4	20,080	29.1*
32	0	0.09	4,579	29.7*
−25	−31	0.005	250	29.8*
−40	−40	0.002	97	29.9*
−60	−51	0.0005	25	29.92*

*Too small a change to be seen on a gauge manifold set.

E. Note: Table 30-2-1 not only demonstrates the comparison in units of measure but dramatically shows the changes in the boiling point of water as the evacuation approaches the perfect vacuum.

F. This is the main purpose of evacuation – to reduce the pressure enough to boil or vaporize the water and then pump it out of the system.

Step 2
Prior to evacuating a system all of the refrigerant should be recovered.

    A. Also new systems will need to be evacuated prior to charging.

    B. The evacuation process can also be an indicator for any leaks in the system prior to charging.

Step 3
Connect the vacuum pump, gauge manifold, and the micron gauge as shown in the illustration below.

    A. The center hose is connected to a vacuum manifold assembly. This is simply a three valve operation for attaching the vacuum pump and a separate micron gauge, each with a shutoff valve.

    B. Prior to the evacuation, you can check the operation of the vacuum pump by pulling a vacuum on the micron gauge only as shown below.

    C. If the vacuum reaches 200 microns or lower, then the vacuum pump is operating properly.

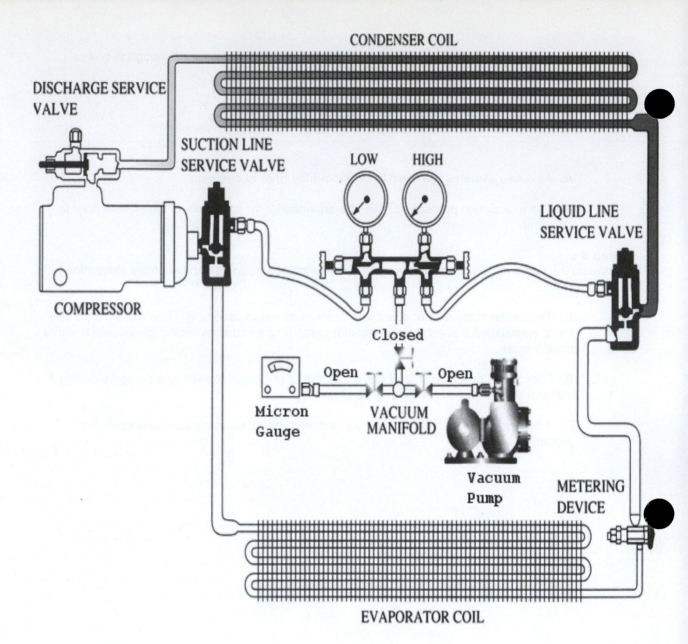

CONDENSER COIL

DISCHARGE SERVICE VALVE

SUCTION LINE SERVICE VALVE

LOW   HIGH

LIQUID LINE SERVICE VALVE

COMPRESSOR

Closed

Micron Gauge

Open   Open

VACUUM MANIFOLD

Vacuum Pump

METERING DEVICE

EVAPORATOR COIL

Figure 30-2-2

Step 4
Begin evacuation as follows:

A. Open the valves to the vacuum pump making sure to follow the pump manufacturer's instructions for the pump suction line size, oil, and calibration.

B. Open wide both valves on the gauge manifold and mid seat both equipment service valves as shown in the illustration below.

C. Start the vacuum pump and evacuate the system until a vacuum of at least 500 microns is achieved.

D. Close the vacuum pump isolation valve and shut it off.

252

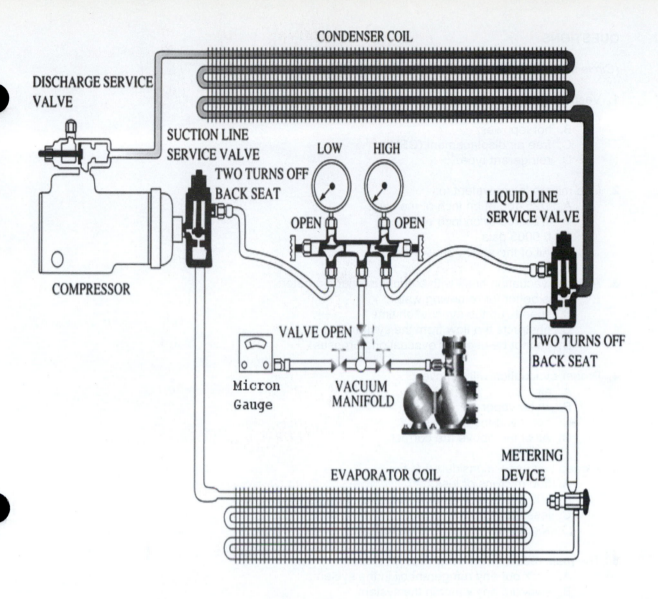

Figure 30-2-3

Step 5
The micron gauge should still be exposed to the system.

    A. If the system maintains fewer than 500 microns for a 10 minute vacuum pressure drop test, the system is considered to be dry, and leak free.

    B. If 800 to 1200 microns is maintained, moisture is present in the oil. A leak will cause the micron gauge to rise steadily.

    C. A large leak will cause a rapid rise while a small leak will cause a slow rise.

    D. Free water in the system will cause a rise to about 20,000 microns.

QUESTIONS

(Circle the letter that indicates the correct answer.)

1. Vacuum pumps are rated according to:
   A. amperage.
   B. horsepower.
   C. free air displacement (CFM).
   D. refrigerant type.

2. One micron is equivalent to:
   A. 1/25,400 of an inch of mercury.
   B. 1/25,400 of an inch of water.
   C. 0.0005 psia.
   D. All of the above are correct.

3. A short evacuation hose with a larger interior diameter:
   A. is better for removing water.
   B. speeds up the evacuation time.
   C. obstructs the flow from the system.
   D. cannot be used for evacuation purposes.

4. Proper evacuation will remove:
   A. air.
   B. water vapor.
   C. non condensable gases.
   D. All of the above are correct.

5. A deep vacuum is considered to be:
   A. 500 microns or less.
   B. 500 microns or more.
   C. less than 200 microns.
   D. slightly more than 200 microns.

6. The main reason for a deep vacuum is to:
   A. suck out any refrigerant oil in the system.
   B. suck out any water in the system.
   C. Both A & B are correct.
   D. allow any water to flash to vapor so that it can be removed.

7. A system is considered to be dry and leak free if it can maintain a vacuum of less than 500 microns with the pump shut off for a period of:
   A. twenty four hours.
   B. ten minutes.
   C. three hours and fifteen minutes.
   D. about one half hour.

8. A reading of 0 microns would be:
   A. atmospheric pressure.
   B. 0 psig.
   C. a perfect vacuum.
   D. All of the above are correct.

9. At a pressure of 250 microns, water will boil at a temperature of:
   A. 212 °F.
   B. 100 °C.
   C. 0 °F.
   D. -25 °F.

10. At a pressure of 25,400 microns, water will boil at a temperature of:

    A. 212 °F.
    B. 26 °C.
    C. 0 °F.
    D. -25 °F.

11. If the micron gauge begins to rise steadily once the vacuum pump is shut off this usually indicates:

    A. a leak.
    B. that everything is normal.
    C. that the micron gauge is faulty.
    D. All of the above are correct.

12. Free water in the system will cause a rise to about 20,000 microns:

    A. True.
    B. False.

13. When recovering refrigerant from an appliance containing less than 200 pounds of R-22 refrigerant, the system must be pulled down to:

    A. a pressure of 10 inches of mercury.
    B. a pressure of 0 psig.
    C. a pressure of 0 psia.
    D. a pressure of 10 psig.

14. When recovering refrigerant from an appliance containing <u>more</u> than 200 pounds of R-22 refrigerant, the system must be pulled down to:

    A. a pressure of 10 inches of mercury.
    B. a pressure of 0 psig.
    C. a pressure of 0 psia.
    D. a pressure of 10 psig.

15. When recovering refrigerant from very high pressure equipment, the system must be pulled down to:

    A. a pressure of 10 inches of mercury.
    B. a pressure of 0 psig.
    C. a pressure of 0 psia.
    D. a pressure of 10 psig.

LAB 30.3 TRIPLE EVACUATION

LABORATORY OBJECTIVE
The student will demonstrate the correct procedure for the triple evacuation method of a refrigeration system.

LABORATORY NOTES
This lab is intended to help students practice the evacuation of a refrigeration system. Proper evacuation of a system will remove non condensable gases (air, water vapor, and inert gases) and assure a tight, dry system before charging.

The triple evacuation procedure consists of three consecutive evacuations spaced by two dilutions of a dry gas. Nitrogen is preferred but helium and CO_2 can also be used. The clean dry gas will act as a carrier, mixing with system contamination (air & water) and carry it out with the subsequent evacuation. It is a time consuming procedure but effective in obtaining a clean dry system.

FUNDAMENTALS OF HVACR TEXT REFERENCE
Unit 30 Refrigerant System Evacuation

REQUIRED TOOLS AND EQUIPMENT
Gloves & Goggles
Vacuum Pump
Service Valve Wrench
Gauge Manifold
Cylinder of nitrogen gas with regulator
Refrigeration System

SAFETY REQUIREMENTS
Wear safety goggles and gloves when working on refrigeration systems. Oil in the system can become acidic over time and can cause acid burns to the skin and eyes.

PROCEDURE
Step 1
In this lab, you will be using a vacuum pump that does not pull a deep vacuum in microns. Therefore the time required for evacuation is critical.

> **A.** The assumption is made that 1 hour of evacuation will remove system contamination without actually measuring the degree of remaining contamination.

> **B.** The time is sometimes varied to fit the time available on site. An evacuation of one half hour might be good enough if the system is known to be clean. One hour or longer would be even better, no matter what the system problems.

> **C.** In moisture removal processes overnight evacuations are common.

Step 2
Prior to evacuating a system all of the refrigerant should be recovered. Also new systems will need to be evacuated prior to charging. The evacuation process can also be an indicator for any leaks in the system prior to charging.

Step 3

When evacuating any system using Schrader valves, the cores are sometimes removed. The purpose is to get a better quality evacuation faster. If the cores are removed, they must be put back in. Some technicians feel that the time and effort required to put the cores back in and the danger of contamination back into the system make it an ineffective practice.

Step 4

The triple evacuation procedure consists of three consecutive evacuations spaced by two dilutions of a dry gas. You will record the time for each evacuation, the dilution time and the dilution pressure.

Step 5

Connect the vacuum pump, gauge manifold, and the gas cylinder as shown in Figure 30-3-1.

A. The center hose is connected to a vacuum manifold assembly. This is simply a three-valve operation for attaching the vacuum pump and the gas cylinder, each with a shutoff valve.

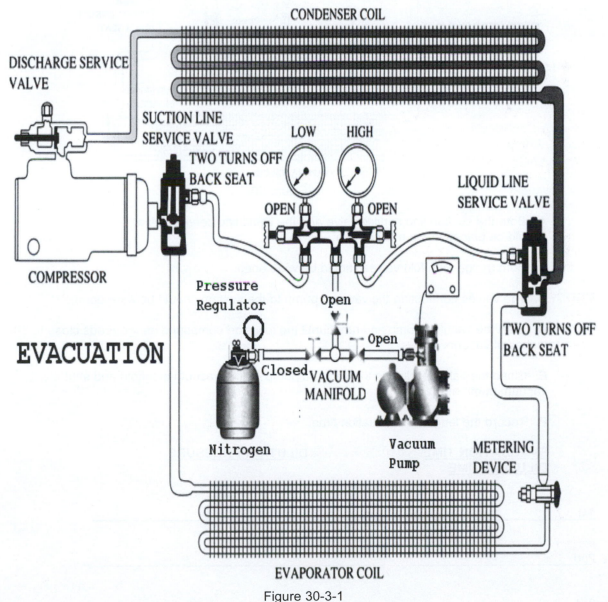

Figure 30-3-1

Step 5
Begin the first evacuation as follows:

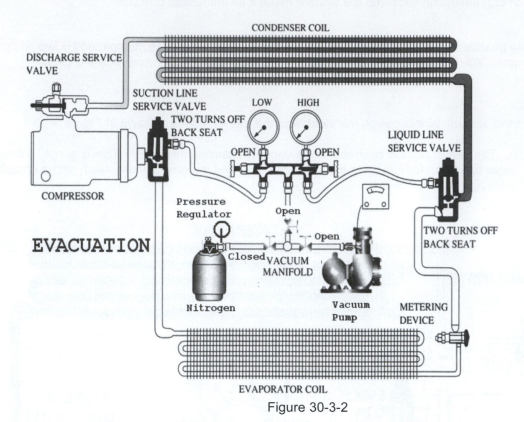

Figure 30-3-2

A. Both the suction line service valve and the liquid line service valve should be placed in the mid position.

B. Both gauge manifold valves should be wide open.

C. The valves connecting the vacuum pump to the system should be wide open.

D. Start the vacuum pump and run it until the low side compound gauge reads close to 30 inches of vacuum.

E. Close the valve leading to the vacuum pump on the vacuum manifold and shut the vacuum pump off.

F. Record the length of evacuation time.

<u>**EVACUATION TIME**</u>   <u>**DILUTION PRESSURE**</u>
<u>**DILUTION TIME**</u>

<u>1st</u>

<u>2nd</u>

<u>3rd</u>

258

Step 7
Open the nitrogen cylinder and adjust the pressure regulator as follows:

      A. Make sure that the nitrogen cylinder pressure regulator is turned all the way out (counterclockwise).

      B. Slowly open the cylinder valve fully open to backseat it. The tank pressure should register on the regulator high pressure gauge. The pressure in the tank can be in excess of 2,000 psi. <u>Do not stand in front of the regulator "T" handle</u>.

      C. Slowly turn the regulator "T" handle inward (clockwise) until the regulator adjusted pressure reaches approximately 50 psig.

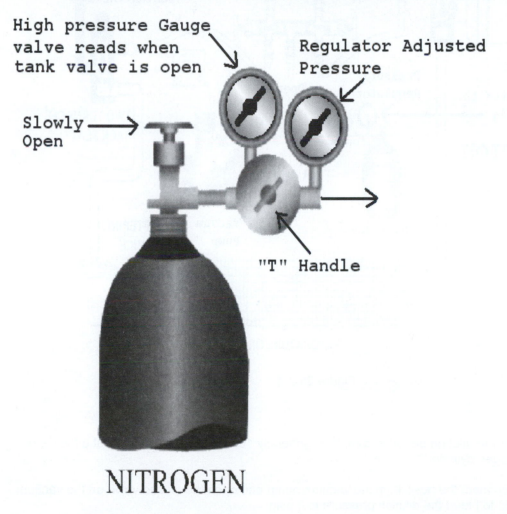

Figure 30-3-3

Step 8
Begin the first dilution as follows:

      A. Close both valves on the gauge manifold.
      B. Open the vacuum manifold isolation valve leading to the gauge manifold and purge any air in the hose between the cylinder pressure regulator and the gauge manifold.
      C. Open both gauge manifold valves slowly and obtain the desired dilution Pressure (generally 10 psig).

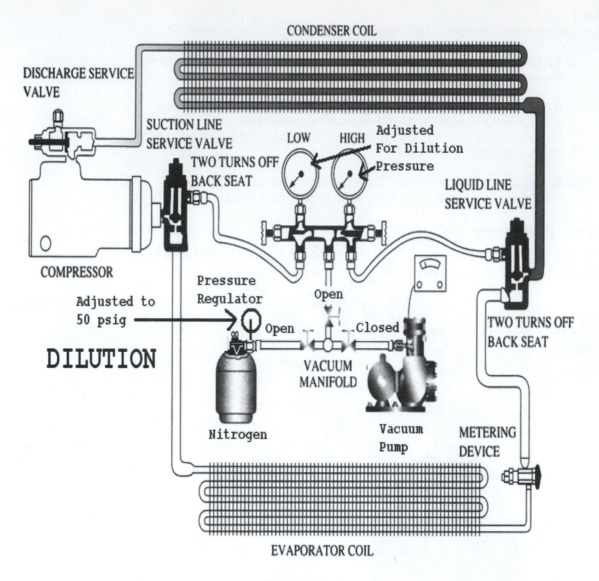

Figure 30-3-4

Step 9

A. After the dilution period is complete (generally 10 minutes), close the shut off valve on the nitrogen cylinder.

B. Disconnect the hose from the vacuum pump and open the closed valve on the vacuum manifold to bleed the dilution pressure to 0 psig.

C. Once the pressure has been bled, you may back all the way off on the "T" handle (counterclockwise) for the pressure regulator.

Step 10

A. Repeat the process over again starting with Step 5 by performing an evacuation.

B. Then follow that with a second dilution and then another final evacuation.

QUESTIONS

(Circle the letter that indicates the correct answer.)

1. Triple evacuation refers to:
   - A. using three separate vacuum pumps.
   - B. evacuating a refrigeration system three consecutive times.
   - C. an evacuation level that increases by thirds.
   - D. an evacuation level that decreases by thirds.

2. Dry gases that can be used in a triple evacuation process include:
   - A. nitrogen.
   - B. helium.
   - C. CO_2.
   - D. All of the above are correct.

3. Nitrogen cylinder pressures:
   - A. are never above 150 psig.
   - B. are generally between 350 to 400 psig.
   - C. can be as high as 2,000 psig.
   - D. can sometimes be negative (vacuum).

4. Proper evacuation will remove:
   - A. air.
   - B. refrigerant.
   - C. oil.
   - D. All of the above are correct.

5. A triple evacuation may be performed:
   - A. because it is very fast.
   - B. only on systems containing CFC refrigerants.
   - C. if a deep vacuum pump is unavailable.
   - D. on ammonia systems only.

6. The reason for introducing a clean dry gas between evacuations is to:
   - A. mix with the system contaminants.
   - B. destroy the system contaminants.
   - C. ensure a tight dry system.
   - D. check for leaks.

7. Nitrogen cylinder tank valves should be:
   - A. never opened fully.
   - B. opened fully to backseat them.
   - C. partially throttled.
   - D. normally closed as there is a relief hole drilled through the seat of the valve.

8. A number of different dry gases may be used for a triple evacuation such as:
   - A. nitrogen, helium and CO_2.
   - B. nitrogen, helium and O_2.
   - C. nitrogen, helium and H_2O.
   - D. All of the above are correct.

9. Nitrogen cylinder tank pressures are relatively low:
   - A. True.
   - B. False.

LAB 30.4 VACUUM PRESSURE DROP TEST

LABORATORY OBJECTIVE
The student will demonstrate the correct procedure for the vacuum pressure drop test to check for leaks in a refrigeration system.

LABORATORY NOTES
To prepare for this lab, students must complete the triple evacuation procedure first.

FUNDAMENTALS OF HVACR TEXT REFERENCE
Unit 30 Refrigerant System Evacuation

REQUIRED TOOLS AND EQUIPMENT
Gloves & Goggles
Vacuum Pump
Service Valve Wrench
Gauge Manifold
Cylinder of inert gas
Refrigeration System

SAFETY REQUIREMENTS
Wear safety goggles and gloves when working on refrigeration systems. Oil in the system can become acidic over time and can cause acid burns to the skin and eyes.

PROCEDURE
Step 1
The timed vacuum pressure drop test is typically done at the conclusion of a timed evacuation or a triple evacuation and used as an additional leak testing procedure.

   A. If the system holds approximately 30 inches of vacuum overnight or over a weekend, there can be no leaks. This is a valid procedure and should be done as time permits at the discretion of the service technician and company policy.

   B. Remember that this procedure will only help determine if there is a leak in the system and does not pinpoint the location of the leak.

Step 2
Upon the final evacuation of a triple evacuation procedure or at the end of a timed evacuation, close both gauge manifold valves with the vacuum pump still operating and then shut off the pump.

Step 3
The unit will be periodically checked to determine if the system is tight and holding a vacuum for a specified time period.

QUESTIONS

(Circle the letter that indicates the correct answer.)

1. A vacuum pressure drop test:
   - A. will be able to pinpoint the location of a leak.
   - B. will be able to provide an approximate location of a leak.
   - C. indicates that there is a leak somewhere.
   - D. will verify the accuracy of the vacuum pump.

2. A vacuum leak can be detected by:
   - A. an electronic leak detector.
   - B. a halide torch.
   - C. bubble test.
   - D. a ultrasonic type leak detector.

3. A triple evacuation:
   - A. is normally performed with a deep vacuum pump.
   - B. will evacuate the system three times.
   - C. can be done with refrigerant still remaining in the system.
   - D. will verify the accuracy of the gauge manifold.

4. When testing for a leak with an electronic leak detector:
   - A. remember that most refrigerants are heavier than air and will settle to the bottom.
   - B. remember that most refrigerants are lighter than air and will settle to the top.
   - C. Move the tip of the detector across the suspected leak area very quickly.
   - D. Always test the outlet of the vacuum pump first.

5. High vacuum indicators have accuracy approaching:
   - A. 500 microns.
   - B. 50 microns.
   - C. 10 microns.
   - D. 2 microns.

6. When a system is evacuated:
   - A. water in the system will vaporize.
   - B. water vapor in the system will condense.
   - C. Both A & B are correct.
   - D. None of the above is correct.

7. Most compound gauges:
   - A. can read pressure and vacuum.
   - B. can read vacuum in microns.
   - C. Both A & B are correct.
   - D. None of the above is correct.

8. If the Schrader valve core is pulled from a service port:
   - A. a vacuum can be pulled much quicker.
   - B. This should be done without opening the system to atmosphere.
   - C. Both A & B are correct.
   - D. None of the above is correct.

LAB 31.1 CHARGING CYLINDER

LABORATORY OBJECTIVE
The student will demonstrate how to properly use a refrigerant charging cylinder.

LABORATORY NOTES
For this lab exercise there should be a typical refrigerant charging cylinder for the student to inspect.

FUNDAMENTALS OF HVACR TEXT REFERENCE
Unit 31 Refrigerant System Charging

REQUIRED TOOLS AND EQUIPMENT
Charging Cylinder

SAFETY REQUIREMENTS:
None

PROCEDURE
Step 1
Locate a charging cylinder and examine it carefully so that you may complete the following exercise.

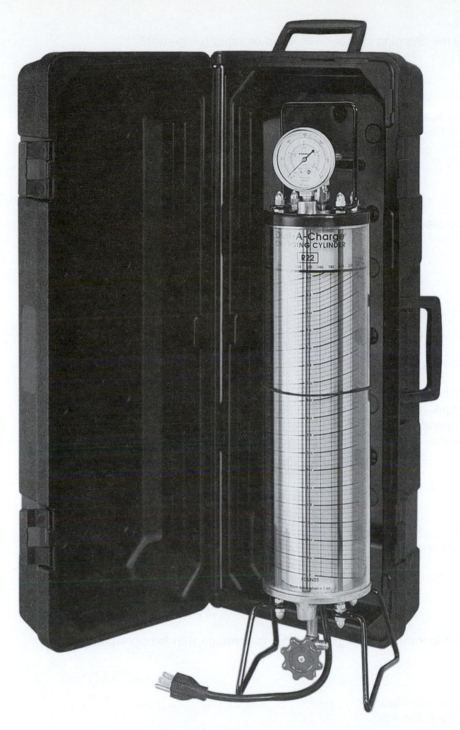

Figure 31-1-1

A. What types of refrigerants can the cylinder be used for?

B. Would you fill the charging cylinder with liquid or vapor refrigerant?

C. What does the scale on the charging cylinder measure and what effect, if any, does room temperature have on its measurement?

D. Why would you electrically heat a charging cylinder?

QUESTIONS

(Circle the letter that indicates the correct answer.)

1. A refrigerant charging cylinder is calibrated:
  A. in units of weight.
  B. in units of temperature.
  C. in units of volume.
  D. in units of pressure.

2. Charging cylinders should always have a:
  A. heater.
  B. relief valve.
  C. a sliding scale to adjust for temperature.
  D. All of the above are correct.

3. A charging cylinder can be used for any type of refrigerant.
  A. True.
  B. False.

4. The volume of liquid refrigerant in a charging cylinder will change with temperature.
  A. True.
  B. False.

5. A charging cylinder is usually filled:
  A. with vapor through the top.
  B. with liquid through the top.
  C. with vapor through the bottom.
  D. with liquid through the bottom.

6. Charging cylinder sizes generally range from:
  A. 8 to 10 ounces of refrigerant.
  B. 16 to 32 ounces of refrigerant.
  C. 2.5 to 10 lb of refrigerant.
  D. 1 to 15 lb of refrigerant.

LAB 31.2 VAPOR CHARGING WITH CHARGING CYLINDER

LABORATORY OBJECTIVE
The student will demonstrate the correct procedure for performing a vapor charge using a charging cylinder on a small refrigeration unit.

LABORATORY NOTES
Prior to charging, the system must be leak tested and evacuated. In this way, when the charging is started, the system is under vacuum so that when the refrigerant enters the system it will be drawn into the unit due to the difference in pressure.

When charging refrigerants that fractionate (zeotropes), such as R-401a, they must be charged as a liquid to prevent separation of the refrigerant as it enters the system.

FUNDAMENTALS OF HVACR TEXT REFERENCE
Unit 31 Refrigerant System Charging

REQUIRED TOOLS AND EQUIPMENT
Gloves & Goggles
Service Valve Wrench
Gauge Manifold
Charging Cylinder
Cylinder of Refrigerant
Operating Refrigeration System

SAFETY REQUIREMENTS
A. Wear safety goggles and gloves when working with refrigerants. Liquid refrigerant can cause frostbite when in contact with eyes and skin.

B. Use low loss hose fittings, or wrap cloth around hose fittings before removing the fittings from a pressurized system or cylinder. Inspect all fittings before attaching hoses.

PROCEDURE
Step 1
The refrigerant type and required amount of charge can be found on the refrigeration system nameplate as shown in Figure 31-2-1.

F.L.A. ☐   F.L.A. ☐

L.R.A. ☐   L.R.A. ☐

H.P. ☐   H.P. ☐

VOLTS ☐   VOLTS ☐

HERTZ ☐   HERTZ ☐
PHASE   PHASE

REFRIGERANT | 5.0. LB. | | KG. |
22

Figure 31-2-1

Step 2

A. Connect a charging cylinder, a vacuum pump, and the refrigerant cylinder with a gauge manifold as shown in Figure 31-2-2.

B. Start the vacuum pump to draw air from the charging cylinder and the hoses. The vacuum in the charging cylinder will also allow for a better flow from the refrigerant cylinder to the charging cylinder.

C. When vacuum has been achieved in the charging cylinder, close the low side gauge manifold valve and secure the vacuum pump and remove it.

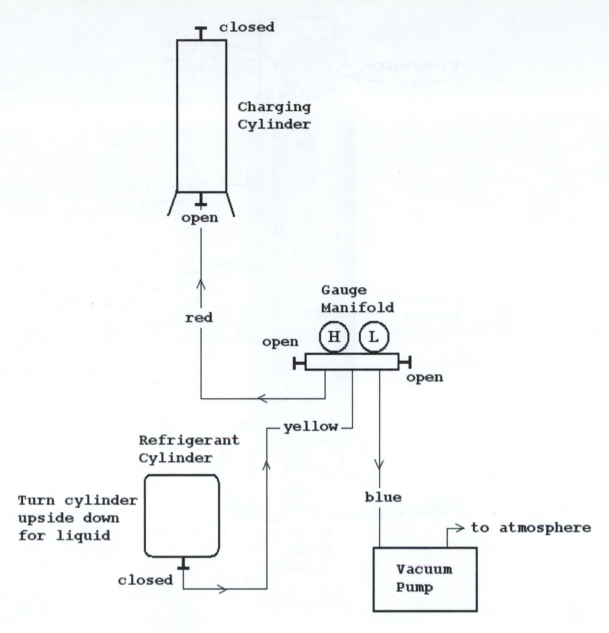

Figure 31-2-2

Step 3
You must be able to accurately read the refrigerant level on a charging cylinder.

    A. Refrigerant is always measured by weight in pounds and ounces or kilograms and grams in SI units. The measured volume on the charging cylinder is calibrated to an equivalent weight on the sliding scale.

    B. Since the temperature and pressure of the refrigerant will have a direct affect on its volume, a sliding weight scale is used. Some charging cylinders have scales for more than one type of refrigerant, so make sure that you are using the proper scale.

    C. The scale should be turned to adjust for the pressure registered by the pressure gauge on the top of the charging cylinder.

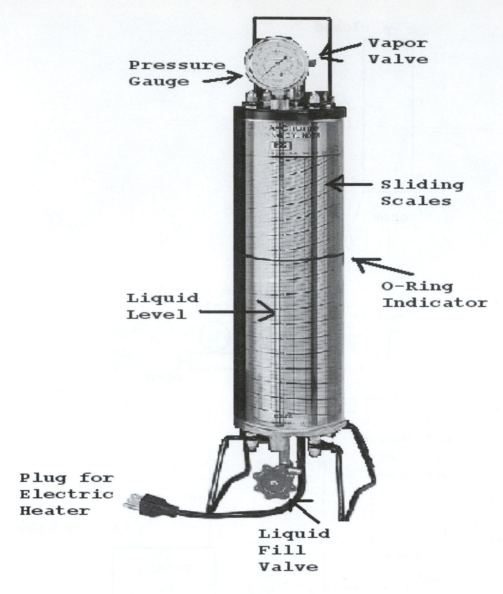

Pressure Gauge →

Vapor Valve

← Sliding Scales

← O-Ring Indicator

Liquid Level →

Plug for Electric Heater

Liquid Fill Valve

Figure 31-2-3

Step 4

    A. Close the high side gauge manifold valve as shown in Figure 31-2-4.

    B. Open the valve on the refrigerant cylinder and then turn it upside Down

    C. Slowly crack and throttle the high side gauge manifold valve to begin filling the charging cylinder with the correct amount of liquid refrigerant.

    D. Read the weight registered on the scale for the volume of refrigerant that you have added to the cylinder and record it here.

_____ lbs. _____ ounces

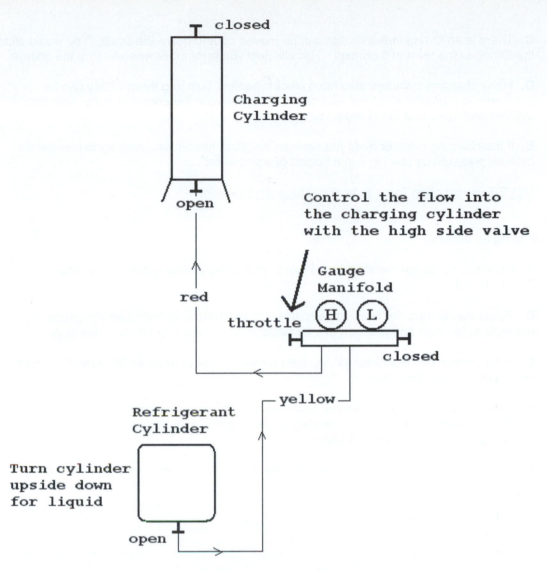

closed

Charging
Cylinder

open

Control the flow into
the charging cylinder
with the high side valve

red

Gauge
Manifold

throttle (H) (L)

closed

yellow

Refrigerant
Cylinder

Turn cylinder
upside down
for liquid

open

Figure 31-2-4

Step 5

A. When the charging cylinder is full with the proper amount of
refrigerant, close the filling valve on the bottom of the charging
cylinder and also the valve on the refrigerant cylinder.

B. Drain any excess pressure in the hoses through the gauge manifold low
pressure hose and then carefully disconnect the charging cylinder and
refrigerant cylinder from the gauge manifold.

Step 6
You should have already determined the amount of charge required for the system.

A. Subtract the amount of charge required from the amount measured in the cylinder and
you will be able to determine when the charge is complete.

B. As an example, assume you have filled the cylinder to a level of 22 ounces after dialing
the scale to match the appropriate pressure. Let's assume the required charge for the
refrigeration system is 16 ounces.

You would charge until the charging cylinder reached a level of 22 – 16 = 6 ounces.

C. There is an O-ring indicator that can be moved up and down the scale. You would slide the O-ring to the level of 6 ounces. This will help you remember when to stop the charge.

D. Many charging cylinders also have electric heaters built into them. They can be plugged into a regular electrical wall outlet. This will raise the pressure in the charging cylinder and speed up the charging process.

E. If the charging cylinder does not have an electrical heater, you may try to elevate the cylinder pressure by placing it in a bucket of warm water.

NEVER use an open flame to heat a refrigerant cylinder!!

Step 7
Begin charging as follows:

A. Connect the gauge manifold to the vapor side of the charging cylinder as shown in Figure 31-2-5

B. Purge the air from the lines while connecting the remaining hoses on the gauge manifold to the high and low side of the refrigeration system that you are charging.

C. If the system has Schrader valves, then be careful not to allow air to enter the system when you connect the hoses.

D. The system should already have been evacuated and in a vacuum and the refrigerant should flow freely as a vapor from the top of the charging cylinder into both the high side and low side of the system as a vapor.

E. Use the electric heater on the charging cylinder to the increase speed of the charge.

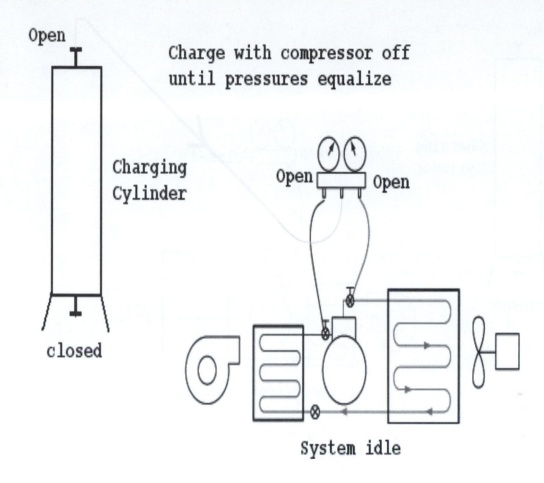

Open

Charge with compressor off
until pressures equalize

Charging
Cylinder

Open   Open

closed

System idle

Figure 31-2-5

Step 8

At some point the pressure in the refrigeration system will equalize with the pressure in the charging cylinder and no more refrigerant will flow. This may happen even though the full charge is not yet complete. At this point you will need to run the system to draw the remaining refrigerant from the charging cylinder into the low side.

    A. To do this, first close the high side valve on the gauge manifold before running the system.

    B. When the system is turned on, the refrigerant should once again begin to flow, this time into the low side (suction) only.

    C. Although there is no flow through the high side of the gauge manifold, you will still be able to monitor the system discharge pressure on the high side gauge.

    D. When the proper amount of refrigerant has been added to the system, close the gauge manifold low side valve.

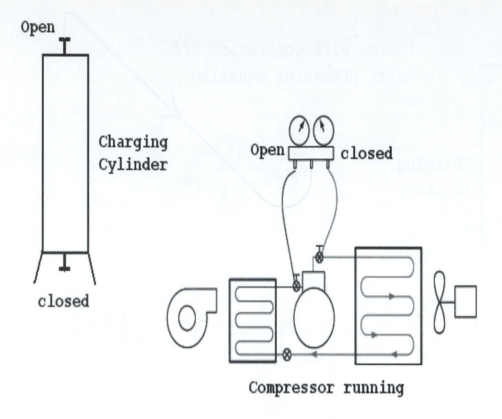

Open

Charging
Cylinder

Open closed

closed

Compressor running

Figure 31-2-6

E. Even though the high side and low side gauge manifold valves are now both closed, you will still be able to monitor the refrigeration system pressures due to the configuration of the gauge manifold.

F. Keep the gauge manifold connected long enough to allow the system time to stabilize.

G. Verify that the suction and discharge pressures of the refrigeration system agree with the manufacturer's recommendations and record the pressures below.

SUCTION PRESSURE _____ DISCHARGE PRESSURE _____

H. After the system has been charged and the operating pressures verified, backseat the suction and discharge service valves and close the vapor valve on the charging cylinder.

I. Carefully disconnect the hoses.

J. Any excess refrigerant remaining in the charging cylinder can be used for subsequent charges or transferred to a recovery cylinder through a recovery unit.

QUESTIONS

(Circle the letter that indicates the correct answer.)

1. It is usually considered good practice to charge with vapor:
    A. to prevent any danger of slugging the compressor with liquid.
    B. because it is faster than a liquid charge.
    C. because zeotropes must be charged as vapors.

D. to ensure accurate measurement of the charge.

2. A color coded disposable refrigeration cylinder in the upright position:
   A. has vapor at the top and liquid on the bottom.
   B. has liquid at the top and vapor at the bottom.
   C. would release liquid if the cylinder valve was opened.
   D. would release a mixture of liquid and vapor if the cylinder valve was opened.

3. The amount of refrigerant charge required for a system is measured by:
   A. volume.
   B. temperature.
   C. pressure.
   D. weight.

4. Zeotropic blends such as R-401a must be:
   A. charged as a vapor because they fractionate
   B. charged as a liquid because they fractionate.
   C. charged only into the suction side of the system.
   D. fractionated before charging.

5. A properly charged system:
   A. should have slightly extra refrigerant added to account for leaks.
   B. should have slightly less refrigerant added to prevent freeze up.
   C. should have exactly the amount of refrigerant required.
   D. should always have a 20% extra capacity reserve.

6. The charging cylinder measures:
   A. refrigerant volume.
   B. refrigerant weight.
   C. refrigerant temperature.
   D. All of the above are correct.

7. The sliding scale on a charging cylinder:
   A. corrects for ambient temperature.
   B. corrects for ambient pressure.
   C. is seldom of any use.
   D. serves as a refrigerant saturation table.

8. Charging cylinders may have built in electric heaters:
   A. so they don't freeze.
   B. to drive off any moisture mixed with the refrigerant.
   C. to reduce fractionation.
   D. speed up the rate of charge.

9. A liquid charge is always must faster than a vapor charge:
   A. True.
   B. False.

LAB 31.3 VAPOR CHARGING WITH DIGITAL SCALE

LABORATORY OBJECTIVE
The student will demonstrate the correct procedure for performing a vapor charge on a small refrigeration unit using a digital scale.

LABORATORY NOTES
Prior to charging, the system must be leak tested and evacuated. In this way, when the charging is started, the system is under vacuum so that when the refrigerant enters the system it will be drawn into the unit due to the difference in pressure.

When charging refrigerants that fractionate (zeotropes), such as R-401a, they must be charged as a liquid to prevent separation of the refrigerant as it enters the system.

FUNDAMENTALS OF HVACR TEXT REFERENCE
Unit 31 Refrigerant System Charging

REQUIRED TOOLS AND EQUIPMENT
Gloves & Goggles
Service Valve Wrench
Gauge Manifold
Digital Scale
Cylinder of Refrigerant
Operating Refrigeration System

SAFETY REQUIREMENTS
A. Wear safety goggles and gloves when working with refrigerants. Liquid refrigerant can cause frostbite when in contact with eyes and skin.

B. Use low loss hose fittings, or wrap cloth around hose fittings before removing the fittings from a pressurized system or cylinder. Inspect all fittings before attaching hoses.

PROCEDURE
Step 1
The refrigerant type and required amount of charge can be found on the refrigeration system nameplate as shown in Figure 31-3-1.

F.L.A. [   ]   F.L.A. [   ]

L.R.A. [   ]   L.R.A. [   ]

H.P. [   ]   H.P. [   ]

VOLTS [   ]   VOLTS [   ]

HERTZ [   ]   HERTZ [   ]
PHASE   PHASE

REFRIGERANT   [5.0.   LB.]   [KG.]
22

Figure 31-3-1

Step 2
Begin charging as follows:

    A. Place a refrigerant cylinder on a digital scale in the upright position and connect the gauge manifold as shown in Figure 31-3-2.

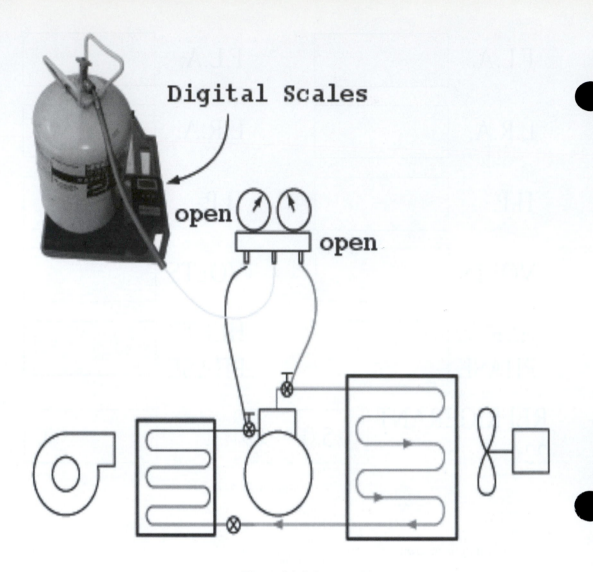

Figure 31-3-2

B. Purge the air from the lines while connecting the remaining hoses from the gauge manifold to the high and low side of the refrigeration system that you are charging.

C. If the system has Schrader valves, then be careful not to allow air to enter the system when you connect the hoses.

D. Set the digital readout on the scale to zero.

E. After the lines have been purged you are ready to begin charging. The service valves on the refrigeration system can be opened, the gauge manifold valves can be opened and then the refrigerant cylinder can be opened.

F. The system should already have been evacuated and in a vacuum. The refrigerant should flow freely into both the high side and low side of the system as a vapor.

G. Watch the digital readout as you charge and the weight of the cylinder should be decreasing by a negative amount. As an example, Lets assume the system requires a 12 ounce charge. You would continue to charge until the readout measured – 12 ounces (minus 12 ounces).

H. It is unlikely that you will overcharge the system at this point because normally only 50% to 75% of the charge will flow before the charging stops and the pressures equalize.

I. You can try to speed up the rate of charge by placing the cylinder in a bucket of warm water. If you choose to do this, then the bucket should be in place before you set the digital scale reading to zero or the charge quantity will be incorrect. Digital scales are very sensitive and the slightest movement can alter the reading.

NEVER let the temperature of the water bath exceed 125 °F.

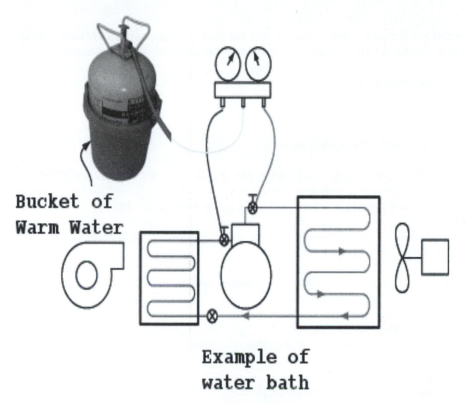

Bucket of
Warm Water

Example of
water bath

Figure 31-3-3

Step 3
At some point the pressure in the refrigeration system will equalize with the pressure in the charging cylinder and no more refrigerant will flow. This may happen even though the full charge is not yet complete. At this point you will need to run the system to draw the remaining refrigerant from the refrigerant cylinder into the low side.

A. To do this, first close the high side valve on the gauge manifold before running the system.

B. When the system is turned on, the refrigerant should once again begin to flow, this time into the low side (suction) only.

C. Although there is no flow through the high side of the gauge manifold, you will still be able to monitor the system discharge pressure on the high side gauge.

D. When the proper amount of refrigerant has been added to the system, close the gauge manifold low side valve.

E. Even though the high side and low side gauge manifold valves are now both closed, you will still be able to monitor the refrigeration system pressures due to the configuration of the gauge manifold.

F. Keep the gauge manifold connected long enough to allow the system time to stabilize.

G. Verify that the suction and discharge pressures of the refrigeration system agree with the manufacturer's recommendations and record the pressures below.

SUCTION PRESSURE _____ DISCHARGE PRESSURE _____

H. After the system has been charged and the operating pressures verified, backseat the suction and discharge service valves and close the valve on the refrigerant cylinder.

I. Carefully disconnect the hoses.

QUESTIONS

(Circle the letter that indicates the correct answer.)

1. When using a digital scale it is good practice to place it on a sturdy and level platform because:
   A. any slight movement can alter the readings.
   B. if the refrigerant cylinder is stable, you will get a more accurate reading.
   C. the digital scale is very accurate and therefore somewhat sensitive.
   D. All of the above are correct.

2. When using a warm water bath to heat a refrigeration cylinder:
   A. make sure that distilled water is used.
   B. never exceed 125 °F.
   C. always use a porcelain container.
   D. temperatures as high as 250 °F are often obtained.

3. A digital scale would be:
   A. much faster to use than a charging cylinder.
   B. much slower to use than a charging cylinder.
   C. used to measure the weight of a charging cylinder.
   D. used to measure the weight of the gauge manifold.

4. Prior to charging:
   A. the system should have a press charge of nitrogen added.
   B. the system should be flushed with air.
   C. the system must be leak tested and evacuated.
   D. the system must be vented to atmosphere.

5. Charging with a vapor:
   A. is cleaner than charging with a liquid.
   B. is not as clean as charging with a liquid.
   C. is faster than charging with a liquid.
   D. may damage the compressor due to a liquid slug.

6. The digital scale is set to zero:
   A. allowing it to calibrate.
   B. so that an accurate measurement of refrigerant can be obtained.
   C. halfway through the charging process.
   D. three quarters of the way through the charging process.

7. An insufficient charge could lead to:
    A. flooding of the evaporator.
    B. flooding of the compressor.
    C. flooding of the condenser.
    D. starving of the evaporator.

8. About a 1% change in refrigerant charge will change the superheat:
    A. 5 °F or more.
    B. 15 °F or more.
    C. 3 °F or more.
    D. 25 °F or more.

9. A vapor charge is always must faster than a liquid charge:
    A. True.
    B. False.

LAB 31.4 LIQUID CHARGING WITH COMPRESSOR OFF

LABORATORY OBJECTIVE
The student will demonstrate the correct procedure for performing a liquid charge on a small refrigeration unit using a digital scale.

LABORATORY NOTES
Prior to charging, the system must be leak tested and evacuated. In this way, when the charging is started, the system is under vacuum so that when the refrigerant enters the system it will be drawn into the unit due to the difference in pressure.

When charging refrigerants that fractionate (zeotropes), such as R-401a, they must be charged as a liquid to prevent separation of the refrigerant as it enters the system.

***FUNDAMENTALS OF HVACR* TEXT REFERENCE**
Unit 31 Refrigerant System Charging

REQUIRED TOOLS AND EQUIPMENT
Gloves & Goggles
Service Valve Wrench
Gauge Manifold
Digital Scale
Cylinder of Refrigerant
Operating Refrigeration System

SAFETY REQUIREMENTS
A. Wear safety goggles and gloves when working with refrigerants. Liquid refrigerant can cause frostbite when in contact with eyes and skin.

B. Use low loss hose fittings, or wrap cloth around hose fittings before removing the fittings from a pressurized system or cylinder. Inspect all fittings before attaching hoses.

PROCEDURE
Step 1
The refrigerant type and required amount of charge can be found on the refrigeration system nameplate as shown below.

F.L.A.		F.L.A.	
L.R.A.		L.R.A.	
H.P.		H.P.	
VOLTS		VOLTS	
HERTZ PHASE		HERTZ PHASE	
REFRIGERANT 22	5.0. LB.		KG.

Figure 31-4-1

Step 2

    A. Liquid charging is always much faster than charging with vapor.

    B. Liquid is charged on the high side of the system.

    C. On small systems the charging is done with the compressor off and the full charge is seldom completed.

    D. Therefore the vapor method must be used to complete the process.

    E. You must also make sure that no liquid enters the compressor.

Step 3

    A. Place a refrigerant cylinder on a digital scale in the upright position and connect the gauge manifold as shown in Figure 31-4-2.

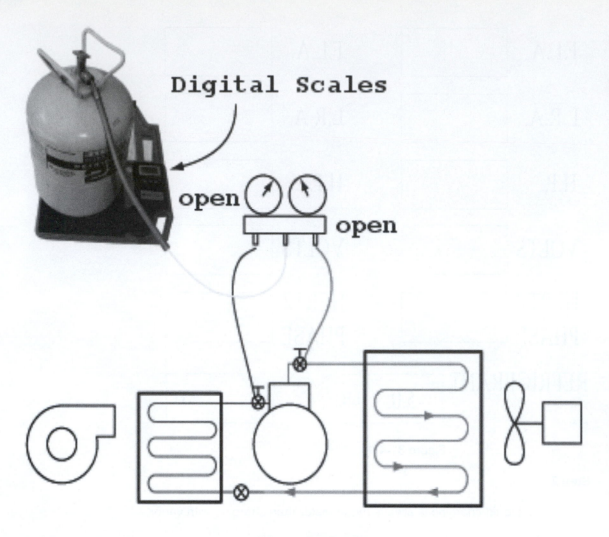

Figure 31-4-2

B. Purge the air from the lines while connecting the remaining hoses from the gauge manifold to the high and low side of the refrigeration system that you are charging.

C. If the system has Schrader valves, then be careful not to allow air to enter the system when you connect the hoses.

Step 4
After the lines have been purged you are ready to begin charging as follows:

Refrigerant cylinder inverted

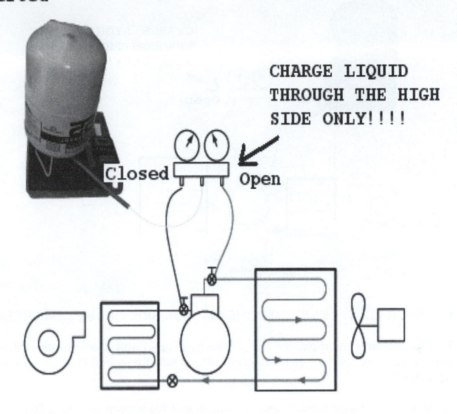

CHARGE LIQUID THROUGH THE HIGH SIDE ONLY!!!!

Closed   Open

Figure 31-4-3

A. Normally the refrigerant cylinder must be turned upside down (inverted) for liquid.

B. It is always good practice to open the cylinder valve all the way open prior to turning it over.

C. Once the bottle is inverted and balanced on the scale, set the digital readout on the scale to zero.

D. Prior to opening the service valves on the refrigeration system, make sure that ONLY THE HIGH SIDE OF THE GAUGE MANIFOLD IS OPEN AND THE LOW SIDE GAUGE MANIFOLD VALVE IS CLOSED as shown in Figure 31-4-4.

Refrigerant cylinder inverted

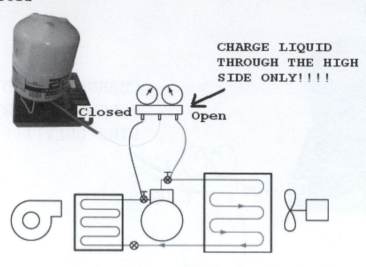

CHARGE LIQUID THROUGH THE HIGH SIDE ONLY!!!!

Closed   Open

Figure 31-4-4

E. NEVER ADMIT LIQUID INTO THE SUCTION SIDE OF THE COMPRESSOR!!

F. The system should already have been evacuated and in a vacuum. When the high side service valve is opened, the refrigerant should flow freely into the high side of the system as a liquid.

G. Watch the digital readout as you charge and the weight of the cylinder should be decreasing by a negative amount. As an example, Lets assume the system requires a 12 ounce charge. You would continue to charge until the readout measured – 12 ounces (minus 12 ounces).

H. It is unlikely that you will overcharge the system at this point because normally only 50% to 75% of the charge will flow before the charging stops and the pressures equalize.

I. At this point close the high side valve on the gauge manifold and record the readout in pounds and ounces from the digital scale in the space provided below.

AMOUNT OF LIQUID CHARGE _____ Lbs. _____ Oz. _____

Step 5
If the full charge is not complete after adding the liquid, then the vapor method must be used to complete the process.

A. Calculate the remaining refrigerant required to complete the charge. Subtract the amount of liquid already added from the total charge required.

Total Charge Required – Liquid Added = Vapor Charge Remaining

B. Turn the cylinder back over so that it is upright for vapor.

C. Start the system in the normal run mode and the discharge pressure should register on the high side of the gauge manifold.

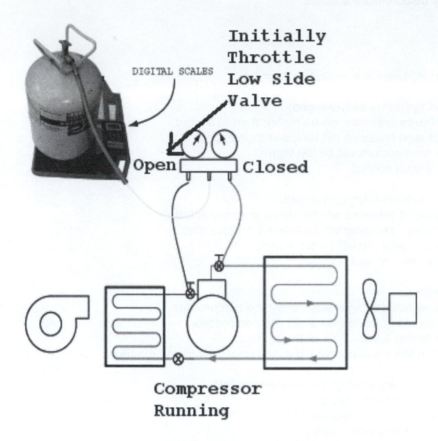

Figure 31-4-5

D. Although there is no flow through the high side of the gauge manifold, you will still be able to monitor the system discharge pressure on the high side gauge.

E. Since there has been liquid in the hoses, you will throttle the gauge manifold low side valve as you begin vapor charging. This will allow any remaining liquid left over in the hose to flash so as not to introduce liquid into the suction side of the compressor.

F. Slowly open the refrigeration system suction service valve while throttling the low side gauge manifold valve.

G. When the proper amount of refrigerant has been added to the system, close the gauge manifold low side valve.

H. Even though the high side and low side gauge manifold valves are now both closed, you will still be able to monitor the refrigeration system pressures due to the configuration of the gauge manifold.

I. Keep the gauge manifold connected long enough to allow the system time to stabilize.
J. Verify that the suction and discharge pressures of the refrigeration system agree with the manufacturer's recommendations and record the pressures below.

SUCTION PRESSURE _____ DISCHARGE PRESSURE _____

K. After the system has been charged and the operating pressures verified, backseat the suction and discharge service valves and close the valve on the refrigerant cylinder.

L. Carefully disconnect the hoses.

QUESTIONS

(Circle the letter that indicates the correct answer.)

1. If the refrigeration system is overcharged:
   A. The discharge pressure will be higher than expected.
   B. The discharge pressure will be lower than expected.
   C. The box temperature will be too warm.
   D. B & C are both correct.

2. If the refrigeration system is <u>under</u>charged:
   A. The discharge pressure will be higher than expected.
   B. The discharge pressure will be lower than expected.
   C. The box temperature will be too warm.
   D. B & C are both correct.

3. When liquid charging:
   A. never allow liquid refrigerant to enter the compressor.
   B. small amounts of liquid may enter the compressor.
   C. charge into the suction side of the system.
   D. charge into the suction side of the system with a gauge manifold.

4. Throttling liquid refrigerant through the gauge manifold:
   A. will increase its temperature.
   B. will allow some of it to flash to vapor.
   C. will damage the gauge manifold.
   D. will damage the gauge manifold valve seats.

5. To speed up the charging process:
   A. heat the cylinder with a propane torch.
   B. heat the cylinder to at least 150 °F.
   C. charge with a liquid.
   D. charge with a vapor.

6. When charging, if the gauge manifold valves are closed (front seated):
   A. no flow will occur through the manifold.
   B. system pressures can still be read.
   C. the system is isolated from the charging cylinder.
   D. All of the above are correct.

7. An insufficient charge could lead to starving of the evaporator:
   A. True.
   B. False.

288

LAB 31.5 LIQUID CHARGING WITH COMPRESSOR RUNNING

LABORATORY OBJECTIVE
The student will demonstrate the correct procedure for performing a liquid charge on a refrigeration unit that has a king valve.

LABORAORY NOTES
The refrigeration unit to be charged according to this procedure must have a manual king valve located after the condenser.

Prior to charging, the system must be leak tested and evacuated. In this way, when the charging is started, the system is under vacuum so that when the refrigerant enters the system it will be drawn into the unit due to the difference in pressure.

When charging refrigerants that fractionate (zeotropes), such as R-401a, they must be charged as a liquid to prevent separation of the refrigerant as it enters the system.

***FUNDAMENTALS OF HVACR* TEXT REFERENCE**
Unit 31 Refrigerant System Charging

REQUIRED TOOLS AND EQUIPMENT
Gloves & Goggles
Service Valve Wrench
Gauge Manifold
Digital Scale
Cylinder of Refrigerant
Operating Refrigeration System

SAFETY REQUIREMENTS
A. Wear safety goggles and gloves when working with refrigerants. Liquid refrigerant can cause frostbite when in contact with eyes and skin.

B. Use low loss hose fittings, or wrap cloth around hose fittings before removing the fittings from a pressurized system or cylinder. Inspect all fittings before attaching hoses.

PROCEDURE
Step 1
The refrigerant type and required amount of charge can be found on the refrigeration system nameplate as shown below.

F.L.A. [ ]   F.L.A. [ ]

L.R.A. [ ]   L.R.A. [ ]

H.P. [ ]   H.P. [ ]

VOLTS [ ]   VOLTS [ ]

HERTZ [ ]   HERTZ [ ]
PHASE   PHASE

REFRIGERANT [5.0. LB.]   [KG.]
22

Figure 31-5-1

Step 2

A. Liquid charging is always much faster than charging with vapor.

B. Liquid is charged on the high side of the system.

C. On small systems the charging is done with the compressor off and the full charge is seldom completed and the vapor method must be used to complete the process. You must also make sure that no liquid enters the compressor. Therefore most small systems are vapor charged.

D. On large systems, a king valve located between the condenser and the metering valve offers a convenient means of charging the system on the high side, with the compressor running.

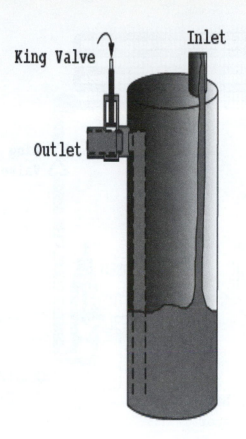

King Valve

Inlet

Outlet

King Valve
Protective
Cap

Charging
Connection

Outlet

Inlet

Figure 31-5-2

Step 3

A. Purge the air from the lines while connecting the hoses from the gauge manifold to the high and low side of the refrigeration system.

B. Purge each line for about 2 seconds.

C. It is always best to purge the lines with vapor and the cylinder in the upright position as shown in Figure 31-5-3.

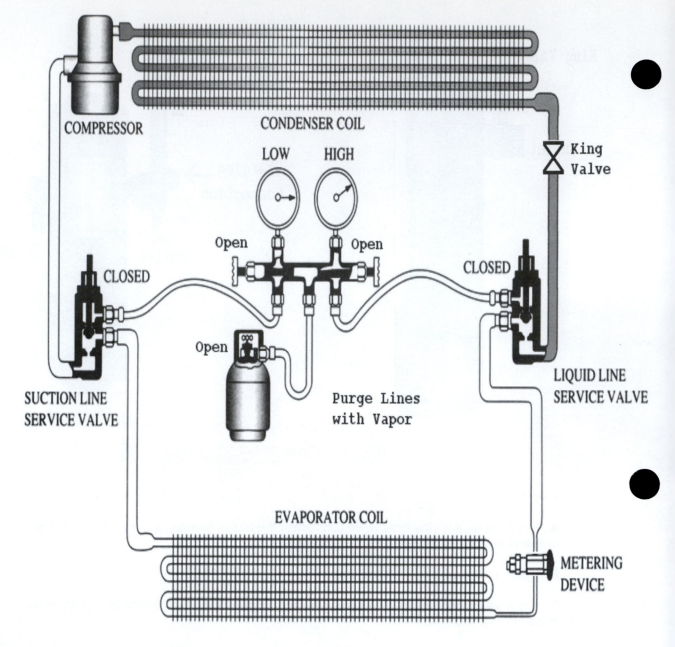

COMPRESSOR

CONDENSER COIL

LOW   HIGH

King
Valve

Open   Open

CLOSED   CLOSED

Open

SUCTION LINE
SERVICE VALVE

Purge Lines
with Vapor

LIQUID LINE
SERVICE VALVE

EVAPORATOR COIL

METERING
DEVICE

Figure 31-5-3

Step 4

A. After the lines have been purged you are ready to begin charging as follows:

B. Make sure that the high and low side gauge manifold valves are closed to start.

C. Open the refrigerant cylinder fully and then turn it over and balance it on a digital scale.

D. Once the bottle is inverted and balanced on the scale, set the digital readout on the scale to zero.

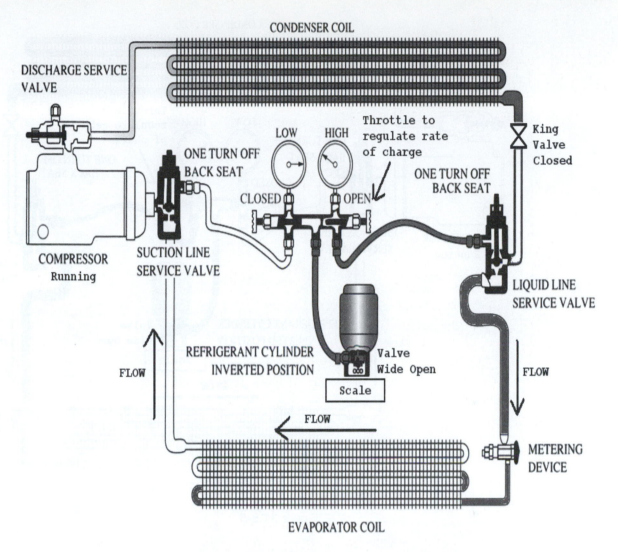

CONDENSER COIL

DISCHARGE SERVICE VALVE

ONE TURN OFF BACK SEAT

LOW   HIGH

Throttle to regulate rate of charge

CLOSED   OPEN

King Valve Closed

ONE TURN OFF BACK SEAT

COMPRESSOR
Running

SUCTION LINE SERVICE VALVE

LIQUID LINE SERVICE VALVE

FLOW

REFRIGERANT CYLINDER INVERTED POSITION

Valve Wide Open

Scale

FLOW

FLOW

METERING DEVICE

EVAPORATOR COIL

Figure 31-5-4

E. Crack the suction and liquid line service valves off their back seats and <u>close the KING VALVE</u>.

F. The system should already have been evacuated and in a vacuum. When the high side service valve is opened, the refrigerant will flow freely into the high side of the system as a liquid as shown in Figure 31-5-5.

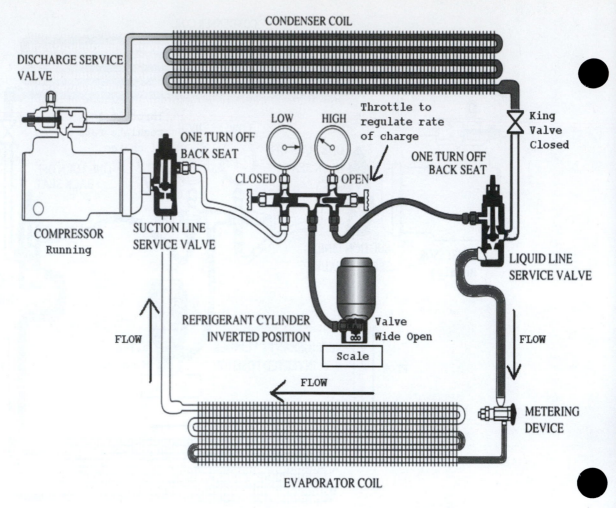

CONDENSER COIL

DISCHARGE SERVICE
VALUE

Throttle to
regulate rate
of charge

LOW   HIGH

King
Valve
Closed

ONE TURN OFF
BACK SEAT

CLOSED   OPEN

ONE TURN OFF
BACK SEAT

COMPRESSOR
Running

SUCTION LINE
SERVICE VALVE

LIQUID LINE
SERVICE VALVE

FLOW

REFRIGERANT CYLINDER
INVERTED POSITION

Valve
Wide Open

FLOW

Scale

FLOW

METERING
DEVICE

EVAPORATOR COIL

Figure 31-5-5

G. The metering device must be in the open position to allow refrigerant to flow into the evaporator. If the system has a box solenoid valve it must be open.

H. As the refrigerant flows through the evaporator, the suction side pressure should begin to rise above the low pressure cutout setting.

I. You can now start the system and run it in the normal operating mode and with the king valve closed, the liquid line service valve pressure will be low enough that refrigerant will be drawn from the cylinder into the system.

J. Throttle the high side gauge manifold valve to control the rate of charge and continue charging until the proper amount of refrigerant has been added as indicated by the reading on the digital scale.

K. Watch the digital readout as you charge and the weight of the cylinder should be decreasing by a negative amount. As an example, Lets assume the system requires a 12 ounce charge. You would continue to charge until the readout measured – 12 ounces (minus 12 ounces).

Step 5

A. When the proper amount of refrigerant has been added to the system, close the gauge manifold high side valve and then <u>open the King Valve</u>.

B. Even though the high side and low side gauge manifold valves are now both closed, you will still be able to monitor the refrigeration system pressures due to the configuration of the gauge manifold.

C. Keep the gauge manifold connected long enough to allow the system time to stabilize.

D. Verify that the suction and discharge pressures of the refrigeration system agree with the manufacturer's recommendations and record the pressures below.

SUCTION PRESSURE _____ DISCHARGE PRESSURE _____

E. After the system has been charged and the operating pressures verified, backseat the suction and discharge service valves and close the valve on the refrigerant cylinder.

F. Carefully disconnect the hoses.

QUESTIONS

(Circle the letter that indicates the correct answer.)

1. When charging through the liquid side with the compressor running:
   A. the box solenoid valve must be open.
   B. the liquid should be added directly into the suction.
   C. the king valve should be open.
   D. All of the above are correct.

2. The king valve should be located:
   A. in the suction line.
   B. right opposite the queen valve.
   C. in the liquid line right after the condenser or receiver.
   D. at the outlet of the evaporator.

3. Closing the king valve while charging:
   A. is bad practice.
   B. allows for the line pressure to drop below charging cylinder pressure.
   C. can damage the compressor.
   D. will trip the unit on the high pressure cutout.

4. When preparing to charge, if the king valve is closed but the refrigerant cylinder is not open:
   A. the discharge pressure will rise rapidly.
   B. the suction pressure will rise rapidly.
   C. both the suction pressure and discharge pressure will rise rapidly.
   D. the unit will stop on the low pressure cut out.

LAB 31.6 LIQUID CHARGING WITH BLENDS (ZEOTROPES)

LABORATORY OBJECTIVE
The student will demonstrate the correct procedure for performing a liquid charge with a blended (zoetrope) refrigerant on a small refrigeration unit using a digital scale.

LABORATORY NOTES
Prior to charging, the system must be leak tested and evacuated. In this way, when the charging is started, the system is under vacuum so that when the refrigerant enters the system it will be drawn into the unit due to the difference in pressure.

When charging refrigerants that fractionate (zeotropes), such as R-401a, they must be charged as a liquid to prevent separation of the refrigerant as it enters the system.

This can be done with a 100% liquid charge as described in Lab 31.5 with a refrigeration system that has a king valve. This can also be accomplished on smaller systems as described in the following procedure.

FUNDAMENTALS OF HVACR TEXT REFERENCE
Unit 31 Refrigerant System Charging

REQUIRED TOOLS AND EQUIPMENT
Gloves & Goggles
Service Valve Wrench
Gauge Manifold
Digital Scale
Cylinder of Refrigerant
Operating Refrigeration System

SAFETY REQUIREMENTS
A. Wear safety goggles and gloves when working with refrigerants. Liquid refrigerant can cause frostbite when in contact with eyes and skin.

B. Use low loss hose fittings, or wrap cloth around hose fittings before removing the fittings from a pressurized system or cylinder. Inspect all fittings before attaching hoses.

PROCEDURE
Step 1
The refrigerant type and required amount of charge can be found on the refrigeration system nameplate as shown below.

F.L.A. [ ]   F.L.A. [ ]

L.R.A. [ ]   L.R.A. [ ]

H.P. [ ]   H.P. [ ]

VOLTS [ ]   VOLTS [ ]

HERTZ [ ]   HERTZ [ ]
PHASE   PHASE

REFRIGERANT
22   5.0. LB. [ ]   KG.

Figure 31-6-1

Step 2

A. Liquid charging is always the method used for charging blended refrigerants (zeotropes).

B. Zeotropic blends generally consist of three different types of refrigerant blended together and they all have different boiling points.

C. Blends can fractionate when allowed to vaporize which means that one of the refrigerants will boil off before the others.

D. If you charge a blend as a vapor, the mixture of refrigerant entering the system will not be representative of the mixture in the blend.

Step 3

A. Liquid is charged on the high side of the system.

B. On small systems the charging is done with the compressor off and the full charge is seldom completed.

C. Therefore additional charging is required to complete the process.

D. You must also make sure that no liquid enters the compressor.

Step 4

    A. Place a refrigerant cylinder on a digital scale in the upright position and connect the gauge manifold as shown in Figure 31-6-2.

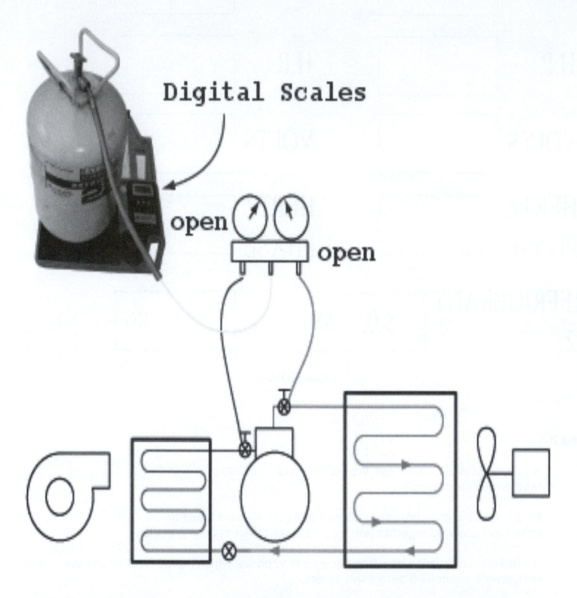

Figure 31-6-2

    B. Purge the air from the lines while connecting the remaining hoses from the gauge manifold to the high and low side of the refrigeration system that you are charging.

    C. If the system has Schrader valves, then be careful not to allow air to enter the system when you connect the hoses.

Step 5
After the lines have been purged you are ready to begin charging as follows:

A. Normally the refrigerant cylinder must be turned upside down (inverted) for liquid.

B. It is always good practice to open the cylinder valve all the way open prior to turning it over.

C. Once the bottle is inverted and balanced on the scale, set the digital readout on the scale to zero.

D. Prior to opening the service valves on the refrigeration system, make sure that ONLY THE HIGH SIDE OF THE GAUGE MANIFOLD IS OPEN AND THE LOW SIDE GAUGE MANIFOLD VALVE IS CLOSED as shown in Figure 31-6-3.

Refrigerant cylinder inverted

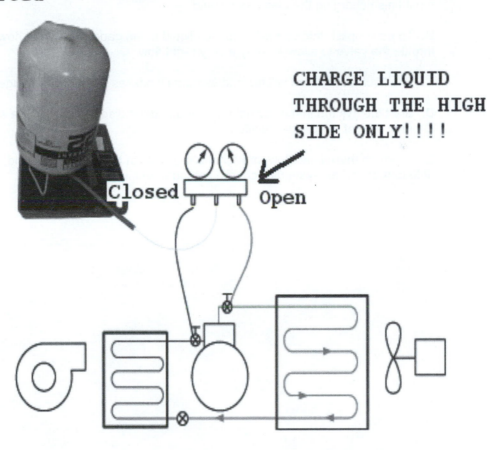

CHARGE LIQUID THROUGH THE HIGH SIDE ONLY!!!!

Closed   Open

Figure 31-6-3

E. NEVER ADMIT LIQUID INTO THE SUCTION SIDE OF THE COMPRESSOR!!

F. The system should already have been evacuated and in a vacuum. When the high side service valve is opened, the refrigerant should flow freely into the high side of the system as a liquid.

G. Watch the digital readout as you charge and the weight of the cylinder should be decreasing by a negative amount. As an example, Lets assume the system requires a 12 ounce charge. You would continue to charge until the readout measured – 12 ounces (minus 12 ounces).

H. It is unlikely that you will overcharge the system at this point because normally only 50% to 75% of the charge will flow before the charging stops and the pressures equalize.

I. At this point close the high side valve on the gauge manifold and record the readout in pounds and ounces from the digital scale in the space provided below.

AMOUNT OF LIQUID CHARGE _____ Lbs. _____ Oz. _____

Step 6
If the full charge is not complete after adding the liquid, then additional charging must be performed to complete the process.

A. You must never allow liquid refrigerant to enter the compressor suction, however you must finish charging the blend as a liquid.

B. To accomplish this you will introduce liquid to the gauge manifold, however you will throttle the valve to allow for only a very slight flow.

C. The liquid refrigerant will flash to vapor as it passes through the gauge manifold valve

D. Start the system in the normal run mode and the discharge pressure should register on the high side of the gauge manifold.

E. Although there is no flow through the high side of the gauge manifold, you will still be able to monitor the system discharge pressure on the high side gauge.

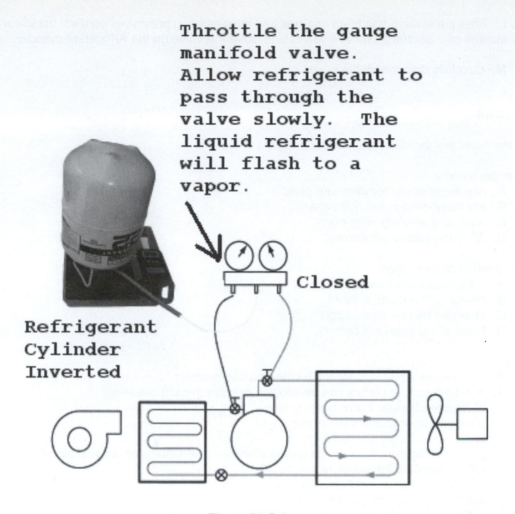

Throttle the gauge manifold valve. Allow refrigerant to pass through the valve slowly. The liquid refrigerant will flash to a vapor.

Closed

Refrigerant Cylinder Inverted

Figure 31-6-4

F. You will throttle the gauge manifold low side valve as you begin vapor charging. This will allow the liquid in the hose to flash so as not to introduce liquid into the suction side of the compressor.

G. Slowly open the refrigeration system suction service valve while throttling the low side gauge manifold valve.

H. When the proper amount of refrigerant has been added to the system, close the gauge manifold low side valve.

I. Even though the high side and low side gauge manifold valves are now both closed, you will still be able to monitor the refrigeration system pressures due to the configuration of the gauge manifold.

J. Keep the gauge manifold connected long enough to allow the system time to stabilize.

K. Verify that the suction and discharge pressures of the refrigeration system agree with the manufacturer's recommendations and record the pressures below.

SUCTION PRESSURE _____ DISCHARGE PRESSURE _____

L. After the system has been charged and the operating pressures verified, backseat the suction and discharge service valves and close the valve on the refrigerant cylinder.

M. Carefully disconnect the hoses.

QUESTIONS

(Circle the letter that indicates the correct answer.)

1. Zeotropic blends:
   A. can experience "temperature glide".
   B. are numbered in the 500 series.
   C. have one single boiling point.
   D. All of the above are correct.

2. Fractionation occurs when:
   A. refrigerants are mixed with oil.
   B. charging blends as a liquid.
   C. charging blends as a vapor.
   D. None of the above is correct.

3. When liquid charging:
   A. never allow liquid refrigerant to enter the compressor.
   B. liquid refrigerant blends can be allowed to enter the compressor.
   C. always fractionate the refrigerant first.
   D. a weight measurement is not required.

4. Throttling liquid refrigerant through the gauge manifold when charging a liquid blend:
   A. will increase its temperature.
   B. will allow some of it to flash to vapor.
   C. will damage the gauge manifold.
   D. will damage the gauge manifold valve seats.

5. What types of refrigerants must be charged as a liquid?
   A. azeotropic blends.
   B. zeotropic blends.
   C. near-azeotropic blends.
   D. Both B & C must be charged as liquids.

6. Zeotropic blends are identified as the:
   A. Z series.
   B. 500 series.
   C. 400 series.
   D. 700 series.

7. The weight of a closed cylinder containing a blend will vary with temperature:
   A. True.
   B. False.

LAB 31.8 CHARGING BY WEIGHT PACKAGED UNIT

LABORATORY OBJECTIVE
You will demonstrate your ability to weigh a charge into an evacuated air conditioning unit.

LABORATORY NOTES
Using the unit that you recovered the refrigerant from in Lab 28.7 and evacuated in Lab 30.2, you will weigh in the manufacturer's specified refrigerant charge.

***FUNDAMENTALS OF HVACR* TEXT REFERENCE**
Unit 31 Refrigerant System Charging

REQUIRED MATERIALS PROVIDED BY STUDENT
Safety Glasses
Gloves
Manifold gauges
Refrigeration wrench

REQUIRED MATERIALS PROVIDED BY SCHOOL
Packaged air conditioning unit
Scale
Refrigerant

PROCEDURE
1. Your gauges should be connected to the low side and high side gauge ports with the gauges closed.
2. The system should be evacuated.
3. Turn on the scale and zero it.
4. Place the refrigerant cylinder on the scale, weigh the cylinder, and record its weight.
5. Connect the center hose to the refrigerant cylinder.
6. Place cylinder on scale in a position to deliver liquid. That is upside down for most non-refillable cylinders. For refrigerant recovery cylinders, that is upright with the hose connected to the liquid valve.
7. Open the valve on the cylinder and purge the hose from the cylinder to the gauges.
8. Zero the scale with the cylinder on the scale.
9. Open the high side manifold gauge.
10. Monitor the scale and close the high side gauge when the correct weight is read on the scale.
11. Close the valve on the charging cylinder.
12. Operate the system.
13. Crack the low side gauge and meter in the refrigerant trapped in the gauge hoses into the low side.
14. Remove you gauges and replace the valve caps.
15. Complete the Data Sheet Below

Charging Refrigerant System by Weight	
Type of Refrigerant	
Amount of Refrigerant	
Cylinder Weight Before Charging	
Cylinder Weight After Charging	
Amount of Refrigerant Charged Into Unit	

LAB 31.9 CHARGING BY WEIGHT SPLIT SYSTEM

LABORATORY OBJECTIVE
You will demonstrate your ability to weigh a charge into an evacuated air conditioning unit.

LABORATORY NOTES
Using the unit that you recovered the refrigerant from in Lab 28.8, you will weigh in the manufacturer's specified refrigerant charge.

***FUNDAMENTALS OF HVACR* TEXT REFERENCE**
Unit 31 Refrigerant System Charging

REQUIRED MATERIALS PROVIDED BY STUDENT
Safety Glasses
Gloves
Manifold gauges
Refrigeration wrench

REQUIRED MATERIALS PROVIDED BY SCHOOL
Split System air conditioning unit
Scale
Refrigerant

PROCEDURE
1. Your gauges should be connected to the low side and high side gauge ports with the gauges closed.
2. The system should be evacuated.
3. Calculate the correct refrigerant charge using the manufacturer's instructions
   A typical formula is:
   Condensing Unit Charge + (0.6 oz x actual line length – 15 ft)
   (Note: this is for units with a 3/8" liquid line and a factory charge for 15 feet of lines)
4. Turn on the scale and zero it.
5. Place the refrigerant cylinder on the scale, weigh the cylinder, and record its weight.
6. Connect the center hose to the refrigerant cylinder.
7. Place cylinder on scale in a position to deliver liquid. That is upside down for most non-refillable cylinders. For refrigerant recovery cylinders, that is upright with the hose connected to the liquid valve.
8. Open the valve on the cylinder and purge the hose from the cylinder to the gauges.
9. Zero the scale with the cylinder on the scale.
10. Open the high side manifold gauge.
11. Monitor the scale and close the high side gauge when the correct weight is read on the scale.
12. Close the valve on the charging cylinder.
13. Operate the system.
14. Crack the low side gauge and meter in the refrigerant trapped in the gauge hoses into the low side.
15. Remove you gauges and replace the valve caps.
16. Complete the Data Sheet Below

Charging Split System by Weight	
Type of Refrigerant	
Amount of Refrigerant	
Cylinder Weight Before Charging	
Cylinder Weight After Charging	
Amount of Refrigerant Charged Into Unit	
Operating Suction Pressure	
Operating Discharge Pressure	

LAB 31.10 PRESSURE-TEMPERATURE METHOD

LABORATORY OBJECTIVE
You will demonstrate your ability to check refrigerant charge in three refrigeration system using a pressure-temperature chart.

LABORATORY NOTES
You will use common temperature relationships and a pressure-temperature chart to determine the correct charge for three different refrigeration systems. You will then check the charge and compare the actual pressures to the desired pressures.

FUNDAMENTALS OF HVACR TEXT REFERENCE
Unit 31 Refrigerant System Charging

REQUIRED MATERIALS PROVIDED BY STUDENT
Safety Glasses
Gloves
Manifold gauges
Refrigeration wrench
Thermometer

REQUIRED MATERIALS PROVIDED BY SCHOOL
Three operating Refrigeration systems

PROCEDURE
1. Examine the unit and record important unit characteristics in the Data Sheet, including:
    a. Type of Unit (air conditioner, freezer, etc.)
    b. Type of Refrigerant
    c. Metering Device (TEV or fixed restriction)
    d. Condenser (air or water)
    e. Evaporator (air or water)
2. Measure and record the relevant operating conditions, including:
    a. Ambient Temperature for air cooled condensers
    b. Inlet water temperature for water cooled condensers
    c. Return Air Temperature for Air Conditioning Units
    d. Return Air Dew Point Temperature for air conditioning units
    e. Box Temperature for Commercial Refrigeration
3. Determine the desired condenser temperature
    a. High Efficiency Air Cooled Condenser Temp = Ambient Temp + 20° (newer units)
    b. Low Efficiency Air Cooled Condenser Temp = Ambient Temp + 30° (older units)
    c. Water Cooled Condenser Temperature = inlet water temperature + 20°
4. Determine the desired condenser pressure using the desired condenser temperature and a PT chart.

5. Determine the desired evaporator temperature.
   a. High Efficiency Air Conditioning (newer units) Use
   b. **Either** Evaporator = Return Air Temp - 35° (easiest because it does not require a psychrometer)
   c. **Or** Evaporator = Return Air Dewpoint - 15° (more accurate because it takes humidity into account)
   d. Low Efficiency Air Conditioning (older units)
   e. **Either** Evaporator = Return Air Temp - 40° (easiest because it does not require a psychrometer)
   f. **Or** Evaporator = Return Dewpoint - 20° (more accurate because it takes humidity into account)
   g. Commercial Refrigeration Evaporator Temperature = Box Temperature - 10°
6. Determine the desired evaporator pressure using the desired evaporator temperature and a PT chart
7. Connect your manifold gauges to the system.
8. Operate the system until the pressures stabilize and the compressor warms up.
9. Record the actual operating condenser and evaporator pressures.
10. Use a PT Chart to determine the Actual Operating Condenser and Evaporator Temperatures
11. Compare the desired temperatures and pressures to the actual temperatures and pressures.
12. Determine if the system is charged correctly, undercharged, overcharge, or has another refrigeration problem.
13. Complete Data Sheet 1.
14. Repeat the entire process on Unit 2 and complete Data Sheet 2.
15. Repeat the Entire Process on Unit 3 and complete Data Sheet 3.

LAB 31.10 PRESSURE–TEMPERATURE CHARGING DATA SHEET 1	
TYPE OF UNIT (*Air conditioner, freezer, etc.*)	
TYPE OF CONDENSER (*Air or Water cooled*)	
TYPE OF EVAPORATOR (*Cools Air or Water*)	
TYPE OF METERING DEVICE (*Expansion valve, cap tube, orifice*)	
TYPE OF REFRIGERANT (*R22, R410A, R134a, etc*)	
OUTDOOR AMBIENT/ WATER TEMPERATURE (*Temp of condenser air or water*)	
DESIRED CONDENSER TEMPERATURE (*Condenser Saturation from formula*)	
DESIRED CONDENSER PRESSURE (*Condenser Pressure from PT chart*)	
RETURN AIR TEMP/ BOX TEMP (*Temp of air in house or refrigerated box*)	
RETURN AIR DEWPOINT (*Measured with digital psychrometer*)	
DESIRED EVAPORATOR TEMPERATURE (*Evaporator saturation using formula*)	
DESIRED EVAPORATOR PRESSURE (*Evaporator Pressure from PT chart*)	
ACTUAL CONDENSER PRESSURE (*High side pressure actually read on gauges*)	
ACTUAL CONDENSER TEMPERATURE (*High side pressure converted to temp by PT*)	
ACTUAL EVAPORATOR PRESSURE (*Low side pressure on gauges*)	
ACTUAL EVAPORATOR TEMPERATURE (*Low side pressure converted to temp by PT*)	
SYSTEM CONDITION (*overcharged, undercharged, or correct*)	

LAB 31.10 PRESSURE–TEMPERATURE CHARGING DATA SHEET 2	
TYPE OF UNIT (*Air conditioner, freezer, etc.*)	
TYPE OF CONDENSER (*Air or Water cooled*)	
TYPE OF EVAPORATOR (*Cools Air or Water*)	
TYPE OF METERING DEVICE (*Expansion valve, cap tube, orifice*)	
TYPE OF REFRIGERANT (*R22, R410A, R134a, etc.*)	
OUTDOOR AMBIENT/ WATER TEMPERATURE (*Temp of condenser air or water*)	
DESIRED CONDENSER TEMPERATURE (*Condenser Saturation from formula*)	
DESIRED CONDENSER PRESSURE (*Condenser Pressure from PT chart*)	
RETURN AIR TEMP/ BOX TEMP (*Temp of air in house or refrigerated box*)	
RETURN AIR DEWPOINT (*Measured with digital psychrometer*)	
DESIRED EVAPORATOR TEMPERATURE (*Evaporator saturation using formula*)	
DESIRED EVAPORATOR PRESSURE (*Evaporator Pressure from PT chart*)	
ACTUAL CONDENSER PRESSURE (*High side pressure actually read on gauges*)	
ACTUAL CONDENSER TEMPERATURE (*High side pressure converted to temp by PT*)	
ACTUAL EVAPORATOR PRESSURE (*Low side pressure on gauges*)	
ACTUAL EVAPORATOR TEMPERATURE (*Low side pressure converted to temp by PT*)	
SYSTEM CONDITION (*overcharged, undercharged, or correct*)	

LAB 31.10 PRESSURE–TEMPERATURE CHARGING
DATA SHEET 3

TYPE OF UNIT (*Air conditioner, freezer, etc.*)	
TYPE OF CONDENSER (*Air or Water cooled*)	
TYPE OF EVAPORATOR (*Cools Air or Water*)	
TYPE OF METERING DEVICE (*Expansion valve, cap tube, orifice*)	
TYPE OF REFRIGERANT (*R22, R410A, R134a, etc*)	
OUTDOOR AMBIENT/ WATER TEMPERATURE (*Temp of condenser air or water*)	
DESIRED CONDENSER TEMPERATURE (*Condenser Saturation from formula*)	
DESIRED CONDENSER PRESSURE (*Condenser Pressure from PT chart*)	
RETURN AIR TEMP/ BOX TEMP (*Temp of air in house or refrigerated box*)	
RETURN AIR DEWPOINT (*Measured with digital psychrometer*)	
DESIRED EVAPORATOR TEMPERATURE (*Evaporator saturation using formula*)	
DESIRED EVAPORATOR PRESSURE (*Evaporator Pressure from PT chart*)	
ACTUAL CONDENSER PRESSURE (*High side pressure actually read on gauges*)	
ACTUAL CONDENSER TEMPERATURE (*High side pressure converted to temp by PT*)	
ACTUAL EVAPORATOR PRESSURE (*Low side pressure on gauges*)	
ACTUAL EVAPORATOR TEMPERATURE (*Low side pressure converted to temp by PT*)	
SYSTEM CONDITION (*overcharged, undercharged, or correct*)	

LAB 31.11 VAPOR CHARGE – PRESSURE/TEMPERATURE

LABORATORY OBJECTIVE

You will demonstrate your ability to safely add refrigerant vapor to an operating unit until the system operating characteristics meet the manufacturer's specifications.

LABORATORY NOTES

You will add refrigerant vapor to the low side with the system operating until the system operating characteristics match the desired pressures determined by common temperature relationships and a pressure-temperature chart.

***FUNDAMENTALS OF HVACR* TEXT REFERENCE**

Unit 31 Refrigerant System Charging

REQUIRED MATERIALS PROVIDED BY STUDENT

Safety Glasses
Gloves
Manifold gauges
Refrigeration wrench
Thermometer

REQUIRED MATERIALS PROVIDED BY SCHOOL

Operating Refrigeration system
Refrigerant
Scale

PROCEDURE

1. Examine the unit and record important unit characteristics in the Data Sheet, including:
    a. Type of Unit (air conditioner, freezer, etc)
    b. Type of Refrigerant
    c. Metering Device (TEV or fixed restriction)
    d. Condenser (air or water)
    e. Evaporator (air or water)
2. Measure and record the relevant operating conditions, including:
    a. Ambient Temperature for air cooled condensers
    b. Inlet water temperature for water cooled condensers
    c. Return Air Temperature for Air Conditioning Units
    d. Return Air Dew Point Temperature for air conditioning units
    e. Box Temperature for Commercial Refrigeration
3. Use the Pressure-Temperature method detailed in Lab 28 to determine the desired system pressures.
4. Connect your gauges to the unit.
5. Purge the air out of the hoses and manifold.
6. Connect the middle gauge hose to the vapor valve on a refrigerant cylinder.
7. *(Note: Make sure that the refrigerant listed on the data plate is the same as the refrigerant in the cylinder)*
8. Open the cylinder valve and purge the middle hose at the manifold.

9. Turn the scale on.
10. Place the cylinder in the upright position on the scale and zero the scale.
11. With the unit operating, open the low side manifold hand wheel.
12. The compound gauge should show a higher pressure – this is the cylinder pressure.
13. Periodically close the low side manifold hand wheel to see what the system operating pressure is.
14. Monitor the low side and high side pressures.
15. When the operating pressures are similar to the desired pressures, you are finished.
16. Record the system operating pressures and amount of refrigerant used.

VAPOR CHARGING DATA SHEET	
TYPE OF UNIT (*Air conditioner, freezer, etc.*)	
TYPE OF CONDENSER (*Air or Water cooled*)	
TYPE OF EVAPORATOR (*Cools Air or Water*)	
TYPE OF METERING DEVICE (*Expansion valve, cap tube, orifice*)	
TYPE OF REFRIGERANT (*R22, R410A, R134a, etc*)	
OUTDOOR AMBIENT/ WATER TEMPERATURE (*Temp of condenser air or water*)	
DESIRED CONDENSER TEMPERATURE (*Condenser Saturation from formula*)	
DESIRED CONDENSER PRESSURE (*Condenser Pressure from PT chart*)	
RETURN AIR TEMP/ BOX TEMP (*Temp of air in house or refrigerated box*)	
RETURN AIR DEWPOINT (*Measured with a psychrometer*)	
DESIRED EVAPORATOR TEMPERATURE (*Evaporator saturation using formula*)	
DESIRED EVAPORATOR PRESSURE (*Evaporator Pressure from PT chart*)	
ACTUAL CONDENSER PRESSURE (*High side pressure actually read on gauges*)	
ACTUAL CONDENSER TEMPERATURE (*High side pressure converted to temp by PT*)	
ACTUAL EVAPORATOR PRESSURE (*Low side pressure on gauges*)	
ACTUAL EVAPORATOR TEMPERATURE (*Low side pressure converted to temp by PT*)	

LAB 31.12 SUPERHEAT CHARGING

LABORATORY OBJECTIVE

You will demonstrate your ability to check refrigerant charge in three refrigeration system using system superheat.

LABORATORY NOTES

You will first determine the manufacturer's recommended superheat. You will check system superheat using the system suction pressure and the suction lines temperature and compare the actual operating superheat to the manufacturer's recommended superheat. Finally, you will determine if the system is overcharged, undercharged, or correctly charged by comparing the recommended superheat with the actual superheat.

FUNDAMENTALS OF HVACR TEXT REFERENCE

Unit 31 Refrigerant System Charging

REQUIRED MATERIALS PROVIDED BY STUDENT

Safety Glasses
Gloves
Manifold gauges
Refrigeration wrench
Thermometer

REQUIRED MATERIALS PROVIDED BY SCHOOL

Three operating Refrigeration systems

PROCEDURE

1. Examine the unit and record important unit characteristics in the Data Sheet, including:
    a. Type of Unit (air conditioner, freezer, etc)
    b. Type of Refrigerant
    c. Metering Device (TEV or fixed restriction) *(Note: superheat charging is only for fixed restriction systems)*
    d. Condenser (air or water)
    e. Evaporator (air or water)
2. Measure and record the relevant operating conditions, including:
    a. Ambient Temperature for air cooled condensers
    b. Inlet water temperature for water cooled condensers
    c. Return Air Temperature for Air Conditioning Units
    d. Return Air Wet Bulb Temperature for air conditioning units
3. Use the manufacturer's information to determine the desired superheat and record on data sheet.
4. Install gauge manifolds on system.
5. Operate system until pressures stabilize and compressor is warm.
6. Read and record the suction line temperature.
7. Read and record the suction pressure.
8. Determine the evaporator saturation temperature using a PT Chart.
9. Operating Superheat = suction Line Temperature - Evaporator saturation temperature
10. Compare the operating superheat to the desired superheat to determine if the unit is overcharged, undercharged, or correct

LAB 31.12 SUPERHEAT CHARGING **DATA SHEET 1**	
TYPE OF UNIT (*Air conditioner, freezer, etc.*)	
TYPE OF CONDENSER (*Air or Water cooled*)	
TYPE OF EVAPORATOR (*Cools Air or Water*)	
TYPE OF METERING DEVICE (*Expansion valve, cap tube, orifice*)	
TYPE OF REFRIGERANT (*R22, R410A, R134a, etc*)	
OUTDOOR AMBIENT/ WATER TEMPERATURE (*Temp of condenser air or water*)	
RETURN AIR DRY BULB TEMP (*Regular return air temperature*)	
RETURN AIR WET BULB TEMP (*Temperature measurement from psychrometer*)	
DESIRED SYSTEM SUPERHEAT (*Superheat from manufacturer's chart*)	
SUCTION LINE TEMPERATURE (*Actual measured suction line temperature*)	
EVAPORATOR PRESSURE (*Low side pressure on gauges*)	
EVAPORATOR SATURATION TEMP (*Low side pressure converted to temp by PT*)	
SYSTEM SUPERHEAT (*Suction line temp minus evaporator saturation temp*)	
SYSTEM CONDITION (*overcharged, undercharged, or correct*)	

LAB 31.12 SUPERHEAT CHARGING DATA SHEET 2	
TYPE OF UNIT (*Air conditioner, freezer, etc.*)	
TYPE OF CONDENSER (*Air or Water cooled*)	
TYPE OF EVAPORATOR (*Cools Air or Water*)	
TYPE OF METERING DEVICE (*Expansion valve, cap tube, orifice*)	
TYPE OF REFRIGERANT (*R22, R410A, R134a, etc*)	
OUTDOOR AMBIENT/ WATER TEMPERATURE (*Temp of condenser air or water*)	
RETURN AIR DRY BULB TEMP (*Regular return air temperature*)	
RETURN AIR WET BULB TEMP (*Temperature measurement from psychrometer*)	
DESIRED SYSTEM SUPERHEAT (*Superheat from manufacturer's chart*)	
SUCTION LINE TEMPERATURE (*Actual measured suction line temperature*)	
EVAPORATOR PRESSURE (*Low side pressure on gauges*)	
EVAPORATOR SATURATION TEMP (*Low side pressure converted to temp by PT*)	
SYSTEM SUPERHEAT (*Suction line temp minus evaporator saturation temp*)	
SYSTEM CONDITION (*overcharged, undercharged, or correct*)	

LAB 31.12 SUPERHEAT CHARGING
DATA SHEET 3

TYPE OF UNIT (*Air conditioner, freezer, etc.*)	
TYPE OF CONDENSER (*Air or Water cooled*)	
TYPE OF EVAPORATOR (*Cools Air or Water*)	
TYPE OF METERING DEVICE (*Expansion valve, cap tube, orifice*)	
TYPE OF REFRIGERANT (*R22, R410A, R134a, etc*)	
OUTDOOR AMBIENT/ WATER TEMPERATURE (*Temp of condenser air or water*)	
RETURN AIR DRY BULB TEMP (*Regular return air temperature*)	
RETURN AIR WET BULB TEMP (*Temperature measurement from psychrometer*)	
DESIRED SYSTEM SUPERHEAT (*Superheat from manufacturer's chart*)	
SUCTION LINE TEMPERATURE (*Actual measured suction line temperature*)	
EVAPORATOR PRESSURE (*Low side pressure on gauges*)	
EVAPORATOR SATURATION TEMP (*Low side pressure converted to temp by PT*)	
SYSTEM SUPERHEAT (*Suction line temp minus evaporator saturation temp*)	
SYSTEM CONDITION (*overcharged, undercharged, or correct*)	

LAB 31.13 VAPOR CHARGE - SUPERHEAT

LABORATORY OBJECTIVE

You will demonstrate your ability to safely add refrigerant vapor to an operating unit until the system operating characteristics meet the manufacturer's specifications.

LABORATORY NOTES

You will connect your gauges, operate a system, and check its charge according to the manufacturer's instructions. You will add refrigerant vapor to the low side with the system operating until the system operating characteristics meet the manufacturer's specification.

FUNDAMENTALS OF HVACR TEXT REFERENCE

Unit 31 Refrigerant System Charging

REQUIRED MATERIALS PROVIDED BY STUDENT

Safety Glasses
Gloves
Manifold gauges
Refrigeration wrench
Thermometer

REQUIRED MATERIALS PROVIDED BY SCHOOL

Operating Refrigeration system
Refrigerant
Scale

PROCEDURE

1. Examine the unit and record important unit characteristics in the Data Sheet, including:
    a. Type of Unit (air conditioner, freezer, etc)
    b. Type of Refrigerant
    c. Metering Device (TEV or fixed restriction) *(Note: superheat charging is only for fixed restriction systems)*
    d. Condenser (air or water)
    e. Evaporator (air or water)
2. Measure and record the relevant operating conditions, including:
    a. Ambient Temperature for air cooled condensers
    b. Inlet water temperature for water cooled condensers
    c. Return Air Temperature for Air Conditioning Units
    d. Return Air Wet Bulb Temperature for air conditioning units
3. Use the manufacturer's information to determine the desired superheat and record on data sheet.
4. Install gauge manifolds on system.
5. Purge the air out of the hoses and manifold.
6. Operate system until pressures stabilize and compressor is warm.
7. Read and record the suction line temperature.
8. Read and record the suction pressure.
9. Determine the evaporator saturation temperature using a PT Chart.
10. Operating Superheat = suction Line Temperature - Evaporator saturation temperature

11. Connect the middle gauge hose to the vapor valve on a refrigerant cylinder.
    (Note: Make sure that the refrigerant listed on the data plate is the same as the refrigerant in the cylinder)
12. Open the cylinder valve and purge the middle hose at the manifold.
13. Turn the scale on.
14. Place the cylinder in the upright position on the scale and zero the scale.
15. With the unit operating, open the low side manifold hand wheel.
16. The compound gauge should show a higher pressure – this is the cylinder pressure.
17. Periodically close the low side manifold hand wheel to see what the system operating pressure is.
18. Monitor the low side and high side pressures and the suction line temperature.
19. When you see a measurable drop in the suction line temperature, close the manifold valve.
20. Recalculate the superheat.
21. As you add charge, the suction pressure(and evaporator saturation temperature) will rise, the suction line temperature will drop. Since both numbers are changing, you must frequently recheck the superheat.
22. When the system operates within the manufacturer's tolerance, you are finished.
23. Record the system operating pressures and amount of refrigerant used.

LAB 31.13 VAPOR CHARGING - SUPERHEAT	
TYPE OF UNIT *(Air conditioner, freezer, etc.)*	
TYPE OF CONDENSER *(Air or Water cooled)*	
TYPE OF EVAPORATOR *(Cools Air or Water)*	
TYPE OF METERING DEVICE *(Expansion valve, cap tube, orifice)*	
TYPE OF REFRIGERANT *(R22, R410A, R134a, etc)*	
OUTDOOR AMBIENT/ WATER TEMPERATURE *(Temp of condenser air or water)*	
RETURN AIR DRY BULB TEMP *(Regular return air temperature)*	
RETURN AIR WET BULB TEMP *(Temperature measurement from psychrometer)*	
DESIRED SYSTEM SUPERHEAT *(Superheat from manufacturer's chart)*	
SUCTION LINE TEMPERATURE *(Actual measured suction line temperature)*	
EVAPORATOR PRESSURE *(Low side pressure on gauges)*	
EVAPORATOR SATURATION TEMP *(Low side pressure converted to temp by PT)*	
SYSTEM SUPERHEAT *(Suction line temp minus evaporator saturation temp)*	

LAB 31.14 SUBCOOLING METHOD

LABORATORY OBJECTIVE
You will demonstrate your ability to check refrigerant charge in three refrigeration system using condenser subcooling.

LABORATORY NOTES
You will first determine the manufacturer's recommended subcooling. You will check system subcooling using the system liquid pressure and the liquid line temperature and compare the actual operating subcooling to the manufacturer's recommended subcooling. Finally, you will determine if the system is overcharged, undercharged, or correctly charged by comparing the recommended subcooling with the actual subcooling.

FUNDAMENTALS OF HVACR TEXT REFERENCE
Unit 31 Refrigerant System Charging

REQUIRED MATERIALS PROVIDED BY STUDENT
Safety Glasses
Gloves
Manifold gauges
Refrigeration wrench
Thermometer

REQUIRED MATERIALS PROVIDED BY SCHOOL
Three operating refrigeration systems

PROCEDURE
1. Examine the unit and record important unit characteristics in the Data Sheet, including:
    a. Type of Unit (air conditioner, freezer, etc)
    b. Type of Refrigerant
    c. Metering Device (TEV or fixed restriction) *(Note: subcooling charging is only for TEV systems)*
2. Use the manufacturer's information to determine the desired subcooling and record on data sheet.
   (Note: Many manufacturers now have the subcooling on their unit data plate.)
3. Install gauge manifolds on system.
4. Operate system until pressures stabilize and compressor is warm.
5. Read and record the liquid line temperature.
6. Read and record the liquid pressure.
7. Determine the condenser saturation temperature using a PT Chart.
8. Operating Subcooling = Condenser Saturation Temperature - Liquid Line Temperature
9. Compare the operating subcooling to the desired subcooling to determine if the unit is overcharged, undercharged, or correct.

LAB 31.14 SUBCOOLING CHARGING
DATA SHEET 1

TYPE OF UNIT (*Air conditioner, freezer, etc.*)	
TYPE OF CONDENSER (*Air or Water cooled*)	
TYPE OF EVAPORATOR (*Cools Air or Water*)	
TYPE OF METERING DEVICE (*Expansion valve, cap tube, orifice*)	
TYPE OF REFRIGERANT (*R22, R410A, R134a, etc*)	
OUTDOOR AMBIENT/ WATER TEMPERATURE (*Temp of condenser air or water*)	
RETURN AIR DRY BULB TEMP (*Regular return air temperature*)	
DESIRED SYSTEM SUBCOOLING (*Subcooling from manufacturer's chart*)	
LIQUID LINE TEMPERATURE (*Actual measured liquid line temperature*)	
CONDENSER PRESSURE (*High side pressure on gauges*)	
CONDENSER SATURATION TEMP (*High side pressure converted to temp by PT*)	
SYSTEM SUBCOOLING (*Condenser saturation temp minus liquid line temp*)	
SYSTEM CONDITION (*overcharged, undercharged, or correct*)	

LAB 31.14 SUBCOOLING CHARGING **DATA SHEET 2**	
TYPE OF UNIT (*Air conditioner, freezer, etc.*)	
TYPE OF CONDENSER (*Air or Water cooled*)	
TYPE OF EVAPORATOR (*Cools Air or Water*)	
TYPE OF METERING DEVICE (*Expansion valve, cap tube, orifice*)	
TYPE OF REFRIGERANT (*R22, R410A, R134a, etc*)	
OUTDOOR AMBIENT/ WATER TEMPERATURE (*Temp of condenser air or water*)	
RETURN AIR DRY BULB TEMP (*Regular return air temperature*)	
DESIRED SYSTEM SUBCOOLING (*Subcooling from manufacturer's chart*)	
LIQUID LINE TEMPERATURE (*Actual measured liquid line temperature*)	
CONDENSER PRESSURE (*High side pressure on gauges*)	
CONDENSER SATURATION TEMP (*High side pressure converted to temp by PT*)	
SYSTEM SUBCOOLING (*Condenser saturation temp minus liquid line temp*)	
SYSTEM CONDITION (*overcharged, undercharged, or correct*)	

LAB 31.14 SUBCOOLING CHARGING DATA SHEET 3	
TYPE OF UNIT (*Air conditioner, freezer, etc.*)	
TYPE OF CONDENSER (*Air or Water cooled*)	
TYPE OF EVAPORATOR (*Cools Air or Water*)	
TYPE OF METERING DEVICE (*Expansion valve, cap tube, orifice*)	
TYPE OF REFRIGERANT (*R22, R410A, R134a, etc*)	
OUTDOOR AMBIENT/ WATER TEMPERATURE (*Temp of condenser air or water*)	
RETURN AIR DRY BULB TEMP (*Regular return air temperature*)	
DESIRED SYSTEM SUBCOOLING (*Subcooling from manufacturer's chart*)	
LIQUID LINE TEMPERATURE (*Actual measured liquid line temperature*)	
CONDENSER PRESSURE (*High side pressure on gauges*)	
CONDENSER SATURATION TEMP (*High side pressure converted to temp by PT*)	
SYSTEM SUBCOOLING (*Condenser saturation temp minus liquid line temp*)	
SYSTEM CONDITION (*overcharged, undercharged, or correct*)	

LAB 31.15 MANUFACTURER'S CHARGING CHART

LABORATORY OBJECTIVE

You will demonstrate your ability to check refrigerant charge using a manufacturer's charging chart in three refrigeration systems.

LABORATORY NOTES

You will inspect the unit information to determine the correct method for checking the refrigerant charge for three different refrigeration systems. You will then check the charge and compare the actual conditions to the desired conditions and determine if the unit is overcharged, undercharged, or correctly charged.

***FUNDAMENTALS OF HVACR* TEXT REFERENCE**

Unit 31 Refrigerant System Charging

REQUIRED MATERIALS PROVIDED BY STUDENT

Safety Glasses
Gloves
Manifold gauges
Refrigeration wrench
Thermometer

REQUIRED MATERIALS PROVIDED BY SCHOOL

Three operating Refrigeration systems

PROCEDURE

1. Examine the unit and record important unit characteristics in the Data Sheet, including:
   a. Type of Unit (air conditioner, freezer, etc)
   b. Type of Refrigerant
   c. Metering Device (TEV or fixed restriction)
   d. Condenser (air or water)
   e. Evaporator (air or water)
2. Examine the system charging information to determine the manufacturer's recommendation
3. Measure and record the relevant operating conditions specified by the manufacturer
4. *(Note: different manufacturers may specify different data. The data sheet contains places for more data than most systems will require. You ONLY need to measure and record data required by the manufacturer.)*
5. Measure the appropriate Data and record in the Data Sheet
6. Compare the measured operating characteristics to the manufacturer's desired operating characteristics to determine if the system is overcharged, undercharged, or charged correctly.

Note: Only the data required by the manufacturer is necessary to collect.

MANUFACTURER'S CHARGING DATA SHEET 1	
TYPE OF UNIT (*Air conditioner, freezer, etc.*)	
TYPE OF CONDENSER (*Air or Water cooled*)	
TYPE OF EVAPORATOR (*Cools Air or Water*)	
TYPE OF METERING DEVICE (*Expansion valve, cap tube, orifice*)	
TYPE OF REFRIGERANT (*R22, R410A, R134a, etc.*)	
OUTDOOR AMBIENT/ WATER TEMPERATURE (*Temp of condenser air or water*)	
RETURN AIR DRY BULB TEMP (*Regular return air temperature*)	
RETURN AIR WET BULB TEMP (*Temperature measurement from psychrometer*)	
DESIRED SUPERHEAT, SUBCOOLING, OR APPROACH (*from manufacturers chart*)	
CONDENSER PRESSURE (*High side pressure on gauges*)	
CONDENSER SATURATION TEMP (*High side pressure converted to temp by PT*)	
LIQUID LINE TEMPERATURE (*Actual measured liquid line temperature*)	
CONDENSER APPROACH TEMP (*liquid line temp minus outdoor ambient*)	
SYSTEM SUBCOOLING (*Condenser saturation temp minus liquid line temp*)	
SYSTEM SUPERHEAT (*Suction line temp minus evaporator saturation temp*)	
DESIRED EVAPORATOR PRESSURE (*Evaporator Pressure from Manufacturers chart*)	
DESIRED CONDENSER PRESSURE (*Condenser Pressure from Manufacturer's chart*)	
SYSTEM CONDITION (*overcharged, undercharged, or correct*)	

Note: Only the data required by the manufacturer is necessary to collect.

MANUFACTURER'S CHARGING DATA SHEET 2	
TYPE OF UNIT (*Air conditioner, freezer, etc.*)	
TYPE OF CONDENSER (*Air or Water cooled*)	
TYPE OF EVAPORATOR (*Cools Air or Water*)	
TYPE OF METERING DEVICE (*Expansion valve, cap tube, orifice*)	
TYPE OF REFRIGERANT (*R22, R410A, R134a, etc.*)	
OUTDOOR AMBIENT/ WATER TEMPERATURE (*Temp of condenser air or water*)	
RETURN AIR DRY BULB TEMP (*Regular return air temperature*)	
RETURN AIR WET BULB TEMP (*Temperature measurement from psychrometer*)	
DESIRED SUPERHEAT, SUBCOOLING, OR APPROACH (*from manufacturers chart*)	
CONDENSER PRESSURE (*High side pressure on gauges*)	
CONDENSER SATURATION TEMP (*High side pressure converted to temp by PT*)	
LIQUID LINE TEMPERATURE (*Actual measured liquid line temperature*)	
CONDENSER APPROACH TEMP (*liquid line temp minus outdoor ambient*)	
SYSTEM SUBCOOLING (*Condenser saturation temp minus liquid line temp*)	
SYSTEM SUPERHEAT (*Suction line temp minus evaporator saturation temp*)	
DESIRED EVAPORATOR PRESSURE (*Evaporator Pressure from Manufacturers chart*)	
DESIRED CONDENSER PRESSURE (*Condenser Pressure from Manufacturer's chart*)	
SYSTEM CONDITION (*overcharged, undercharged, or correct*)	

Note: Only the data required by the manufacturer is necessary to collect.

MANUFACTURER'S CHARGING DATA SHEET 3	
TYPE OF UNIT (*Air conditioner, freezer, etc.*)	
TYPE OF CONDENSER (*Air or Water cooled*)	
TYPE OF EVAPORATOR (*Cools Air or Water*)	
TYPE OF METERING DEVICE (*Expansion valve, cap tube, orifice*)	
TYPE OF REFRIGERANT (*R22, R410A, R134a, etc.*)	
OUTDOOR AMBIENT/ WATER TEMPERATURE (*Temp of condenser air or water*)	
RETURN AIR DRY BULB TEMP (*Regular return air temperature*)	
RETURN AIR WET BULB TEMP (*Temperature measurement from psychrometer*)	
DESIRED SUPERHEAT, SUBCOOLING, OR APPROACH (*from manufacturers chart*)	
CONDENSER PRESSURE (*High side pressure on gauges*)	
CONDENSER SATURATION TEMP (*High side pressure converted to temp by PT*)	
LIQUID LINE TEMPERATURE (*Actual measured liquid line temperature*)	
CONDENSER APPROACH TEMP (*liquid line temp minus outdoor ambient*)	
SYSTEM SUBCOOLING (*Condenser saturation temp minus liquid line temp*)	
SYSTEM SUPERHEAT (*Suction line temp minus evaporator saturation temp*)	
DESIRED EVAPORATOR PRESSURE (*Evaporator Pressure from Manufacturers chart*)	
DESIRED CONDENSER PRESSURE (*Condenser Pressure from Manufacturer's chart*)	
SYSTEM CONDITION (*overcharged, undercharged, or correct*)	

LAB 31.16 CHECKING CHARGE ON WATER COOLED SYSTEM

LABORATORY OBJECTIVE

You will demonstrate your ability to check refrigerant charge in a system with a water cooled condenser using the manufacturer's charging chart.

LABORATORY NOTES

You will inspect the unit information to determine the correct method for checking the refrigerant charge. You will then check the charge and compare the actual conditions to the desired conditions and determine if the unit is overcharged, undercharged, or correctly charged.

***FUNDAMENTALS OF HVACR* TEXT REFERENCE**

Unit 31 Refrigerant System Charging

REQUIRED MATERIALS PROVIDED BY STUDENT

Safety Glasses
Gloves
Manifold gauges
Refrigeration wrench
Thermometer

REQUIRED MATERIALS PROVIDED BY SCHOOL

Operating water cooled refrigeration system

PROCEDURE

1. Examine the unit and record important unit characteristics in the Data Sheet, including:
    a. Type of Unit (air conditioner, freezer, etc)
    b. Type of Refrigerant
    c. Metering Device (TEV or fixed restriction)
2. Examine the system charging information to determine the manufacturer's recommendation
3. Measure and record the relevant operating conditions specified by the manufacturer
4. *(Note: different manufacturers may specify different data. The data sheet contains places for more data than most systems will require. You ONLY need to measure and record data required by the manufacturer.)*
5. Measure the appropriate Data and record in the Data Sheet
6. Compare the measured operating characteristics to the manufacturer's desired operating characteristics to determine if the system is overcharged, undercharged, or charged correctly.

Note: Only the data required by the manufacturer is necessary to collect.

LAB 31.16 WATER COOLED SYSTEM CHARGING DATA SHEET	
TYPE OF REFRIGERANT (*R22, R410A, R134a, etc*)	
SPECIFIED CONDENSER WATER FLOW (*water flow from manufacturer's chart*)	
ACTUAL CONDENSER WATER FLOW (*water flow from flow meter*)	
RETURN AIR DRY BULB TEMP (*Regular return air temperature*)	
RETURN AIR WET BULB TEMP (*Temperature measurement from psychrometer*)	
SPECIFIED WATER TEMP DIFFERENCE (*water temp difference from manufac chart*)	
ENTERING WATER TEMPERATURE (*Temperature of entering water*)	
LEAVING WATER TEMPERATURE (*Temperature of leaving water*)	
ACTUAL WATER TEMP DIFFERENCE (*leaving water minus entering water*)	
DESIRED CONDENSER PRESSURE (*Condenser Pressure from manufacturer chart*)	
DESIRED EVAPORATOR PRESSURE (*Evaporator Pressure from manufacturer chart*)	
SYSTEM CONDITION (*overcharged, undercharged, or correct*)	

LAB 32.1 USING NON-CONTACT VOLTAGE DETECTORS

LABORATORY OBJECTIVE

You will use a non-contact voltage detector to safely distinguish between energized electrical components and wires and non-energized components and wires.

LABORATORY NOTES

You will hold your non-contact voltage detector near components and wires to determine which are energized and which are not energized.

***FUNDAMENTALS OF HVACR* TEXT REFERENCE**

Unit 32 Electrical Safety

REQUIRED MATERIALS PROVIDED BY STUDENT

Safety glasses
Non-contact voltage detector

REQUIRED MATERIALS PROVIDED BY SCHOOL

Energized electrical components and wires
De-energized electrical components and wires

PROCEDURE

Safety Note: In general, avoid touching things. Do not touch ANYTHING without first testing with your non-contact voltage detector.

Safety Note: A circuit does not have to be operating to be energized with a potentially fatal voltage. Always assume that ALL circuits can kill you!

1. Hold your non-contact voltage detector near the components and wires assigned by the instructor.
2. Do NOT open electrical panels or stick your hand near electrical equipment that the instructor has not assigned you to inspect.
3. Most testers have some type of confidence check, such as a light flash or beep when the button is first pressed. Become familiar with yours and do not use it when it fails to beep or flash when first activated.
4. Non-contact voltage detectors normally do not indicate through grounded metal enclosures or shielded cables.
5. Be prepared to show the instructor the difference between an energized circuit and a de-energized circuit.

LAB 32.2 ELECTRICAL SAFETY PROCEDURES

LABORATORY OBJECTIVE
You will identify the location of the electric disconnect switches in the shop and demonstrate your ability to turn off a disconnect switch.

LABORATORY NOTES
You will identify the disconnect switch for an air conditioning unit and demonstrate how to turn it off. You will then locate the electric panel that feeds that disconnect switch and identify the circuit breaker that feeds the disconnect switch. Finally, you will locate the main switchgear that feeds the electric panel.

FUNDAMENTALS OF HVACR TEXT REFERENCE
Unit 32 Electrical Safety

REQUIRED MATERIALS PROVIDED BY STUDENT
Safety glasses

REQUIRED MATERIALS PROVIDED BY SCHOOL
Electrical disconnects

PROCEDURE
Safety Note: In general, avoid touching things. Do not touch ANYTHING without first testing with your non-contact voltage detector.
1. Locate the power conduit leaving the unit.
2. Follow it to the disconnect switch. It should be within sight of the unit, often right next to it.
3. Check with your non-contact voltage detector to make sure the box is not energized.
4. Demonstrate how to turn the disconnect switch off.
5. The disconnect switch should be labeled, including the name of the electric panel feeding it.
6. Locate that electric panel.
7. Find the circuit breaker that controls the disconnect switch. It should be labeled on the panel.
8. Find the main switch gear by the back shop door.
9. Find the disconnect switch for the panel.

LAB 32.3 CHANGING FUSES

LABORATORY OBJECTIVE

You will demonstrate your ability to safely replace a cartridge fuse in a disconnect switch using non-conductive fuse pullers.

LABORATORY NOTES

You will check the disconnect switch to make sure it is de-energized, turn off the disconnect switch, open the disconnect switch, remove a cartridge fuse using a non-conductive fuse puller, replace the cartridge fuse using a non-conductive fuse puller, and close the disconnect switch.

FUNDAMENTALS OF HVACR **TEXT REFERENCE**

Unit 32 Electrical Safety

REQUIRED MATERIALS PROVIDED BY STUDENT

Safety glasses

Non-contact voltage detector

Non-conductive fuse puller

REQUIRED MATERIALS PROVIDED BY SCHOOL

Electrical Disconnect with cartridge fuses.

PROCEDURE

Safety Note: In general, avoid touching things. Do not touch ANYTHING without first testing with your non-contact voltage detector.

Safety Note: A circuit does not have to be operating to be energized with a potentially fatal voltage. Always assume that ALL circuits can kill you!

1. Check the disconnect switch to make sure it is not energized.
2. Turn the switch off.
3. Open the disconnect switch.
4. Use non-conductive fuse pullers to remove a cartridge fuse from the disconnect switch.
5. Safety Note: Do NOT use pliers or fingers!
6. Replace the cartridge fuse.
7. Close the disconnect switch.

LAB 33.1 SERIES AND PARALLEL CIRCUITS

LABORATORY OBJECTIVE
You will demonstrate your understanding of series and parallel circuits by safely wiring series and parallel circuits to specification.

LABORATORY NOTES
In this lab exercise you will learn the characteristics of series, parallel, and series- parallel circuits by designing, wiring, and operating circuits using a 110-volt source.

FUNDAMENTALS OF HVACR TEXT REFERENCE
Unit 33 Basic Electricity

REQUIRED MATERIALS PROVIDED BY STUDENT
Safety glasses
Multimeter
Wire stripper
Crimp tool
Wire cutter
6-in-1 screwdriver

REQUIRED MATERIALS PROVIDED BY SCHOOL
120-volt power cord
Wire
Electrical terminals
Wire nuts
Three 120-volt light sockets
Three 120-volt lights
Three SPST toggle switches

PROCEDURE
You should first draw each circuit using the following symbols:

Power Supply   Switch   Light

Have the instructor check your drawing before wiring the circuit.
Wire each circuit according to your drawing.
All circuits should be checked by the instructor BEFORE they are energized.

CIRCUIT #1

1. Design and draw a circuit with 1 switch controlling 1 load. The switch should be wired in series with the load.

2. Wire and operate the circuit. (Have instructor check the circuit before operation.)

3. What happens when the switch is opened?

4. What happens when the switch is closed?

CIRCUIT #2

1. Design and draw a circuit with 2 switches controlling 1 load. The switches should be wired in series with each other and in series with the load.

2. Wire and operate the circuit. (Have the instructor check the circuit before operation.)

3. Close both switches. What happens?

4. Open 1 switch. What happens?

5. Close the first switch and open the other switch. What happens?

6. Summarize the operation of this circuit.

CIRCUIT #3

1. Design and draw a circuit with 2 switches controlling 1 load. The switches should be wired in parallel with each other but in series with the load.

2. Wire and operate the circuit. (Have instructor check the circuit before operation.)

3. Open both switches. What happens?

4. Close 1 switch. What happens?

5. Open the first switch and close the other switch. What happens?

6. Summarize the operation of this circuit.

CIRCUIT #4

1. Design and draw a circuit with 1 switch controlling 2 loads. The loads should be wired in parallel with each other but in series with the switch.

2. Wire and operate the circuit. (Have instructor check the circuit before operation.)

3. Summarize the operation of this circuit.

CIRCUIT #5

1. Design and draw a circuit with 2 switches controlling 2 loads. The loads should be wired in parallel with each other. The switches should be wired in series with each other. The switches should be wired in series with the loads.

2. Wire and operate the circuit. (Have instructor check the circuit before operation.)

3. Close both switches. What happens?

4. Open 1 switch. What happens?

5. Close the first switch and open the other switch. What happens?

6. Summarize the operation of this circuit.

CIRCUIT #6

1. Design and draw a circuit with 2 switches controlling 2 loads. The loads should be wired in parallel with each other. The switches should be wired in parallel with each other. The switches should be wired in series with the loads.

2. Wire and operate the circuit. (Have instructor check circuit before operation.)

3. Close both switches. What happens?

4. Open 1 switch. What happens?

5. Close the first switch and open the other switch. What happens?

6. Summarize the operation of the circuit.

CIRCUIT #7

1. Design and draw a circuit with 2 switches controlling 2 loads. Each load should be wired in series with one switch. The switches should be wired so that each switch controls one load and the loads operate independently.

2. Wire and operate the circuit. (Have instructor check circuit before operation.)

3. Close both switches. What happens?

4. Open 1 switch. What happens?

5. Close the first switch and open the other switch. What happens?

6. Summarize the operation of this circuit.

CIRCUIT #8

1. Design and draw a circuit with 1 switch controlling 2 loads. The loads should be wired in series with each other and in series with the switch.

2. Wire and operate the circuit. (Have the instructor check the circuit before operation.)

3. What do you notice about the amount of light the bulbs are producing?

4. Why is this?

5. With the circuit operating, CAREFULLY unscrew one of the light bulbs. What happens? Why?

6. Summarize the operation of this circuit.

CIRCUIT #9

1. Design and draw a circuit with 1 switch controlling 2 loads. The loads should be wired in series with each other and in series with the switch. Wire another switch in parallel with one light.

2. Wire and operate the circuit? (Have instructor check the circuit before operating.)

3. Close both switches. What happens?

4. Open the switch that is wired in parallel with one light. What happens?

5. Explain the operation of the parallel-wired switch.

6. How does this contrast to the operation of the series wired switch?

CIRCUIT #10

1. Design and draw a circuit with 3 loads in series. Two of the loads should have a switch wired in parallel to them. Each switch should control 1 load. In this circuit, 1 load will operate continuously while the other 2 loads will each be controlled by their own switch. This circuit is tricky.

2. Wire and operate the circuit. (Have instructor check circuit before operation.)

3. Close both switches. What happens?

4. Open 1 switch. What happens?

5. Close the first switch and open the other switch. What happens?

6. Summarize the operation of this circuit.

SUMMARY

1. Explain how switches wired in series with loads control the loads.

2. Explain how switches wired in parallel with loads control the loads.

3. Explain how loads wired in series with each other behave.

4. Explain how loads wired in parallel with each other behave.

5. Explain the operation of a circuit with one load wired in series with several switches wired in series with each other.

6. Explain the operation of a circuit with one load wired in series with two switches that are wired in parallel with each other.

7. How are MOST loads in air conditioning systems wired with respect to each other?

8. How are MOST loads in air conditioning systems wired with respect to switches?

9. How are MOST switches in air conditioning systems wired with respect to each other?

LAB 33.2 APPLY OHMS LAW TO SERIES CIRCUIT VOLTAGE CHANGES

LABORATORY OBJECTIVE
You will demonstrate your understanding of ohms law in series circuits by safely wiring series circuits to specification and measuring the circuit characteristics.

LABORATORY NOTES
In this lab exercise you will build a simple SERIES circuit using a variable power supply and two electric strip heaters. You will measure the resistance, voltage, and current at different voltage levels and compare your measurements using Ohm's law.

FUNDAMENTALS OF HVACR TEXT REFERENCE
Unit 33 Basic Electricity

REQUIRED MATERIALS PROVIDED BY STUDENT
Safety glasses
Multimeter
Ammeter
Wire stripper
Crimp tool
Wire cutter
6-in-1 screwdriver

REQUIRED MATERIALS PROVIDED BY SCHOOL
120-volt power cord
Adjustable power supply
Wire
Electrical terminals
Wire nuts
Two electric heaters

PROCEDURE
Step 1
WITH THE CIRCUIT DISCONNECTED FROM THE POWER SUPPLY, measure the resistance of each heater and the total circuit resistance. Record your measurements.

Note: the power cord must be unplugged from the adjustable power supply to get a correct reading because the coil in the power supply will affect the reading.

Step 2
Use ohms law to calculate the total circuit resistance, the circuit current and voltage drop across each heater for the voltages listed in the chart. Record your calculations.

Step 3
Adjust the voltage source to produce the first voltage shown on the chart below.

Step 4
Measure the circuit current with a clamp on amp meter. Use a volt meter to measure the voltage drop across each heater and across the entire circuit. Record your measurements.

NOTE: You may need to wrap the wire around the jaw of the amp meter several times to get a reading.

Step 5
Repeat steps 3 and 4 for each of the other voltages listed in the chart.

LAB 33.2 SERIES HEATERS DATA TABLE				
	100 Volt Setting Calculated	**100 Volt Setting Measured**	**50 Volt Setting Calculated**	**50 Volt Setting Measured**
Heater 1 Resistance				
Heater 2 Resistance				
Total Circuit Resistance				
Heater 1 Volts				
Heater 2 Volts				
Total Circuit Volts	100 volts		50 volts	
Heater 1 Amps				
Heater 2 Amps				
Total Circuit Amps				

LAB 33.3 OHMS LAW & PARALLEL CIRCUIT VOLTAGE CHANGES

LABORATORY OBJECTIVE
You will demonstrate your understanding of ohms law in parallel circuits by safely wiring parallel circuits to specification and measuring the circuit characteristics.

LABORATORY NOTES
In this lab exercise you will build a simple PARALLEL circuit using a variable power supply and two electric strip heaters. You will measure the resistance, voltage, and current at different voltage levels and compare your measurements using Ohm's law.

FUNDAMENTALS OF HVACR TEXT REFERENCE
Unit 33 Basic Electricity

REQUIRED MATERIALS PROVIDED BY STUDENT
Safety glasses
Multimeter
Ammeter
Wire stripper
Crimp tool
Wire cutter
6-in-1 screwdriver

REQUIRED MATERIALS PROVIDED BY SCHOOL
120-volt power cord
Adjustable power supply
Wire
Electrical terminals
Wire nuts
Two electric heaters

PROCEDURE
Step 1

BEFORE CONSTRUCTING THE CIRCUIT, measure the resistance OF EACH HEATER.

Note: you cannot read the resistance of an individual component in a parallel circuit without removing it from the circuit. If the device is still connected in parallel, you will read the total resistance of the entire parallel circuit, not just the individual device.

Step 2

Wire the two heaters in parallel and read the total circuit resistance BEFORE connecting the circuit to the adjustable power supply.

Note: the power cord must be unplugged from the adjustable power supply to get a correct reading because the coil in the power supply will affect the reading.

Step 3

Use ohms law to calculate the total circuit resistance, circuit current and current draw of each heater for the voltages listed in the chart. Record your calculations.

Step 4

Adjust the voltage source to produce the first voltage shown on the chart below.

Step 5

Measure the total circuit current and the individual heater currents with a clamp on amp meter. Measure the voltage across the total circuit and across each heater with a volt meter. Record your measurements.

Step 6

Repeat steps 4 and 5 for each of the other voltages listed in the chart.

LAB 33.3 PARALLEL HEATERS DATA TABLE				
	100 Volt Setting Calculated	100 Volt Setting Measured	50 Volt Setting Calculated	50 Volt Setting Measured
Heater 1 Resistance				
Heater 2 Resistance				
Total Circuit Resistance				
Heater 1 Volts				
Heater 2 Volts				
Total Circuit Volts	100 volts		50 volts	
Heater 1 Amps				
Heater 2 Amps				
Total Circuit Amps				

LAB 34.1 ALTERNATING CURRENT PRINCIPLES

LABORATORY OBJECTIVE
You will demonstrate the effects of capacitive and inductive reactance in alternating current circuits.

LABORATORY NOTES
In this portion of the lab exercise you will wire a 20 microfarad capacitor in series with a 120-volt light, measure the circuit current and the voltage drop across the light and the capacitor. You will then repeat this using a 10 microfarad capacitor and compare the results.

FUNDAMENTALS OF HVACR TEXT REFERENCE
Unit 34 Alternating Current Fundamentals

REQUIRED MATERIALS PROVIDED BY STUDENT
Safety glasses
Multimeter
Ammeter
Wire cutter
Wire crimper/stripper
6-in-1 screwdriver

REQUIRED MATERIALS PROVIDED BY SCHOOL
120-volt power cord
60-watt incandescent light
20 microfarad 370 volt run capacitor
10 microfarad 370 volt run capacitor
120-volt coil and iron center
120-volt shaded pole motor

PROCEDURE
CAPACATIVE REACTANCE

1. Wire a 20 microfarad run capacitor in series with a 60-watt light. Connect the circuit to a 110-volt A.C. source.

2. Measure the circuit amp draw. _____

3. Measure the voltage across the light. _____

4. Measure the voltage across the capacitor. _____

5. Measure the source voltage. _____

6. Discharge the capacitor by unplugging the circuit and shorting the plug ends across a piece of metal or wire. Test the capacitor charge by measuring the voltage across the capacitor terminals with the DC scale on your volt meter. If there is no charge, proceed to the next step. If there is a voltage, wait until the meter discharges the capacitor and the voltage is close to 0 volts DC. What we are doing here is removing the charge on the capacitor so that it is safe to

handle.

7. Replace the 20 MFD capacitor with a 10 MFD capacitor and operate the circuit.

8. Measure the circuit amp draw. _____

9. Measure the voltage across the light. _____

10. Measure the voltage across the capacitor. _____

11. Measure the source voltage. _____

12. Which capacitor produces a higher amp draw?

13. Which capacitor appears to have a higher capacitive reactance, the 20 MFD or the 10 MFD? Why?

OVERVIEW
Inductive reactance of coil

In this portion of the lab exercise you will wire a coil in series with a 120-volt light, measure the circuit current and the voltage drop across the light and the coil. You will then insert an iron center, measure the voltages and current again, and compare the results with the iron center in and out of the coil.

INDUCTIVE REACTANCE - COIL

1. Wire a 120-volt solenoid coil in series with a 60-watt light. The solenoid should have its core removed.

2. Measure the amp draw of the circuit. _____

3. Measure the voltage across the light. _____

4. Measure the voltage across the solenoid coil _____

5. Slowly slide an iron or steel core into the solenoid while the circuit is operating.

6. What happens to the light?

7. Measure the amp draw of the circuit. _____

8. Measure the voltage across the light. _____

9. Measure the voltage across the solenoid coil _____

10. Explain you results.

11. When does the solenoid appear to have a higher inductive reactance? Why?

OVERVIEW
Inductive reactance of motor

In this portion of the lab exercise you will wire a shaded pole motor in series with a 120-volt light, measure the circuit current and the voltage drop across the light and the motor. You will then stall the motor, measure the voltages and current again, and compare the results with the motor spinning and the motor stalled.

INDUCTIVE REACTANCE - MOTOR

1. Replace the solenoid with a small 110-volt shaded pole motor and operate the circuit.

2. Measure the amp draw of the circuit. _____

3. Measure the voltage across the light. _____

4. Measure the voltage across the motor _____

5. Explain your results.

6. Hold the rotor of the motor so that it can't turn and operate the circuit.

7. NOTE: You should be using a LOW TORQUE MOTOR. This would NOT BE SAFE with a high torque motor. If you do not know the difference, ASK THE INSTRUCTOR!

8. Now, release the motor.

9. What happens?

10. Does the motor appear to have more inductive reactance when it is turning or when it is stationary? Why?

LAB 34.2 ALTERNATING CURRENT DEMONSTRATION

LABORATORY OBJECTIVE
(This lab is demonstrated by the instructor.)
The instructor will demonstrate the effects of capacitive and inductive reactance in a resonant alternating current circuit.

LABORATORY NOTES
The instructor will wire a circuit with a light, capacitor, and motor in series. One SPST switch is wired in parallel with the capacitor and another SPST switch is wired in parallel with the motor. With both switches off, all three devices are in the circuit. Turning either switch on shorts the component it is wired in parallel with out of the circuit. This allows the instructor to show the effects of pure resistance, inductive reactance, capacitive reactance, and resonance where the inductive and capacitive reactance offset each other. A dual trace oscilloscope will be connected to the incoming voltage and to the voltage across the light. The voltage across the light will be in phase with the circuit current, so the current phase can be seen with it. Comparing the two waveforms you will see the waveforms shift as the capacitive and inductive reactance is added or subtracted from the circuit.

FUNDAMENTALS OF HVACR **TEXT REFERENCE**
Unit 34 Alternating Current Fundamentals

REQUIRED MATERIALS PROVIDED BY SCHOOL
Safety glasses
Multimeter
Ammeter
Wire cutter
Wire crimper/stripper
6-in-1 screwdriver
120-volt power cord
60 watt incandescent light
10 microfarad 370 volt run capacitor
Small 120-volt shaded pole motor (8 watt unit bearing motor)
Two SPST switches

PROCEDURE
RESONANCE

1. Wire a small 230-volt shaded pole motor in series with a 20-MFD capacitor and a 60-watt light bulb.

2. Wire a toggle switch in parallel to the capacitor.

3. Connect the circuit to a 110 volt A.C. power source.

4. While operating the circuit, open and close the toggle switch.

5. Explain the operation of the circuit.

6. Turn the toggle switch OFF

7. Measure the voltage across the motor. _____

8. Measure the voltage across the light. _____

9. Measure the voltage across the capacitor. _____

10. Turn the toggle switch ON

11. Measure the voltage across the motor. _____

12. Measure the voltage across the light. _____

13. Measure the voltage across the capacitor. _____

14. Explain your results.

15. With a dual-trace oscilloscope, place lead A on the incoming voltage. This lead will show the phase angle of the voltage in the circuit.

16. Place lead B between the light bulb socket and the motor. This lead will measure the phase angle of the current in the circuit.

17. Operate the circuit. Switch the toggle switch on and off and observe the effect on the oscilloscope.

EXPLANATION

An oscilloscope is a meter that reads voltage. However, rather than giving a number which corresponds to the effective voltage of an AC circuit, it shows the entire waveform. We see the entire sine wave. Lead A, on the incoming voltage, will show us the sine wave of the voltage being supplied to the circuit. Lead B will read the sine wave of the voltage drop across the light, our resistive load in this circuit. Remember, in resistive circuits the current and voltage are in phase with each other. Therefore, by reading the voltage across the resistive load, we are also showing the phase of the current in the circuit. By comparing the two sine waves produced, we can visually see the phase shift between voltage and current in an alternating current circuit.

LAB 34.3 IDENTIFICATION OF RELAY & CONTACTOR PARTS

LABORATORY OBJECTIVE
The purpose of this lab is to learn how relays and contactors work.

LABORATORY NOTES
You will disassemble a contactor and a relay, identify their parts, and discuss their operation.

***FUNDAMENTALS OF HVACR* TEXT REFERENCE**
Unit 34 Alternating Current Fundamentals

REQUIRED MATERIALS PROVIDED BY STUDENT
Electrical hand tools

REQUIRED MATERIALS PROVIDED BY SCHOOL
Relay to disassemble
Contactor to disassemble

PROCEDURE
Inspect the sealed relay and identify parts.

Identify the coil.

Where are the electrical connections to the coil?

Identify the armature.

What type of armature does it have – swinging or sliding?

Identify the contacts.

How many contacts does it have?

Are the contacts normally open or normally closed?

Where are the electrical connections for the contacts?

Disassemble a contactor and identify the parts.

Identify the coil.

Where are the electrical connections to the coil?

Identify the armature.

What type of armature does it have – swinging or sliding?

Identify the contacts.

How many contacts does it have?

Are the contacts normally open or normally closed?

Where are the electrical connections for the contacts?

Reassemble the contactor and relay.

LAB 35.1 ELECTRICAL MULTIMETERS

LABORATORY OBJECTIVE
The student will demonstrate how to properly use an electrical multimeter.

LABORATORY NOTES
For this lab exercise there should be a typical multimeter and a live electrical circuit that can be used for test purposes.

Testing electrical circuits is an important skill that each technician needs to develop. Although practice and experience are significant, a high degree of success is obtainable by following a proven procedure such as the following:

    A. Know the unit electronically. This means understanding the proper function of each control and the sequence of the control operation.

    B. Be able to read schematic wiring diagrams and have them available.

    C. Be able to use the proper electrical test instruments. Know the instrument. Read instructions carefully before using.

***FUNDAMENTALS OF HVACR* TEXT REFERENCE**
Unit 35 Electrical Measuring and Test Instruments

REQUIRED TOOLS AND EQUIPMENT
Digital & analog multimeters
Clamp-on ammeter
Live electrical circuit & resistor

SAFETY REQUIREMENTS

A. Check all circuits for voltage before doing any service work.

B. Stand on dry nonconductive surfaces when working on live circuits.

C. Never bypass any electrical protective devices.

PROCEDURE
Step 1
Familiarize yourself with the operation of an analog multimeter by checking the resistance of a known resistor.

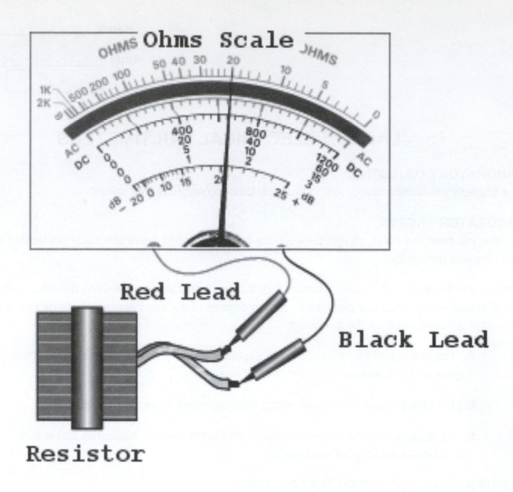

Figure 35-1-1

A. The ohmmeter uses a battery to furnish the current needed for resistance measurements. The higher the resistance the lower the current flow.

B. Whenever checking resistance, the power to the circuit being tested must be shut off.

C. Set the meter to measure resistance (Ohms).

D. Zero the meter by touching the red and black leads together. Since they are touching the resistance should read zero. When they are apart the resistance should be infinite.

E. Touch the red lead to one end of the resistor and the black lead to the other end and measure the resistance. Record the reading.

Resistance = _____ <u>Ohms</u>

F. Repeat the same measurement using a digital multimeter and record the reading. How do the two readings compare?

Resistance = _____ Ohms

Step 2
Check the voltage of a circuit using a multimeter.

A. Voltmeters are connected in parallel with the load to read the voltage drop. A knob in the center face of the meter adjusts the meter to the scale being used. Always start to measure voltage using the highest range on the meter.

B. Examine the wiring diagram for the circuit to be tested and determine how the leads from the meter should be connected.

C. Once familiar with circuit, turn on the power and measure the voltage. Record the reading.

Voltage = _____ Volts

Step 3
Check the amperage of a circuit using a clamp-on ammeter.

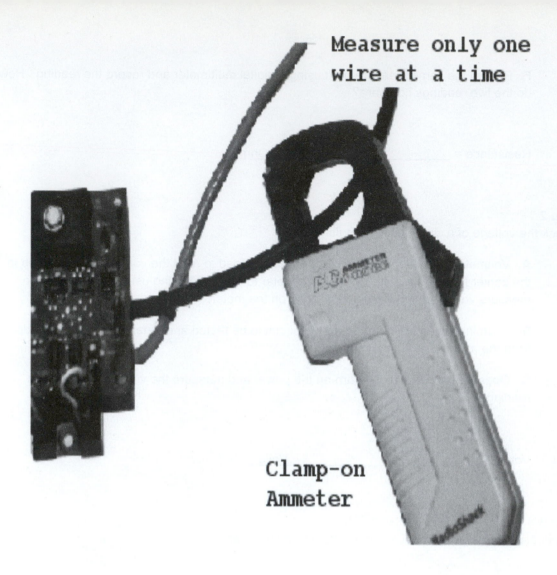

Measure only one wire at a time

Clamp-on Ammeter

Figure 35-1-2

A. When taking a reading always start with the highest possible scale and then work down to the most appropriate scale.

B. Never put the clamp around two different wires at the same time.

C. Examine the wiring diagram for the circuit to be tested and determine where to clamp the ammeter.

D. Once familiar with circuit, turn on the power and measure the amperage. Record the reading.

Amperage = _____ Amps

QUESTIONS

(Circle the letter that indicates the correct answer.)

1. When measuring voltage:
    - A. attach the lead in series with the load.
    - B. always use the highest scale on the meter first.
    - C. always use the lowest scale on the meter first.
    - D. use a clamp-on ammeter.

2. Electrical potential is measured with a/an.
    - A. ohmmeter.
    - B. wattmeter.
    - C. megohm meter.
    - D. voltmeter.

3. Always remember that when using an ohmmeter:
    - A. start with the lowest scale reading first.
    - B. make sure that the circuit is de-energized.
    - C. make sure that the circuit is energized.
    - D. the dial must be tapped gently to calibrate the needle.

4. A clamp-on ammeter can be used to measure current flow through:
    - A. a single wire only.
    - B. multiple wires.
    - C. multiple wires of different polarity.
    - D. motor foundations.

5. Using an ohmmeter on a live circuit:
    - A. is always correct.
    - B. will provide a voltage reading.
    - C. must be done very carefully.
    - D. may destroy the meter.

6. When the two leads of an ohmmeter are touched together:
    - A. there will be a spark.
    - B. there will be zero resistance indicated.
    - C. there will be infinite resistance indicated.
    - D. The meter will short out.

7. Most clamp-on ammeters:
    - A. cannot accurately read low amp draws.
    - B. can accurately read low amp draws.
    - C. will measure the resistance of the wire along with the current.
    - D. will measure the voltage of the wire along with the current.

8. Digital meters:
    - A. are less accurate than analog meters.
    - B. are generally cheaper than comparable analog meters.
    - C. are more easily damaged than analog meters.
    - D. can be accurate to three decimal places.

9. When checking the voltage of a DC circuit:
    A. only one lead from the meter is required.
    B. verify the correct polarity of the probes that are to be used before connecting the meter to the circuit.
    C. start with the lowest scale.
    D. All of the above are correct.

10. When measuring the resistance of a circuit, the reading is zero ohms:
    A. this indicates an open circuit.
    B. this means that the meter is faulty.
    C. this indicates a short circuit.
    D. Any of the above could be correct.

11. Solid state circuits:
    A. may be tested with any type ohmmeter.
    B. may only be tested with an analog ohmmeter.
    C. may only be tested with a digital ohmmeter.
    D. should normally not be tested with an ohmmeter.

12. Voltage is normally measured:
    A. in series with the load.
    B. across the load.
    C. Both A & B are correct.
    D. None of the above is correct.

LAB 35.2 USING VOLT METERS

LABORATORY OBJECTIVE
You will demonstrate your ability to safely read voltages using a multimeter.

LABORATORY NOTES
You will use a multimeter to check voltage at the different locations indicated by the chart.

FUNDAMENTALS OF HVACR TEXT REFERENCE
Unit 35 Electrical Measuring and Test Instruments

REQUIRED MATERIALS PROVIDED BY STUDENT
Safety glasses
Non-contact voltage detector
Multimeter
Electrical tools

REQUIRED MATERIALS PROVIDED BY SCHOOL
Equipment with voltage to read

PROCEDURE
Safety Note: In general, avoid touching things. Do not touch ANYTHING without first testing with your non-contact voltage detector.

Safety Note: A circuit does not have to be operating to be energized with a potentially fatal voltage. Always assume that ALL circuits can kill you!

Read and record the voltage of the following applications using both digital and analog volt meters.

Read Volts on	Volt Reading
Voltage of Wall Receptacle	
A/C Condensing Unit Disconnect	
A/C Condensing Unit Contactor	
Voltage Across Run Capacitor while operating	
Control Transformer Primary	
Control Transformer Secondary	
Leg to Leg on Single Phase Box	
Leg to Ground on Single Phase Box	
Leg to Leg on Three Phase Box	
Leg to Ground on Three Phase Box	

LAB 35.3 USING AMP METERS

LABORATORY OBJECTIVE

You will demonstrate your ability to safely read current (amps) using a clamp on ammeter.

LABORATORY NOTES

You will use a ammeter to check current at the different locations indicated by the chart.

***FUNDAMENTALS OF HVACR* TEXT REFERENCE**

Unit 35 Electrical Measuring and Test Instruments

REQUIRED MATERIALS PROVIDED BY STUDENT

Safety glasses

Non-contact voltage detector

Ammeter

Electrical tools

REQUIRED MATERIALS PROVIDED BY SCHOOL

Operating equipment for reading current.

PROCEDURE

Safety Note: In general, avoid touching things. Do not touch ANYTHING without first testing with your non-contact voltage detector.

Safety Note: A circuit does not have to be operating to be energized with a potentially fatal voltage.

Always assume that ALL circuits can kill you!

A circuit does have to be operating to read current.

Read and record the current (amps) of 10 operating electric loads.

Load	Amps
Power wire of operating air conditioner	
Black wire on operating compressor	
Red wire on operating compressor	
Yellow wire on operating compressor	
Red secondary transformer wire on operating unit	
Wire to operating furnace fan motor	
Wire to operating condenser fan motor	
Wire to operating electric strip heater	
Any operating electric load	
Any operating electric load	

LAB 35.4 USING OHM METERS

LABORATORY OBJECTIVE
You will demonstrate your ability to safely read the resistance of air conditioning components using a multimeter.

LABORATORY NOTES
You will use a multimeter to read the resistance of three variable resistors and record your readings in the chart below. You will then read the resistance of several air conditioning components and record the readings the chart.

FUNDAMENTALS OF HVACR TEXT REFERENCE
Unit 35 Electrical Measuring and Test Instruments

REQUIRED MATERIALS PROVIDED BY STUDENT
Multimeter

REQUIRED MATERIALS PROVIDED BY SCHOOL
3 variable resistors

PROCEDURE
1. Safety Note: Never check resistance on an energized circuit.
2. Use a non-contact voltage detector to check the circuit before disconnecting any wires or ohming any part of the circuit.
3. Components in an electrical circuit should be isolated from the circuit before ohming. This is normally done by removing the wires from them.
4. Avoid touching the tips of the leads while checking resistance. Touching the meter leads while ohming devices with a high resistance will cause an inaccurate reading because your body has a measurable resistance to most digital ohm meters.
5. Pay attention to the scale, especially on auto ranging and auto selecting meters.
   The Ω symbol stands for ohms. A reading of 2.5 means 2.5 ohms
   The K or KΩ stands for thousand ohms. A reading of 2.5 means 2500 ohms
   The M or MΩ stands for million ohms. A reading of 2.5 means 2500000 ohms.
   The V symbol stands for volts. If you see this, the circuit is energized!
6. Obtain 3 variable resistors from your instructor.
7. Read from the left to the middle, from the right to the middle, and from end to end.
8. Record three readings from each resistor in the following table.

Device	Numeric Reading	Scale	Actual Ω Reading
Resistor 1 Left to Middle			
Resistor 1 Right to Middle			
Resistor 1 End to End			
Resistor 2 Left to Middle			
Resistor 2 Right to Middle			
Resistor 2 End to End			
Resistor 3 Left to Middle			
Resistor 3 Right to Middle			
Resistor 3 End to End			
RBM 90-63 Relay Term 2 to 5			
RBM 90-64 Relay Term 2 to 5			
RBM 90-65 Relay Term 2 to 5			
RBM 90-66 Relay Term 2 to 5			
RBM 90-67 Relay Term 2 to 5			
RBM 90-68 Relay Term 2 to 5			
Transformer Yellow to Yellow			
Transformer Black to White			
Transformer Black to Red			
Transformer Black to Orange			
24-volt relay coil			
120-volt relay coil			
208/230-volt relay coil			
Electric strip heater			
Fuse			
Shaded pole motor			

LAB 35.5 CHECKING RESISTORS

LABORATORY OBJECTIVE
You will demonstrate your ability to safely read the resistance of fixed resistors.

LABORATORY NOTES
You will determine the resistor rating using the color bands on the resistors. Then, you will use a multimeter to read the resistance of fixed resistors and compare the actual resistance to the rating.

FUNDAMENTALS OF HVACR TEXT REFERENCE
Unit 35 Electrical Measuring and Test Instruments

REQUIRED MATERIALS PROVIDED BY STUDENT
Multimeter

REQUIRED MATERIALS PROVIDED BY SCHOOL
Fixed resistors

PROCEDURE
Write down the color bands on the resistor.

Use the guide on the following page to determine the rating of the resistor and record it in the data table.

Read the resistance of the resistor and compare to the rating.

	Resistor 1	Resistor 2	Resistor 3	Resistor 4
Band 1 Color				
First Digit				
Band 2 Color				
Second Digit				
Band 3 Color				
Multiplier				
Band 4 Color				
Tolerance				
Rating				
Reading				

Reading Resistor Color Bands

Resistors "resist" the flow of electrical current. Resistors are color coded to identify their resistance value. Each color is assigned a number. The table below lists the colors used and the number that each color represents.

Digits		Tolerances
Black	0	
Brown	1	1%
Red	2	2%
Orange	3	
Yellow	4	
Green	5	0.5%
Blue	6	0.25%
Violet	7	0.1%
Grey	8	0.05%
White	9	
Gold		5%
Silver		10%
None		20%

Most resistors use a 4 band color code. The first two digits represent digits, the third band represents a power of tem multiplier, and the fourth band represents tolerance. The tolerance band is usually separated some from the other bands. A few resistors use a 5 band scheme that adds an extra digit band. They have 3 digit color bands, a multiplier band, and a tolerance band. For example, a resistor with bands of orange, yellow, red, and silver:

The first digit is 3 and the second digit is 4. The power of 10 is 2, for a multiplier of 10^2, which is 100. Silver indicates a 10% tolerance.

$34 \times 10^2 = 3400$ ohms with a 10% tolerance. The 10% tolerance means that the actual value could be anywhere from 3060 to 3740 ohms.

Because resistors are not the exact value as indicated by the color bands, manufactures have included a tolerance color band to indicate the accuracy of the resistor. The tolerance band is usually gold or silver, but other colors are used for very close tolerance resistors. Some resistors may have no tolerance color. Gold band indicates the resistor is within 5% of what is indicated. Silver = 10% and None = 20%.

LAB 36.1 SETTING THE HIGH PRESSURE CUT-OUT

LABORATORY OBJECTIVE

The student will demonstrate how to properly set the cut-out and cut-in points on a high pressure cutout switch.

LABORATORY NOTES

For this lab exercise there needs to be an operating refrigeration system that has a high pressure cut-out switch. It can be set to stop the compressor before excessive pressures are reached. Such conditions might occur because of a water supply failure in water cooled condensers or because of a fan motor stoppage on air cooled condensers.

***FUNDAMENTALS OF HVACR* TEXT REFERENCE**

Unit 36 Electrical Components

REQUIRED TOOLS AND EQUIPMENT

Operating Refrigeration Unit with High Pressure Cutout Switch
Gauge Manifold

SAFETY REQUIREMENTS

A. Always read the equipment manual to become familiarized with the refrigeration system and its accessory components prior to start up.

B. Wear safety goggles and gloves when working with refrigerants. Liquid refrigerant can cause frostbite when in contact with eyes and skin.

C. Use low loss hose fittings, or wrap cloth around hose fittings before removing the fittings from a pressurized system or cylinder. Inspect all fittings before attaching hoses.

PROCEDURE

Step 1

Familiarize yourself with the major components in the refrigeration system including the condenser, compressor, evaporator, and metering device. Determine where the high side and low side connections for the system are located.

> **A.** The operating manual with the refrigeration specifications should indicate what the expected discharge pressure should be. If not, then you must calculate the expected discharge pressure based upon the refrigerant type for the system you are working on.
>
> **B.** Air cooled condensers – The refrigerant condensing temperature is typically 30°F higher than ambient (room) temperature. If it is 90°F outside, then the refrigerant condensing temperature will equal
>
> 90°F + 30°F = 120°F.

You should be able to determine the condenser pressure (high side gauge) from the P-T chart, Table 47-1-1, depicted in Lab 47.1 *Basic Refrigeration System Startup*.

For R-22 it would be expected to be approximately 260 psig. For another type of refrigerant such as R-134a the expected condenser pressure would be somewhat lower at approximately 172 psig.

C. Water cooled condensers – The refrigerant condensing temperature is typically 10 degrees higher than the water leaving the condenser. If the water temperature is 85°F entering the condenser and 95°F leaving the condenser, then the refrigerant condensing temperature will equal

95°F + 10°F = 105°F.

You should be able to determine the condenser pressure (high side gauge) from the P-T chart, Table 47-1-1, depicted in Lab 47.1 *Basic Refrigeration System Startup*.

For R-22 it would be expected to be approximately 211 psig. For another type of refrigerant such as R-134a the expected condenser pressure would be somewhat lower approximately at 135 psig.

Step 2
Determine the High pressure cut-out setting as follows:

A. The high pressure cut-out is typically set at 125% of normal discharge pressure.

B. For an air cooled condenser using R-22 the setting would be approximately 1.25 x 260 psig = 325 psig. For R-134a it would be

1.25 x 172 psig = 215 psig.

C. For a water cooled condenser using R-22 the setting would be approximately 1.25 x 211 psig = 264 psig. For R-134a it would be

1.25 x 135 psig = 169 psig.

D. Depending on the type of control switch, the HP cut-out may reset automatically once the pressure reaches the cut-in value (this would be at some point below the normal discharge pressure – in this example about 100 psig).

E. Remember - cycling on the HP cut-out is <u>harmful</u> to the unit!

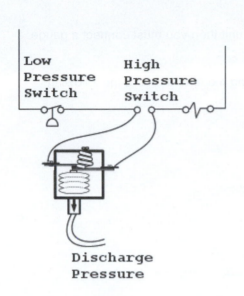

Low Pressure Switch   High Pressure Switch

Discharge Pressure

Figure 36-1-1

Thermostat

High Pressure Cutout Opens on High Pressure

Starter

L₁   L₂

Figure 36-1-2

Step 3

A. Adjust the High pressure cut-out setting by turning the adjusting screw to the correct setting as determined from Step 2.

B. Also adjust the cut-in setting if the high pressure cutout is so equipped.

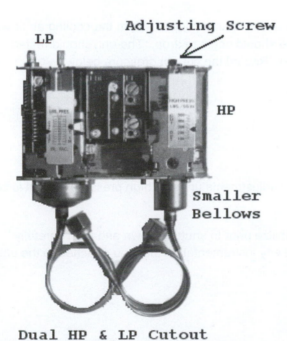

LP   Adjusting Screw

HP

Smaller Bellows

Dual HP & LP Cutout

Figure 36-1-3

Step 4

A. If there are no pressure gauges currently on the unit then you must connect a gauge manifold to the high and low side of the system.

B. Start the refrigeration system in the normal cooling mode and record the following:

RUNNING

High Side Pressure _____

Low Side Pressure _____

Step 5

A. After the system has stabilized, turn off the cooling air or water and allow the system to cut out on high pressure. Record the cut-out pressure below.

High Side Cut-out Pressure _____

B. Check the cut-in pressure if the switch is so equipped. Turn the cooling air or water back on and the discharge pressure should begin to drop. The unit should restart at the high pressure cutout – cut-in setting. Record the cut - in pressure below.

High Side Cut-in Pressure _____

Step 5

A. Continue making any necessary adjustments until the high pressure cut-out is correctly set.

B. Allow the system to run and stabilize prior to shutting down and then carefully disconnect the gauge manifold and any instrumentation that you attached to the unit.

QUESTIONS

(Circle the letter that indicates the correct answer.)

1. The high pressure cut-out switch:
   - A. is a safety device.
   - B. is a device to cycle the compressor on and off with load.
   - C. contacts are normally open.
   - D. contacts close on high pressure.

2. The cut-in setting on a high pressure cut-out switch.
   - A. bypasses the cutout portion.
   - B. restarts the compressor when the pressure is back to normal.
   - C. shuts off the compressor on high pressure.
   - D. restarts the compressor on high pressure.

3. The high pressure cut-out should be set to stop the compressor at:
   - A. a pressure that is 75% above normal discharge pressure.
   - B. a pressure that is 25% above normal discharge pressure.
   - C. a pressure that is 125% above normal discharge pressure.
   - D. a pressure that is 150% above normal discharge pressure.

4. Cycling of the compressor on the high pressure cut-out:
   - A. is always desirable.
   - B. is unavoidable.
   - C. is normal.
   - D. is harmful to the compressor.

5. A refrigeration system with a normal discharge pressure of 120 psig:
   - A. would have a high pressure cut-out setting of 125 psig.
   - B. would have a high pressure cut-out setting of 150 psig.
   - C. would have a high pressure cut-out setting of 135 psig.
   - D. would have a high pressure cut-out setting of 185 psig.

6. The cut-out setting on a high pressure cut-out switch.
   - A. bypasses the cutout portion.
   - B. restarts the compressor when the pressure is back to normal.
   - C. shuts off the compressor on high pressure.
   - D. restarts the compressor on high pressure.

7. Two operating conditions that could lead to the compressor stopping on the high pressure cut-out are:
   - A. lack of cooling air or cooling water.
   - B. high current.
   - C. a shorted winding.
   - D. All of the above are correct.

LAB 36.2 SETTING THE LOW PRESSURE CUT-OUT

LABORATORY OBJECTIVE
The student will demonstrate how to properly set the cut-out and cut-in points on a low pressure cutout switch.

LABORATORY NOTES
For this lab exercise there needs to be an operating refrigeration system that has a low pressure cut-out switch.

The low-pressure switch, which senses compressor suction pressure, opens on a drop in pressure. It is set to cut out at a protective low limit pressure, but remains closed at normal operating pressures. It is also known as a loss of charge or low pressure cut-out switch.

FUNDAMENTALS OF HVACR TEXT REFERENCE
Unit 36 Electrical Components

REQUIRED TOOLS AND EQUIPMENT
Operating refrigeration unit with low pressure cutout switch and king valve
Gauge manifold

SAFETY REQUIREMENTS

A. Always read the equipment manual to become familiarized with the refrigeration system and its accessory components prior to start up.

B. Wear safety goggles and gloves when working with refrigerants. Liquid refrigerant can cause frostbite when in contact with eyes and skin.

C. Use low loss hose fittings, or wrap cloth around hose fittings before removing the fittings from a pressurized system or cylinder. Inspect all fittings before attaching hoses.

PROCEDURE
Step 1
Familiarize yourself with the major components in the refrigeration system including the condenser, compressor, evaporator, and metering device. Determine where the high side and low side connections for the system are located.

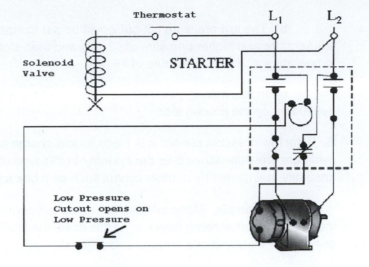

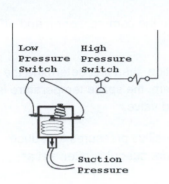

Figure 36-2-1

Figure 36-2-2

A. The operating manual with the refrigeration specifications should indicate what the expected suction pressure should be. Many low pressure control settings exist and different situations call for different settings.

B. If you have no information on the system, you may be able to calculate the expected suction pressure based upon the refrigerant type for the system you are working on.

C. The refrigerant temperature in the evaporator should generally be no more than 15°F colder than the medium being cooled.

D. As an example, if a freeze box is kept at 0°F, then the refrigerant temperature should be about minus 15°F.

You should be able to determine the lowest evaporator pressure from the P-T chart, Table 47-1-1, depicted in Lab 47.1 *Basic Refrigeration System Startup*.

For R-22 it would be expected to be approximately 13 psig. If the space to be cooled was for vegetables to be kept at 40°F, then the corresponding refrigerant temperature would be 40°F - 15°F = 25°F which corresponds to an evaporator pressure of 49 psig.

Step 2
Determine the Low pressure cut-out setting as follows:

A. If the low pressure cut-out is in place to control the space temperature, then first determine the maximum and minimum temperature of the space.

a) Example: The space is to be kept at a temperature of between 40°F to 50°F. As explained in the previous section, the refrigerant temperature would be approximately 15°F colder than the space and would therefore vary between 25°F to 35°F. Using R-22, this would correspond to an evaporator pressure of from 49 to 62 psig.

b) The low pressure cut-out would be set to start the compressor at the warmer temperature and higher pressure of 62 psig and then stop the compressor at the lower temperature and lower pressure of 49 psig.

c) In this manner, the low pressure cut-out would cycle the compressor on and off dependent on the cooling load.

B. If the low pressure cut-out is in place for low charge protection, then first determine the minimum pressure allowed for the system. In this type of system, the space temperature is generally maintained by another control such as a box solenoid valve.

a) Example: Many refrigeration systems operate at positive pressures to reduce the possibility of air being drawn in. In this case, the low pressure cut-out is often set at some point slightly above atmospheric pressure.

b) A common setting for this type of low pressure cut-out would be to stop the compressor at a low load and low pressure of 2 psig and then restart the compressor as the space warms up and the pressure rises to about 8 psig.

c) In this manner, the refrigerant pressure in the system is never allowed to fall below atmospheric pressure.

C. If the system loses its refrigerant charge, the space warms up and the compressor continuously runs. Due to a lack of refrigerant circulating through the compressor and its continuous running, the compressor and motor may be damaged.

a) In this type of undercharged situation, the suction pressure would continue to drop. If the system is equipped with a low pressure cut-out, the system would automatically shut down at the system cut-out pressure.

D. Depending on the type of refrigeration system, a low pressure cut-out can operate as both a temperature control and a low charge protection control.

Step 3
Adjust the Low pressure cut-out setting as determined from Step 2 by turning the adjusting screws as follows:

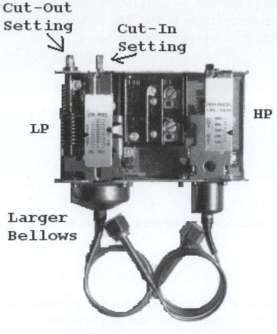

Cut-Out Setting

Cut-In Setting

LP

HP

Larger Bellows

Dual HP & LP Cutout

Figure 36-2-3

A. Range adjustment – This is the difference between the minimum and maximum operating pressures within which the control will function. As an example, a cut-out of 2 psig and a cut-in of 8 psig would indicate a typical range setting of a low pressure cut-out.

B. Differential adjustment – This is the difference between the cut-out and cut-in pressures for the control. The differential of a low-pressure switch set to cut-out at 2 psig and cut-in at 8 psig would be 6 psig.

C. A low pressure cut-out will have two adjusting screws as shown below. In many cases, one adjusting screw will control the range while the other controls the differential.

Be careful as you make adjustments to the control. In some cases, the set range may change as you change the differential and vice versa.

Step 4

A. If there are no pressure gauges currently on the unit then you must connect a gauge manifold to the high and low side of the system.

B. Start the refrigeration system in the normal cooling mode and allow the system to stabilize.

C. Close the king valve which is located in the liquid line directly following the receiver or condenser. The suction pressure will begin to decrease and eventually the compressor will stop. Record the suction pressure at which the compressor stopped below:

Suction Pressure at which compressor stops _____

D. After recording the suction pressure at which the compressor stops, slowly open the king valve and observe the suction pressure begin to rise. The compressor should restart. Record the suction pressure at which the compressor restarts below:

Suction Pressure at which compressor restarts _____

Step 5

A. Continue making any necessary adjustments until the low pressure cut-out is correctly set.

B. Allow the system to run and stabilize prior to shutting down and then carefully disconnect the gauge manifold and any instrumentation that you attached to the unit.

QUESTIONS

(Circle the letter that indicates the correct answer.)

1. The low pressure cut-out switch:
    A. is a safety device.
    B. is a device to cycle the compressor on and off with load.
    C. contacts are normally open.
    D. Both A & B are correct.

2. The cut-in setting on a low pressure cut-out switch.
    A. bypasses the cutout portion.
    B. restarts the compressor when the pressure is low.
    C. shuts off the compressor on a rise in suction pressure.
    D. restarts the compressor on a rise in suction pressure.

3. The low pressure cut-out should be set to stop the compressor:
    A. when the desired box temperature is reached.
    B. before the suction pressure drops below atmospheric.
    C. Both A & B are correct.
    D. None of the above is correct.

4. The differential can be defined as the difference between the cut-out and cut-in points of the control.
    A. True.
    B. False.

LAB 36.3 SETTING A BOX THERMOSTAT CONTROL

LABORATORY OBJECTIVE

The student will demonstrate how to properly set a thermostat that is controlling the temperature of a refrigerated space.

LABORATORY NOTES

For this lab exercise there needs to be an operating refrigeration system that has a thermostat controlling the temperature of the space being cooled.

FUNDAMENTALS OF HVACR TEXT REFERENCE

Unit 36 Electrical Components

REQUIRED TOOLS AND EQUIPMENT

Operating Refrigeration Unit with thermostat Control

SAFETY REQUIREMENTS

A. Always read the equipment manual to become familiarized with the refrigeration system and its accessory components prior to start up.

PROCEDURE

Step 1

Familiarize yourself with the major components in the refrigeration system including the condenser, compressor, evaporator, and metering device.

> A. Temperature motor control – This is the simplest type of control system. For this type of system there is only one space being cooled. The thermostat will cycle the compressor motor on and off dependent on the load. This type of system is shown in Figure 36-3-1 below.

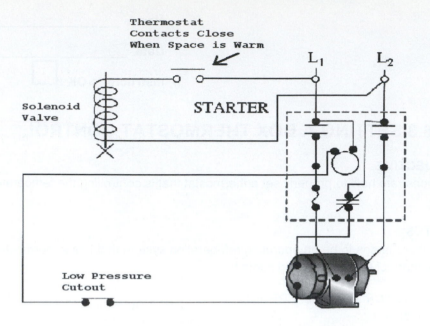

Figure 36-3-1

B. Pressure motor control – This is used on systems that cool multiple spaces at one time. Each individual space will have its own solenoid valve that is controlled by a thermostat. The low pressure cut-out will cycle the compressor motor on and off dependent on the load.

C. The thermostat will have a sensing bulb that should be located in the space being cooled. The sensing bulb should be in such a location that it is sensing the average temperature of the space. If it is in a walk-in cooler, it should be mounted in a central location. Care should be taken not to damage the thin interconnecting tube located between the sensing bulb and the thermostat contacts.

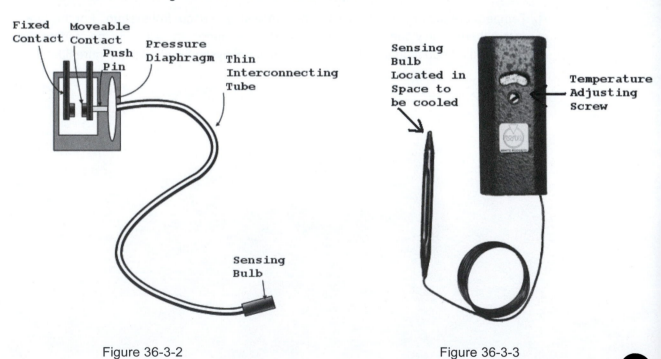

Figure 36-3-2   Figure 36-3-3

Step 2

Set the thermostat as follows:

A. Determine the desired temperature for the space to be cooled and check the thermostat setting.

B. Generally the thermostat can be adjusted with a screwdriver or a small wrench. Some thermostats have a readable scale for temperature. Others are adjusted by counting the number of turns of the screw. As an example, one full turn of the adjusting screw may equal four degrees, etc.

C. Differential adjustment – This is the difference between the cut-out and cut-in pressures for the control. On a temperature type motor control, this setting will reduce the cycling of the compressor motor on and off. On a pressure type motor control, this setting will reduce the cycling of the solenoid valve open and close.

As an Example, assume the space is to be kept at a temperature of 33°F with a 4 degree differential. With a temperature motor control, the compressor will start when the space temperature reaches 35°F and stop again some time later when the space cools back down to 31°F. This provides for an average temperature of 33°F.

D. Some spaces are allowed higher differentials than others. For example, dairy cooler temperatures should be fairly constant, without large variations in temperatures. Therefore they would be set for a smaller differential.

Step 3

A. After setting the thermostat, start the refrigeration system in the normal cooling mode and allow the system to stabilize.

B. Depending on the size of the unit, it may take some time before the set point temperature is reached. Record the temperature of the space when the compressor stops or the solenoid valve closes.

_____ °F

C. Allow the space to warm up and then record the temperature of the space when the compressor starts or the solenoid valve opens.

_____ °F

D. Continue making any necessary adjustments until the thermostat is correctly set.

QUESTIONS

(Circle the letter that indicates the correct answer.)

1. If the thin interconnecting tube between the thermostat contacts and the sensing bulb is damaged:
- A. the compressor will not start.
- B. the space temperature will be too low.
- C. the system will operate as normal.
- D. the compressor will not stop.

2. A high differential setting on a thermostat.
- A. will lead to decreased cycling.
- B. will lead to lower than normal temperatures.
- C. will lead to increased cycling.
- D. will lead to higher than normal temperatures.

3. A temperature type motor control:
- A. cycles the compressor motor on and off dependent on load.
- B. is connected to the low pressure cut-out.
- C. protects the motor from overheating.
- D. unloads the compressor at high space temperatures.

4. Generally the thermostat sensing bulb is located:
- A. right at the entrance to the space.
- B. connected to the evaporator coil.
- C. at a central location to sense average temperature.
- D. at the lowest point in the space.

5. The thermostats for a multiple space system that uses a pressure type motor control:
- A. cycles the compressor motor on and off dependent on load.
- B. is connected to the low pressure cut-out.
- C. protects the motor from overheating.
- D. control individual space solenoid valves.

LAB 36.4 TESTING A CAPACITOR

LABORATORY OBJECTIVE

The student will demonstrate how to properly discharge and test a capacitor used on a motor starting circuit.

LABORATORY NOTES

Many single phase motors require capacitors. Capacitors can be dangerous. Capacitors can hold a high voltage charge even after the power is turned off. Always discharge capacitors before touching them.

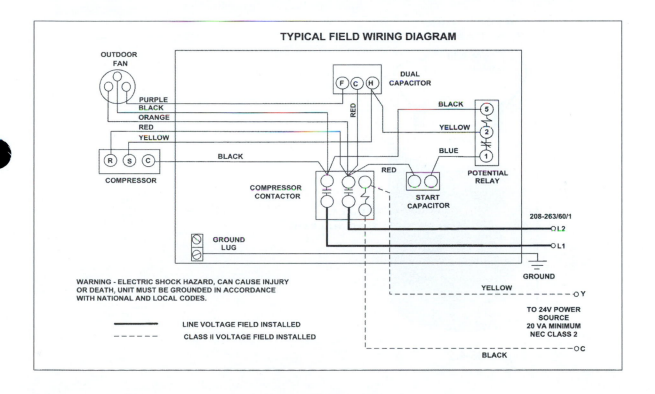

Figure 36-4-1

FUNDAMENTALS OF HVACR TEXT REFERENCE

Unit 36 Electrical Components

REQUIRED TOOLS AND EQUIPMENT

Capacitor
20,000 ohm, 2 W resistor

SAFETY REQUIREMENTS

A. Turn the power off if the capacitor to be tested is installed in an operating system. Lock and tag out the power supply.

B. Confirm the power is secured by testing for zero voltage with you meter.

C. Capacitors should be discharged with a bleed resistor.

D. <u>Never</u> discharge a capacitor with a screwdriver or other metal device touching the terminals. The spark that is generated externally can also occur internally, damaging the compressor.

PROCEDURE
Step 1
Discharge the capacitor as follows.

        A. Place the capacitor in a protective case and connect a 20,000 ohm,

2 W resistor across the terminals.

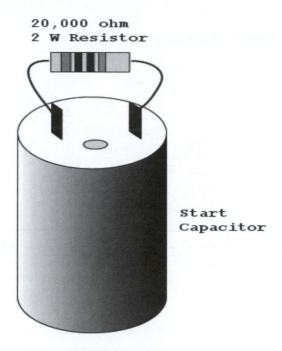

20,000 ohm
2 W Resistor

Start
Capacitor

Figure 36-4-2

        B. Most start capacitors have a bleed resistor; however it is good practice to make sure the charge has been bled off.

        C. To test the capacitor, disconnect it from the wiring and place the ohmmeter leads on the terminals as shown in Figure 36-4-2. The ohmmeter should have at least an R x 100 scale. Familiarize yourself with electrical meters.

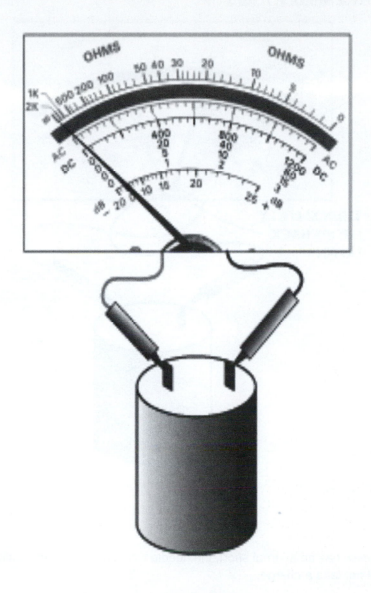

Figure 36-4-3

D. If the capacitor is good, the needle will make a rapid swing toward zero and slowly return to infinity as shown in Figure 36-4-3.

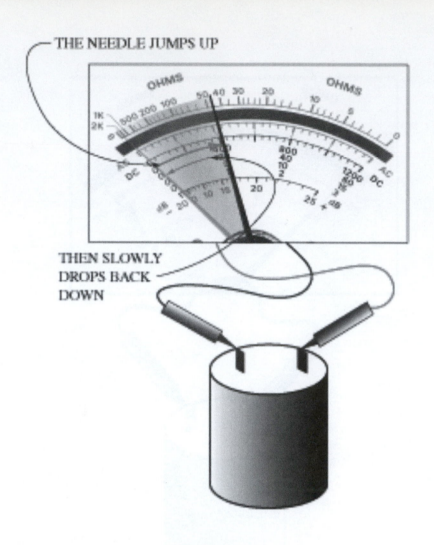

THE NEEDLE JUMPS UP

THEN SLOWLY DROPS BACK DOWN

Figure 36-4-4

E. If the capacitor has an internal short, the needle will stay at zero, indicating the instrument will not take a charge.

F. In replacing a capacitor, it is desirable to use an exact replacement – a capacitor with the same mfd rating and voltage limit rating.

QUESTIONS
(Circle the letter that indicates the correct answer.)

1. Start capacitors:
- A. are high capacity (100-800 mfd).
- B. are low capacity (2-40 mfd).
- C. are only charged when the circuit is live.
- D. can be discharged with a screwdriver.

2. Capacitors rated for continuous duty are:
- A. start capacitors.
- B. run capacitors.
- C. Both A & b are correct.
- D. None of the above s correct.

3. In regard to capacitors:
    A. start and run capacitors are never interchangeable.
    B. start capacitors are always rated for continuous duty.
    C. start capacitors are low capacity (2-40 mfd).
    D. All of the above are correct.

4. A red dot on a run capacitor terminal indicates:
    A. the high side of the system.
    B. the free end.
    C. it should be connected to the run terminal.
    D. its mfd rating.

5. If a capacitor has a bleed resistor:
    A. it must be removed before testing.
    B. the capacitor can still be tested with the bleed resistor in place.
    C. it will fail open.
    D. Both B & C are correct.

6. When testing a capacitor:
    A. the ohmmeter should have at least an R x 100 scale.
    B. the ohmmeter should have at least an R x 1000 scale.
    C. the ohmmeter should have at least an R x 10 scale.
    D. None of the above is correct.

7. Discharging a capacitor by shorting its terminals:
    A. is always the preferred method.
    B. will cause violent sparking.
    C. melt the tip of a screwdriver.
    D. can damage the capacitor.

8. If the capacitor is good, the ohmmeter needle will stay at zero.
    A. True.
    B. False.

LAB 36.5 IDENTIFY COMPONENTS ON AIR CONDITIONING DIAGRAM

LABORATORY OBJECTIVE

You will identify common HVACR electrical components on an air conditioning unit and locate their symbols on the unit electrical diagram.

LABORATORY NOTES

In this lab exercise you will identify common electrical components on an air conditioner. You will then identify the symbols in the unit diagram that represent the components.

FUNDAMENTALS OF HVACR TEXT REFERENCE

Unit 36 Electrical Components / Unit 38 Electrical Diagrams

REQUIRED MATERIALS PROVIDED BY STUDENT

Safety glasses
Non-contact voltage detector
6-in-1 screwdriver

REQUIRED MATERIALS PROVIDED BY SCHOOL

Air Conditioner with electrical components and an electrical diagram

PROCEDURE

Turn off the power to the packaged air conditioner assigned by the instructor.

Remove the electrical panel to the unit.

Check with the non-contact voltage detector to insure that the panel is not energized.

DO NOT TOUCH ANYTHING YOU DON'T HAVE TO!

You may use the unit component location diagram or pictorial diagram to locate the components.

You should be able to locate the component on the diagram and on the unit.

You must point out the following components and their symbols to the instructor.

- Condenser fan motor
- (outdoor fan motor)
- Evaporator fan motor
- (indoor blower motor)
- Contactor
- Transformer
- Compressor
- Capacitor
- Indoor fan relay
- Thermostat

LAB 36.6 IDENTIFY COMPONENTS ON HEAT PUMP DIAGRAM

LABORATORY OBJECTIVE
You will identify common HVACR electrical components on a heat pump and locate their symbols on the unit electrical diagram.

LABORATORY NOTES
In this lab exercise you will identify common electrical components on a heat pump. You will then identify the symbols in the unit diagram that represent the components.

FUNDAMENTALS OF HVACR TEXT REFERENCE
Unit 36 Electrical Components / Unit 38 Electrical Diagrams

REQUIRED MATERIALS PROVIDED BY STUDENT
Safety glasses
Non-contact voltage detector
6-in-1 screwdriver

REQUIRED MATERIALS PROVIDED BY SCHOOL
Heat pump with electrical components and an electrical diagram

PROCEDURE
Turn off the power to the packaged heat pump assigned by the instructor.
Remove the electrical panel to the unit.
Check with the non-contact voltage detector to insure that the panel is not energized.
DO NOT TOUCH ANYTHING YOU DON'T HAVE TO!
You may use the unit component location diagram or pictorial diagram to locate the components.
You should be able to locate the component on the diagram and on the unit.
You must point out the components and their symbols to the instructor.

Compressor	High pressure switch
Supplementary heater	Thermal overload for electric heater
Indoor fan motor	Indoor fan relay
Transformer	Contactor
Defrost control	Defrost thermostat
Reversing Valve	Outdoor fan motor
High pressure switch	

LAB 36.7 IDENTIFYING START AND RUN CAPACITORS

LABORATORY OBJECTIVE

The purpose of this lab is to learn how to distinguish between start and run capacitors and identify their important electrical characteristics.

LABORATORY NOTES

You will identify capacitors as either starting capacitors or running capacitors. You will also record their microfarad and voltage ratings.

FUNDAMENTALS OF HVACR TEXT REFERENCE

Unit 36 Electrical Components
Unit 37 Electric Motors

REQUIRED MATERIALS PROVIDED BY SCHOOL

Assortment of starting and running capacitors.

PROCEDURE

Examine the assortment of capacitors, separating the run capacitors from the starting capacitors. Record their microfarad and voltage ratings in the table below.

SAFETY NOTE: Capacitors can hold a potentially fatal electric charge. Always discharge capacitors before touching their Electrical terminals. Better yet, do NOT touch the Electrical terminals.

SUMMARY

Capacitor Type	Voltage Rating	Microfarad Rating

LAB 36.8 CHECKING START AND RUN CAPACITORS USING MULTIMETER

LABORATORY OBJECTIVE

The purpose of this lab is to learn how to check start and run capacitors using the capacitor test function on a multimeter.

LABORATORY NOTES

You will check both starting capacitors and running capacitors. You will also record their microfarad and voltage ratings and compare their rated values to the value displayed by the meter to determine if the capacitor is good or bad.

FUNDAMENTALS OF HVACR TEXT REFERENCE

Unit 36 Electrical Components
Unit 37 Electric Motors

REQUIRED MATERIALS PROVIDED BY STUDENT

Multimeter with capacitor test function

REQUIRED MATERIALS PROVIDED BY SCHOOL

Assortment of starting and running capacitors.

PROCEDURE

1. For each capacitor being tested, you should:
2. Discharge the capacitor.
3. Record their microfarad and voltage ratings in the table below.
4. Use the multimeter to check the capacitance and record that.
5. Compare the capacitor rating to the tested value and determine if the capacitor is good or bad.

SAFETY NOTE: Capacitors can hold a potentially fatal electric charge. Always discharge capacitors before touching their Electrical terminals. Better yet, do NOT touch the Electrical terminals.

Capacitor Type	Voltage Rating	Microfarad Rating	Microfarad Test

LAB 36.9 IDENTIFYING OVERLOADS

LABORATORY OBJECTIVE
The purpose of this lab is to learn the types of overloads and become familiar with their specifications.

LABORATORY NOTES
You will examine an assortment of overloads and record their specifications.

FUNDAMENTALS OF HVACR TEXT REFERENCE
Unit 36 Electrical Components

Unit 37 Electric Motors

REQUIRED MATERIALS PROVIDED BY STUDENT
Multimeter
Electrical hand tools

REQUIRED MATERIALS PROVIDED BY SCHOOL
Overloads

PROCEDURE

Examine four pressure switches and record the following specifications:

1. Type of overload – Magnetic, thermal, or current
2. Line duty or pilot duty
3. Rating – this can be different depending upon the type of overload. Some common overload ratings are
   a. Amperage
   b. Must Hold Current
   c. Must trip current
   d. Temperature

Type - Magnetic, thermal, or current	Line duty or Pilot Duty	Rating	Rating

LAB 36.10 TROUBLESHOOTING THERMAL OVERLOADS

LABORATORY OBJECTIVE
The purpose of this lab is to learn to troubleshoot line duty thermal overloads.

LABORATORY NOTES
You will check line duty thermal overloads using volt, ohms, and amps.

FUNDAMENTALS OF HVACR TEXT REFERENCE
Unit 36 Electrical Components
Unit 37 Electric Motors

REQUIRED MATERIALS PROVIDED BY STUDENT
Multimeter
Amp meter
Electrical hand tools

REQUIRED MATERIALS PROVIDED BY SCHOOL
Line duty thermal Overloads

PROCEDURE
1. On open overloads, physically look at the overload element and contacts and check to see if the contacts or element of the overload are burned out.
2. WITH POWER OFF check the contacts of the overload by ohming out the points. They should have a resistance of close to 0 ohms (closed).
3. If the contacts have a measurable resistance of more than a few tenths of an ohm, the overload is bad and should be changed.
4. If the overload is open, an infinite ohm reading, check:
a. the voltage
b. compressor windings
c. run capacitor
d. starting components
5. If all of the above components are good, attempt to start the motor after the overload closes and check the amp draw when the motor tries to start.
6. Compare the amp draw when the overload opens to its rating.
7. If amp draw is below the overload rating, change the overload.
8. If the operating amp draw is at or above the overload rating, it is just doing its job and is good. Further checks of the compressor and compressor circuit are required.
9. If the overload is bad, change with correct replacement.

LAB 36.11 PILOT DUTY OVERLOAD OPERATION

LABORATORY OBJECTIVE
The purpose of this lab is to learn how pilot duty overloads relays work.

LABORATORY NOTES
You will build a circuit containing a pilot duty magnetic overload, a 24-volt contactor, and a 24-volt lock out relay to demonstrate how a pilot duty overload works.

***FUNDAMENTALS OF HVACR* TEXT REFERENCE**
Unit 36 Electrical Components
Unit 37 Electric Motors

REQUIRED MATERIALS PROVIDED BY STUDENT
Multimeter   Electrical hand tools
Amp meter

REQUIRED MATERIALS PROVIDED BY SCHOOL
Pilot duty Overloads   24-volt contactor
Heat Strips   24-volt lock out relay
24-volt Transformer   Toggle switch

PROCEDURE
1. Use the diagram below to wire a pilot duty magnetic overload protecting a set of electric strip heaters controlled by a contactor.
2. Have an instructor check your wiring.
3. Turn the power on and operate the unit.

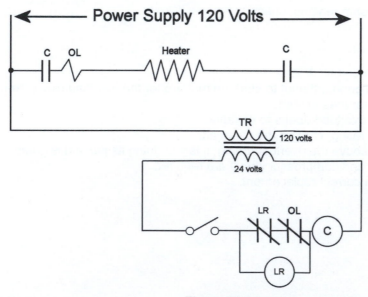

Figure 36-11-1

LAB 36.12 MAGNETIC STARTER FAMILIARIZATION

LABORATORY OBJECTIVE
The purpose of this lab is to learn how magnetic starters work.

LABORATORY NOTES
You will disassemble a magnetic starter, identify the parts, and discuss its operation.

FUNDAMENTALS OF HVACR TEXT REFERENCE
Unit 36 Electrical Components

REQUIRED MATERIALS PROVIDED BY STUDENT
Electrical hand tools

REQUIRED MATERIALS PROVIDED BY SCHOOL
Magnetic starter to disassemble

PROCEDURE
Inspect the parts.
1. Identify the coil.

2. Where are the electrical connections to the coil?

3. Identify the armature.

4. What type of armature does it have – swinging or sliding?

5. Identify the contacts.

6. How many contacts does it have?

7. Are the contacts normally open or normally closed?

8. Where are the auxiliary contacts?

9. Where are the electrical connections for the contacts?

10. Identify the overloads.

11. Where are the electrical connections for the overloads?

12. Remove one of the overload heaters and discuss how to change them.

13. Where are the electrical connections for the contacts?

Reassemble the magnetic starter.

LAB 36.13 WIRING MAGNETIC STARTER WITH A TOGGLE SWITCH

LABORATORY OBJECTIVE
The purpose of this lab is to learn how to design and wire a simple magnetic starter circuit.

LABORATORY NOTES
You will design, wire, and operate a basic magnetic starter circuit using a toggle switch will control the coil.

The magnetic starter will control a 3 phase motor.

FUNDAMENTALS OF HVACR TEXT REFERENCE
Unit 36 Electrical Components

REQUIRED MATERIALS PROVIDED BY STUDENT
Electrical hand tools
Multimeter
Amp meter

REQUIRED MATERIALS PROVIDED BY SCHOOL
Magnetic starter with 120-volt coil
Toggle switch
3 phase motor

PROCEDURE
1. Use the below to wire a toggle switch controlling the magnetic starter. Make sure the circuit to the coil passes through the normally closed overload contacts. The starter will control a three-phase motor.
2. Have instructor check your drawing.
3. Wire the circuit you have drawn.
4. Have an instructor check it.
5. Operate your circuit.

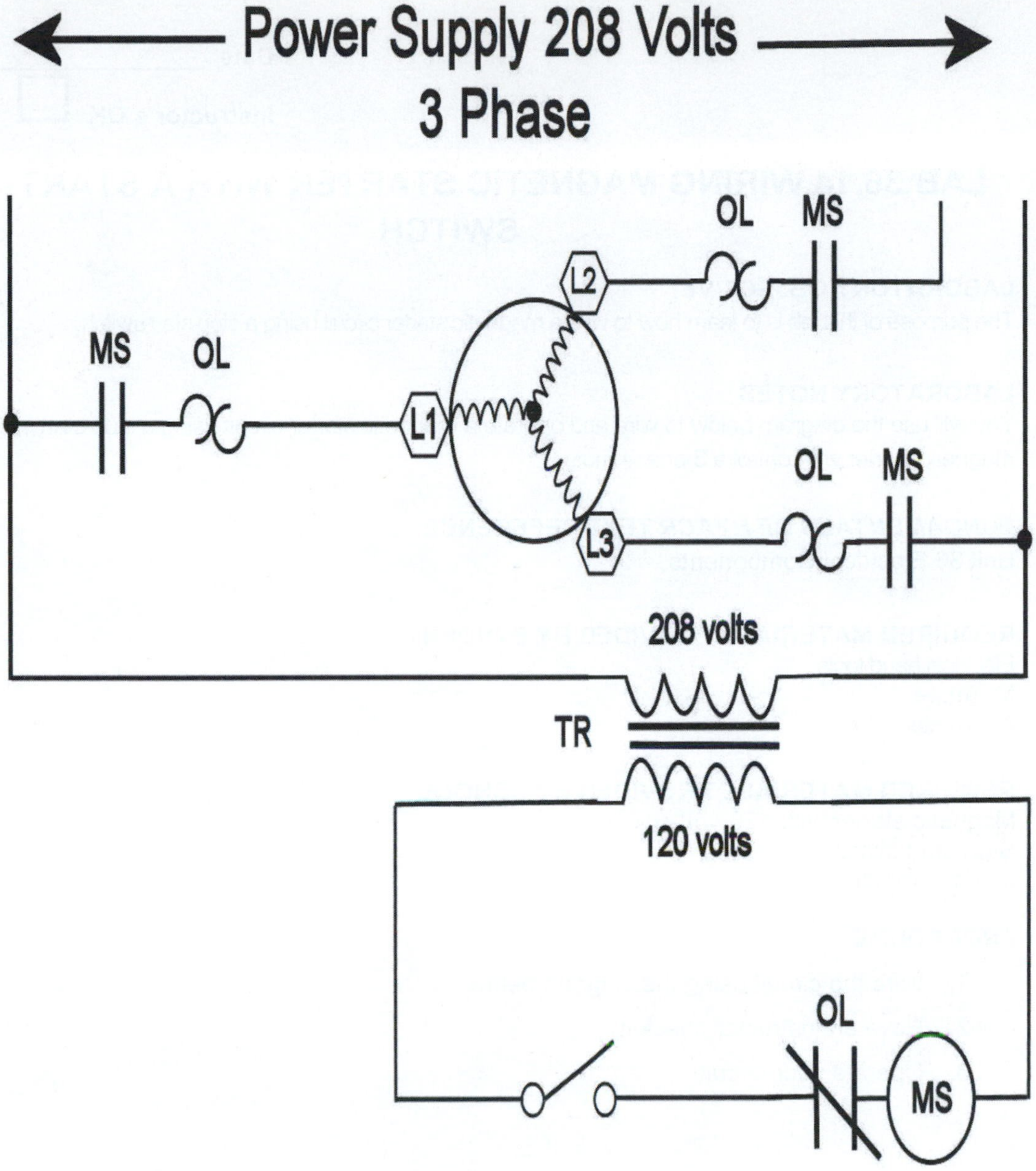

Power Supply 208 Volts
3 Phase

Figure 36-13-1

LAB 36.14 WIRING MAGNETIC STARTER WITH A START-STOP SWITCH

LABORATORY OBJECTIVE
The purpose of this lab is to learn how to wire a magnetic starter circuit using a stop-start switch.

LABORATORY NOTES
You will use the diagram below to wire and operate a magnetic starter circuit using a stop-start switch. The magnetic starter will control a 3 phase motor.

FUNDAMENTALS OF HVACR TEXT REFERENCE
Unit 36 Electrical Components

REQUIRED MATERIALS PROVIDED BY STUDENT
Electrical hand tools
Multimeter
Amp meter

REQUIRED MATERIALS PROVIDED BY SCHOOL
Magnetic starter with 120-volt coil
Stop-start switch
3 phase motor

PROCEDURE
1. Wire the circuit using the diagram below.

2. Have an instructor check it.

3. Operate your circuit.

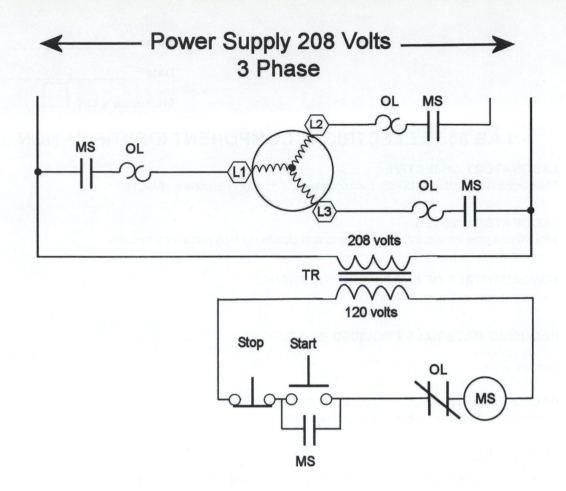

Figure 36-14-1

389

LAB 36.15 ELECTRICAL COMPONENT IDENTIFICATION

LABORATORY OBJECTIVE

The purpose of this lab is to identify common electrical components used in HVACR.

LABORATORY NOTES

You will examine the electrical components and identify each by name and function.

FUNDAMENTALS OF HVACR TEXT REFERENCE

Unit 36 Electrical Components

REQUIRED MATERIALS PROVIDED BY STUDENT

Safety glasses
Gloves

REQUIRED MATERIALS PROVIDED BY SCHOOL

Assortment of Common Electrical Components

PROCEDURE

Examine each of the numbers electrical components and write down their name and function.

\#	Name	Function
\multicolumn{3}{c}{**Lab 36.15 Electrical Component Identification**}		

\#	Name	Function
1		
2		
3		
4		
5		
6		
7		
8		
9		
10		

LAB 36.16 IDENTIFICATION OF RELAY & CONTACTOR PARTS

LABORATORY OBJECTIVE
The purpose of this lab is to learn how relays and contactors work.

LABORATORY NOTES
You will disassemble a contactor and a relay, identify their parts, and discuss their operation.

FUNDAMENTALS OF HVACR TEXT REFERENCE
Unit 34 Alternating Current Fundamentals
Unit 36 Electrical Components

REQUIRED MATERIALS PROVIDED BY STUDENT
Electrical hand tools

REQUIRED MATERIALS PROVIDED BY SCHOOL
Relay to disassemble
Contactor to disassemble

PROCEDURE

Inspect the sealed relay and identify parts.

Identify the coil.

Where are the electrical connections to the coil?

Identify the armature.

What type of armature does it have – swinging or sliding?

Identify the contacts.

How many contacts does it have?

Are the contacts normally open or normally closed?

Where are the electrical connections for the contacts?

Disassemble a contactor and identify the parts.

Identify the coil.

Where are the electrical connections to the coil?

Identify the armature.

What type of armature does it have – swinging or sliding?

Identify the contacts.

How many contacts does it have?

Are the contacts normally open or normally closed?

Where are the electrical connections for the contacts?

Reassemble the contactor and relay.

LAB 36.17 EXAMINING LOW VOLTAGE THERMOSTATS

LABORATORY OBJECTIVE
The purpose of this lab is to learn the types of low voltage thermostats and become familiar with their electrical details.

LABORATORY NOTES
You will examine an assortment of low voltage thermostats and record their specifications.

FUNDAMENTALS OF HVACR TEXT REFERENCE
Unit 36 Electrical Components

REQUIRED TOOLS AND MATERIALS
Low voltage thermostats
Multimeter
Electrical hand tools

PROCEDURE
Examine four low voltage thermostats and record the following specifications:

1. Voltage –what is the maximum voltage rating for the thermostat?

2. Amperage – what is the maximum amperage rating for the thermostat?

3. Operation – Is the thermostat a mechanical or digital?

4. Heat -Cool or Heat Pump? Is the thermostat designed to work with a furnace and air conditioner or a heat pump?

5. Stages – How many stages of heating and cooling does the thermostat have?

6. Terminal Identification – what are the terminal markings and what is each terminal's function?

Low Voltage Thermostat Data				
	Thermostat 1	Thermostat 2	Thermostat 3	Thermostat 4
Volt Rating				
Amp Rating				
Operation				
Heat-Cool or Heat Pump				
Terminal				
Terminal				
Terminal				
Terminal				
Terminal				
Terminal				
Terminal				
Terminal				
Terminal				
Terminal				

LAB 36.18 EXAMINING LINE VOLTAGE THERMOSTATS

LABORATORY OBJECTIVE

The purpose of this lab is to learn the types of line voltage thermostats and become familiar with their electrical details.

LABORATORY NOTES

You will examine an assortment of line voltage thermostats, record their specifications, and test them with an ohm meter.

***FUNDAMENTALS OF HVACR* TEXT REFERENCE**

Unit 36 Electrical Components

REQUIRED TOOLS AND MATERIALS

Line voltage thermostats
Multimeter
Electrical hand tools

PROCEDURE

Examine three line voltage thermostats and record the following specifications:

1. Voltage –what is the maximum voltage rating for the thermostat?
2. Amperage – what is the maximum amperage rating for the thermostat?
3. Switching Action – Close on temperature rise, open on temperature rise, or both
4. Terminal Identification – what are the terminal markings and what is each terminal's function? (Where does power connect, where do the heating and cooling circuits connect?)
5. Temperature Range – At what temperatures can this thermostat function?
6. Check the thermostat with an ohm meter by adjusting the temperature while measuring the resistance across the contacts.

	Thermostat 1	Thermostat 2	Thermostat 3
Volt Rating			
Amp Rating			
Switching Action			
Temperature Range			
Terminal			
Terminal			
Terminal			

LAB 36.19 IDENTIFY PRESSURE SWITCHES

LABORATORY OBJECTIVE

The purpose of this lab is to learn the types of pressure switches and become familiar with their electrical details.

LABORATORY NOTES

You will examine an assortment of pressure switches and record their specifications.

***FUNDAMENTALS OF HVACR* TEXT REFERENCE**

Unit 36 Electrical Components

REQUIRED TOOLS AND MATERIALS

Pressure switches
Multimeter
Electrical hand tools

PROCEDURE

Examine four pressure switches and record the following specifications:

1. Voltage –what is the maximum voltage rating for the pressure switch?

2. Amperage – what is the maximum amperage rating for the pressure switch?

3. Switching Action – Close on rise, open on rise?

4. Pressure Range – At what pressures can this switch function?

5. Cut in pressure – what is the cut in pressure setting?

6. Cut out setting – what is the cut out setting?

7. Differential – what is the differential?

	Switch 1	Switch 2	Switch 3	Switch 4
Voltage				
Amperage				
Close or Open on Rise				
Range				
Cut In				
Cut Out				
Differential				

LAB 36.20 TESTING TRANSFORMERS

LABORATORY OBJECTIVE
The purpose of this lab is to learn how to check transformers for faults.

LABORATORY NOTES
You will check the resistance of the primary and secondary windings on five transformers assigned by the instructor. You will wire the transformers that ohm out correctly and check their voltage output.

FUNDAMENTALS OF HVACR TEXT REFERENCE
Unit 36 Electrical Components

REQUIRED TOOLS AND MATERIALS
Transformers
Volt-Ohm meter
Electrical hand tools

PROCEDURE

  SAFETY NOTE: Do not touch BOTH transformer leads when checking ohms.

  A shock can occur as the test leads are disconnected.

1. Make sure the transformer is disconnected from any power source.
2. Check the resistance of the primary and secondary windings on the transformers.
    a. A reading of infinity (OL) indicates an open motor winding.
    b. A reading less than 1.0 ohm on the primary indicates a shorted winding.
    c. A reading less than 0.7 ohms on the secondary indicates a shorted winding.
3. Record the readings in the data table.
4. If both windings are good, apply the correct voltage to the primary winding and read the voltage on the secondary winding.
5. Record the applied primary voltage and induced secondary voltage in the data table.

Transformer VA Rating	Primary Resistance	Secondary Resistance	Primary Voltage	Secondary Voltage

LAB 36.21 CHECKING LOW VOLTAGE THERMOSTATS

LABORATORY OBJECTIVE
The purpose of this lab is to learn how to test low voltage thermostats using an ohm meter.

LABORATORY NOTES
You will wire a low voltage thermostat to a short set of control wires and test the thermostat operation with an ohm meter.

***FUNDAMENTALS OF HVACR* TEXT REFERENCE**
Unit 36 Electrical Components

REQUIRED TOOLS AND MATERIALS
Low voltage thermostats
Multimeter
Electrical hand tools

PROCEDURE

1. Connect a set of 18 gauge, solid control wires to the terminals on the thermostat or thermostat sub base.
2. Connect the thermostat to the sub base.
3. Set the fan switch to "Auto" and the system switch to "OFF"
4. Check resistance between the red wire and all the others. It should be infinite (OL)
5. Set the fan switch to "On"
6. Check the resistance between R and G – it should read shorted (less than 1 ohm)
7. Set the fan switch back to "Auto", the system switch to "Cool", and run the temperature setting up above room temperature.
8. Check the resistance between R and O – it should read shorted (less than 1 ohm)
9. Run the temperature down and check the resistance between the R and Y terminals as well as the R and G terminals. They should read shorted (less than 1 ohm)
10. Move the temperature setting up and recheck the resistance between R and Y and R and G. They should now read infinite (OL)
11. Set the system switch to "Heat", and run the temperature setting down below room temperature.
12. Check the resistance between R and B – it should read shorted (less than 1 ohm)
13. Run the temperature up and check the resistance between the R and W terminals. They should read shorted (less than 1 ohm)
14. Move the temperature setting down and recheck the resistance between R and W.. They should now read infinite (OL)

LAB 36.22 CHECKING RELAYS AND CONTACTORS

LABORATORY OBJECTIVE
The purpose of this lab is to learn how to test relays and contactors.

LABORATORY NOTES
You will test an assortment of relays and contactors by checking the resistance of the coil and contacts. If the relay or contactor ohms correctly, you will apply the correct voltage to the coil and check to see that the contacts open and close by ohming them.

FUNDAMENTALS OF HVACR TEXT REFERENCE
Unit 36 Electrical Components

REQUIRED TOOLS AND MATERIALS
Relays
Contactors
Transformer
Multimeter
Electrical hand tools

PROCEDURE

1. Check the resistance of the coil. The resistance should be measurable – not 0 and not infinite.

2. The coil condition listed as Open, Shorted, or Good. A coil resistance measurement of less than 1 ohm indicates a shorted coil. A measurement of OL (infinite) indicates an open coil. A very rough estimate for coil resistance is 1 ohm per volt coil rating.

3. Ohm all the contacts. With no power to the coil, normally open contacts should be infinite (OL) and normally closed should be shorted (less than 1 ohm)

4. The normal position of all the contacts should be listed and the reading for each set of contacts should be listed.

5. If the coil is good, energize the coil with the correct voltage and check the resistance of the contacts. With the coil energized, normally open contacts should read shorted (less than 1) and normally closed contacts should read infinite (OL).

6. Record the resistance reading of the contacts with the coil energized.

Relay Data					
	Relay 1	Relay 2	Relay 3	Relay 4	Relay 5
Coil Voltage					
Coil Resistance					
Coil Condition					
Contacts Set1 Normal					
Contacts Set1 Coil Energized					
Contacts Set2 Normal					
Contacts Set2 Coil Energized					
Contacts Set3 Normal					
Contacts Set3 Coil Energized					

Contactor Data					
	Contactor 1	Contactor 2	Contactor 3	Contactor 4	Contactor 5
Coil Voltage					
Coil Resistance					
Coil Condition					
Contacts Set1 Normal					
Contacts Set1 Coil Energized					
Contacts Set2 Normal					
Contacts Set2 Coil Energized					
Contacts Set3 Normal					
Contacts Set3 Coil Energized					

LAB 36.23 SETTING LOW PRESSURE CUTOUT

LABORATORY OBJECTIVE
The purpose of this lab is to learn how to set the cutout on a low-pressure switch used for compressor safety.

LABORATORY NOTES
You will set the low-pressure safety cutout switch on the system assigned by the instructor and test the operation of the pressure switch.

FUNDAMENTALS OF HVACR TEXT REFERENCE
Unit 36 Electrical Components

REQUIRED TOOLS AND MATERIALS
Refrigeration system with low-pressure switch
Gauges
Multimeter
Electrical hand tools

PROCEDURE
1. Refer to the information in the text and available with the unit to determine the correct cutout setting.
2. Set the switch.
3. Operate the unit.
4. Force the system pressure to drop below the cutout by blocking evaporator airflow or front seating the suction service valve.
5. Observe the pressure and note where the system cuts out.
6. Observe the pressure rise and observe where the system cuts back on.
7. Record the data in the chart below.

High Pressure Switch Setting and Operation			
	Recommended	Setting	Operation
Cut In Pressure			
Differential			
Cut Out Pressure			

LAB 36.24 SETTING HIGH PRESSURE CUTOUTS

LABORATORY OBJECTIVE
The purpose of this lab is to learn how to set the cutout on a high pressure switch used for compressor safety.

LABORATORY NOTES
You will set the high pressure safety cutout switch on the system assigned by the instructor and test the operation of the pressure switch.

FUNDAMENTALS OF HVACR **TEXT REFERENCE**
Unit 36 Electrical Components

REQUIRED TOOLS AND MATERIALS
Refrigeration system with high pressure switch
Gauges
Multimeter
Electrical hand tools

PROCEDURE

1. Refer to the information in the text and available with the unit to determine the correct cutout setting.

2. Set the switch.

3. Operate the unit.

4. Force the system pressure to rise above the cutout by blocking condenser airflow.

5. Observe the pressure rise and note where the system cuts out.

6. Observe the pressure drop and observe where the system cuts back on.

High Pressure Switch Setting and Operation			
	Recommended	**Setting**	**Operation**
Cut Out Pressure			
Differential			
Cut In Pressure			

LAB 37.1 TESTING WINDINGS ON A SINGLE PHASE MOTOR

LABORATORY OBJECTIVE
The student will demonstrate how to properly test a single phase motor winding.

LABORATORY NOTES
Many single phase motors require capacitors. Capacitors can be dangerous. Capacitors can hold a high voltage charge even after the power is turned off. Always discharge capacitors before touching them.

FUNDAMENTALS OF HVACR TEXT REFERENCE
Unit 37 Electric Motors

REQUIRED TOOLS AND EQUIPMENT
Single phase motor
Ohmeter

SAFETY REQUIREMENTS
A. Turn the power off if the winding to be tested is installed in an operating system. Lock and tag out the power supply.

B. Confirm the power is secured by testing for zero voltage with you meter.

C. Capacitors should be discharged before you work on the system. <u>Never</u> discharge a capacitor with a screwdriver or other metal device touching the terminals.

D. Occasionally the terminal has been damaged and on a pressurized system the terminal can blow out. To avoid possible injury, unless the charge has been removed, use terminal points some distance away from the compressor.

PROCEDURE
Step 1
The first thing to do in testing is to check the motor windings.

        A. It may be necessary to remove a guard that encloses the terminals on a hermetic unit.

        B. The terminals should be marked C, S, and R.

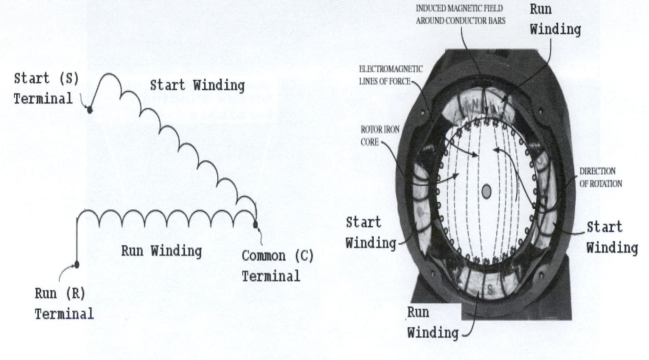

Figure 37-1-1

C. Use an ohmmeter to read the resistance in both the start and the run winding. Familiarize yourself with electrical meters.

D. If the terminals are not marked, they can be identified by a simple test of measuring the resistance across each pair of terminals.

The greatest resistance reading would be across both the start(S) and the run (R)windings leaving the remaining terminal not being touched as the common terminal (C).

The higher of the two readings from C would be across the start winding thereby identifying terminal S

Step 2
Measure the winding resistance and record the values.

Start Winding S to C _____Ohms

Run Winding R to C _____Ohms

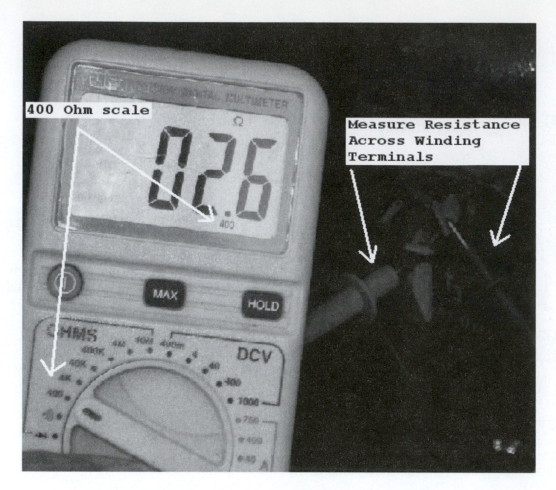

400 Ohm scale

Measure Resistance Across Winding Terminals

Figure 37-1-2

A. A resistance of zero would indicate what?

B. An infinite resistance would indicate what?

C. A low resistance would indicate what?

Note: Large motors (5 to 10 Hp and higher) have heavy copper windings to carry the motor current. They may therefore indicate a reading very close to a short circuit when winding resistance is read, depending on the ohmmeter.

Step 3 Check for a grounded winding as follows:

A. The ohmmeter needs to be capable of measuring very high resistance (R x 100,000).

B. When testing for a grounded winding, one test lead is placed on the terminal and the other on the outer shell of the compressor.

C. Be careful to make good contact with the motor shell as a coat of paint or a layer of dirt can hide a grounded winding.

D. The temperature is important, so run the compressor for about five minutes before testing.

E. For an ungrounded winding, the resistance is generally one to three mega ohms.

Not Grounded - High Resistance

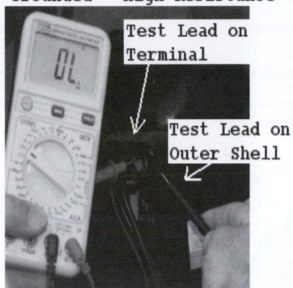

Test Lead on Terminal

Test Lead on Outer Shell

Grounded - Low Resistance

Figure 37-1-3

QUESTIONS

(Circle the letter that indicates the correct answer.)

1. Motor windings are insulated:
   A. with heavy copper wire.
   B. with red latex paint.
   C. with rubber.
   D. with a thin coat of varnish.

2. A shorted motor winding would indicate:
   A. an infinite resistance.
   B. a low resistance.
   C. no resistance.
   D. exactly 100 mega ohms.

3. A open winding would indicate:
   A. an infinite resistance.
   B. a low resistance.
   C. no resistance.
   D. exactly 100 mega ohms.

4. As motor windings get old they turn from a tan color to black:
   A. True.
   B. False.

LAB 37.2 MOTOR INSULATION TESTING

LABORATORY OBJECTIVE
The student will demonstrate how to properly test for leakage resistance from motor windings.

LABORATORY NOTES
The wiring that is used fro motor windings is insulated with a thin coat of varnish. As the motor is used and the windings are heated, this material will slowly carbonize. As it carbonizes, it will slightly change to a darker tan and eventually on to black. When it is completely carbonized, it does not provide any insulating capacity.

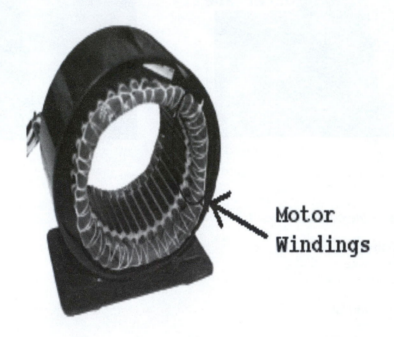

Figure 37-2-1

An instrument referred to as a megger tests the degree to which carbonization has taken place within the motor winding insulation.

FUNDAMENTALS OF HVACR TEXT REFERENCE
Unit Electric Motors

REQUIRED TOOLS AND EQUIPMENT
Motor
Megger

SAFETY REQUIREMENTS
A. Turn the power off if the winding to be tested is installed in an operating system. Lock and tag out the power supply.

B. Confirm the power is secured by testing for 0 voltage with a multimeter.

PROCEDURE
Step 1

Familiarize yourself with the operation of a megger.

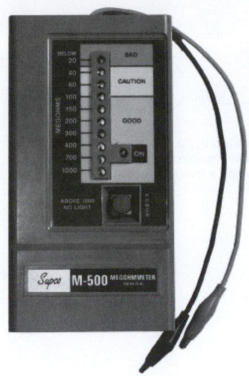

Figure 37-2-2

A. Meggers can test leakage at high voltage. 500 volts for 208 to 240 volt motors and 1000 volts for 480 volt motors.

B. Meggers may be battery operated or have a built in hand cranked generator.

C. They may detect insulation faults, while an ordinary multimeter using a few volts DC would show a satisfactory reading.
D. The megger shown below has a scale that reads from 20 to 1000 megohms.

E. Below 30 megohms would be considered bad insulation.

F. 30 to 100 megohms would indicate that the windings are questionable.

G. 100 megohms or greater would be considered good.

Step 2
Prepare to test the motor insulation as follows:

A. Turn the power off if the winding to be tested is installed in an operating system. Lock and tag out the power supply.

B. Confirm the power is secured by testing for 0 voltage with a multimeter.

C. Attach one lead of the Megger to the motor terminal and the other lead to ground (the motor or compressor frame is suitable).

D. Record your reading below.

Run Winding _____ Ohms

Start Winding _____ Ohms

Common Terminal _____ Ohms

E. Larger motors are checked for insulation breakdown by having megger readings taken on a regular basis in accordance with the preventative maintenance schedule. This provides a record of the motor condition over time and a lowering resistance indicates that the windings are becoming increasingly carbonized.

LAB 37.3 TYPES OF ELECTRIC MOTORS

LABORATORY OBJECTIVE
You will learn to identify the types of electric motors used in the HVACR field.

FUNDAMENTALS OF HVACR TEXT REFERENCE
Unit 37 Motors

REQUIRED MATERIALS PROVIDED BY SCHOOL
Selection of Electric Motors

PROCEDURE
The Instructor will assign ten motors to identify according to type.
- SP-Shaded Pole
- PSC-Permanent Split Capacitor
- CS-Capacitor Start
- CSR-Capacitor Start-Capacitor Run

Write down the type of motor in the table below. Use the other columns to check off motor characteristics to help you identify the motor type. You will be expected to explain to the instructor how you determined the motor types.

	Motor Type	Shaded Pole	Start Winding	Centrifugal Switch	Capacitors
1					
2					
3					
4					
5					
6					
7					
8					
9					
10					

LAB 37.4 EXAMINE SHADED POLE MOTOR

LABORATORY OBJECTIVE

You will identify the parts of a shaded pole motor and explain its operation.

LABORATORY NOTES

You will disassemble the shaded pole motor assigned by the instructor, identify its parts, discuss its operation, and reassemble the motor.

***FUNDAMENTALS OF HVACR* TEXT REFERENCE**

Unit 37 Motors

REQUIRED MATERIALS PROVIDED BY STUDENT

Nut drivers

REQUIRED MATERIALS PROVIDED BY SCHOOL

Shaded Pole motor for disassembly

PROCEDURE

1.   Use a marker to mark where the end bells align with the motor frame.

2.   Loosen the nuts on the bolts that hold the motor together.

3.   Disassemble the motor.

4.   Identify the rotor and stator.

5.   Identify the shaded poles.

6.   Use the shaded poles to determine the direction of rotation.

7.   Count the number of poles and calculate the motor speed (RPM)

8.   Discuss your findings with the instructor.

9.   Reassemble the motor.

10. Be careful to get the end bells aligned and check that the shaft turns freely after assembly.

LAB 37.5 INSPECTING OPEN SPLIT PHASE MOTOR

LABORATORY OBJECTIVE
The purpose of this lab is to identify the parts of an open split phase motor and explain its operation.

LABORATORY NOTES
You will disassemble the open split phase motor assigned by the instructor, identify its parts, discuss its operation, and reassemble the motor.

FUNDAMENTALS OF HVACR TEXT REFERENCE
Unit 37 Motors

REQUIRED MATERIALS PROVIDED BY STUDENT
Nut drivers

REQUIRED MATERIALS PROVIDED BY SCHOOL
Open split phase motor for disassembly

PROCEDURE
1. Use a marker to mark where the end bells align with the motor frame.
2. Loosen the nuts on the bolts that hold the motor together.
3. Disassemble the motor.
4. Identify the rotor and stator.
5. Identify the start and run windings.
6. Identify the centrifugal switch.
7. Count the number of poles and calculate the motor speed (RPM)
8. Discuss your findings with the instructor.
9. Reassemble the motor.
10. Be careful to get the end bells aligned and check that the shaft turns freely after assembly.

LAB 37.6 INSPECTING OPEN CAPACITOR START MOTOR

LABORATORY OBJECTIVE
The purpose of this lab is to identify the parts of an open capacitor start motor and explain its operation.

LABORATORY NOTES
You will disassemble the open capacitor start motor assigned by the instructor, identify its parts, discuss its operation, and reassemble the motor.

FUNDAMENTALS OF HVACR TEXT REFERENCE
Unit 37 Motors

REQUIRED MATERIALS PROVIDED BY STUDENT
Nut drivers

REQUIRED MATERIALS PROVIDED BY SCHOOL
Open capacitor start motor for disassembly

PROCEDURE

1. Use a marker to mark where the end bells align with the motor frame.

2. Loosen the nuts on the bolts that hold the motor together.

3. Disassemble the motor.

4. Identify the rotor and stator.

5. Identify the start and run windings.

6. Identify the centrifugal switch.

7. Examine how the start capacitor is wired to the centrifugal switch

8. Count the number of poles and calculate the motor speed (RPM)

9. Discuss your findings with the instructor.

10. Reassemble the motor.

11. Be careful to get the end bells aligned and check that the shaft turns freely after assembly.

LAB 37.7 INSPECTING PERMANENT SPLIT CAPACITOR (PSC) MOTOR

LABORATORY OBJECTIVE
The purpose of this lab is to identify the parts of a PSC motor and explain its operation.

LABORATORY NOTES
You will disassemble the permanent split capacitor (PSC) motor assigned by the instructor, identify its parts, discuss its operation, and reassemble the motor.

FUNDAMENTALS OF HVACR TEXT REFERENCE
Unit 37 Motors

REQUIRED MATERIALS PROVIDED BY STUDENT
Nut drivers

REQUIRED MATERIALS PROVIDED BY SCHOOL
Permanent split capacitor motor (PSC) for disassembly

PROCEDURE

1. Use a marker to mark where the end bells align with the motor frame.

2. Loosen the nuts on the bolts that hold the motor together.

3. Disassemble the motor.

4. Identify the rotor and stator.

5. Identify the start and run windings.

6. Count the number of poles and calculate the motor speed (RPM)

7. Discuss your findings with the instructor.

8. Reassemble the motor.

9. Be careful to get the end bells aligned and check that the shaft turns freely after assembly.

LAB 37.8 WIRE AND OPERATE SHADED POLE MOTORS

LABORATORY OBJECTIVE
You will learn how to wire and operate shaded pole motors used in HVACR.

LABORATORY NOTES
You will wire shaded pole motors, operate them, and measure their amp draw.

FUNDAMENTALS OF HVACR TEXT REFERENCE
Unit 37 Motors

REQUIRED MATERIALS PROVIDED BY STUDENT
Volt-ohm meter
Amp meter
Alectrical hand tools

REQUIRED MATERIALS PROVIDED BY SCHOOL
Shaded pole motors

PROCEDURE

1. Use the wiring diagram on the motor and/or the diagrams in the book to wire the shaded pole motors assigned by the instructor.

2. Be sure to check the motor nameplate voltage and compare it to the voltage you are connecting to the motor.

3. Operate the motor.

4. Use a clamp on amp meter to read the motor current.

5. Write down your reading in the chart below.

6. Discus your results with the instructor.

7. Put up all the motors.

Voltage	Speeds	RPM	Nameplate FLA	Running Amps (for each speed)

LAB 37.9 TROUBLESHOOTING SHADED POLE MOTORS

LABORATORY OBJECTIVE
The purpose of this lab is to learn how to check shaded pole motors for faults.

LABORATORY NOTES
You will check the shaded pole motors assigned by the instructor. You will wire and try to operate motors that ohm out correctly.

FUNDAMENTALS OF HVACR TEXT REFERENCE
Unit 37 Motors

REQUIRED MATERIALS PROVIDED BY STUDENT
Volt-Ohm meter
Amp meter
Electrical hand tools

REQUIRED MATERIALS PROVIDED BY SCHOOL
Shaded pole motors

PROCEDURE
1. Check to see that the shaft turns easily by hand. If it does not turn easily by hand, the motor has a mechanical problem and should not be wired up.
2. Ohm the motor windings and record your readings.
3. A reading of infinity (OL) indicates an open motor winding.
4. A reading less than 1.0 ohm indicates a shorted winding.
5. A reading from the motor winding to the motor frame indicates a grounded winding.
6. If the motor ohms out correctly, wire and operate it.
7. Be sure to check the motor nameplate voltage and compare it to the voltage you are connecting to the motor.
8. Operate the motor.
9. Use a clamp on amp meter to read the motor current.
10. Write down your analysis in the table below.
11. Discus your results with the instructor.
12. Put up all the motors.

SUMMARY

Voltage	Speeds	RPM	Nameplate FLA	Running Amps	Condition (Good, open, shorted, grounded, bad bearings)

LAB 37.10 WIRE AND OPERATE OPEN SPLIT PHASE AND CAPACITOR START MOTORS

LABORATORY OBJECTIVE
You will learn how to wire and operate open split phase motors used in HVACR.

LABORATORY NOTES
You will wire open split phase motors, operate them, and measure their amp draw.

***FUNDAMENTALS OF HVACR* TEXT REFERENCE**
Unit 37 Motors

REQUIRED MATERIALS PROVIDED BY STUDENT
Volt-Ohm meter
Amp meter
Electrical hand tools

REQUIRED MATERIALS PROVIDED BY SCHOOL
Split phase motors

PROCEDURE

1. Use the wiring diagram on the motor and/or the diagrams in the book to wire the split phase motors assigned by the instructor.

2. Be sure to check the motor nameplate voltage and compare it to the voltage you are connecting to the motor.

3. Operate the motor.

4. Use a clamp on amp meter to read the motor current.

5. Write down your reading in the chart below.

6. Discus your results with the instructor.

7. Put up all the motors.

SUMMARY

Voltage	Speeds	RPM	Nameplate FLA	Running Amps

LAB 37.11 TROUBLESHOOTING OPEN SPLIT PHASE & CAPACITOR START MOTORS

LABORATORY OBJECTIVE
The purpose of this lab is to learn how to check open split phase motors for faults.

LABORATORY NOTES
You will check the open split phase motors assigned by the instructor. You will wire and try to operate motors that ohm out correctly.

FUNDAMENTALS OF HVACR TEXT REFERENCE
Unit 37 Motors

REQUIRED MATERIALS PROVIDED BY STUDENT
Volt-Ohm meter
Amp meter
Electrical hand tools

REQUIRED MATERIALS PROVIDED BY SCHOOL
Open split phase motors and Capacitor Start Motors

PROCEDURE
1. Check to see that the shaft turns easily by hand. If it does not turn easily by hand, the motor has a mechanical problem and should not be wired up.
2. Ohm the motor windings and record your readings.
    a. A reading of infinity (OL) indicates an open motor winding.
    b. A reading less than 1.0 ohm indicates a shorted winding.
    c. A reading from the motor winding to the motor frame indicates a grounded winding.
3. If the motor ohms out correctly, wire and operate it.
4. Be sure to check the motor nameplate voltage and compare it to the voltage you are connecting to the motor.
5. Operate the motor.
6. Use a clamp on amp meter to read the motor current.
7. Write down your analysis in the table below.
8. Discus your results with the instructor.
9. Put up all the motors.

SUMMARY

Voltage	Speeds	RPM	Nameplate FLA	Running Amps	Condition (Good, open, shorted, grounded, bad bearings)

LAB 37.12 WIRE AND OPERATE OPEN PSC MOTORS

LABORATORY OBJECTIVE
You will learn how to wire and operate open PSC motors used in HVACR.

LABORATORY NOTES
You will wire open PSC motors, operate them, and measure their amp draw.

***FUNDAMENTALS OF HVACR* TEXT REFERENCE**
Unit 37 Motors

REQUIRED MATERIALS PROVIDED BY STUDENT
Volt-Ohm meter
Amp meter
Electrical hand tools

REQUIRED MATERIALS PROVIDED BY SCHOOL
PSC motors

PROCEDURE

1. Use the wiring diagram on the motor and/or the diagrams in the book to wire the shaded pole motors assigned by the instructor.
2. Be sure to check the motor nameplate voltage and compare it to the voltage you are connecting to the motor.
3. Operate the motor.
4. Use a clamp on amp meter to read the motor current.
5. Write down your reading in the chart below.
6. Discus your results with the instructor.
7. Put up all the motors.

SUMMARY

Voltage	Speeds	RPM	Nameplate FLA	Running Amps (for each speed)

LAB 37.13 TROUBLESHOOTING OPEN PSC MOTORS

LABORATORY OBJECTIVE
The purpose of this lab is to learn how to check open PSC motors for faults.

LABORATORY NOTES
You will check the open PSC motors assigned by the instructor. You will wire and try to operate motors that ohm out correctly.

FUNDAMENTALS OF HVACR TEXT REFERENCE
Unit 37 Motors

REQUIRED MATERIALS PROVIDED BY STUDENT
Volt-Ohm meter
Amp meter
Electrical hand tools

REQUIRED MATERIALS PROVIDED BY SCHOOL
Open PSC motors

PROCEDURE

1. Check to see that the shaft turns easily by hand. If it does not turn easily by hand, the motor has a mechanical problem and should not be wired up.
2. Ohm the motor windings and record your readings.
    a. A reading of infinity (OL) indicates an open motor winding.
    b. A reading less than 1.0 ohm indicates a shorted winding.
    c. A reading from the motor winding to the motor frame indicates a grounded winding.
3. If the motor ohms out correctly, wire and operate it.
4. Be sure to check the motor nameplate voltage and compare it to the voltage you are connecting to the motor.
5. Operate the motor.
6. Use a clamp on amp meter to read the motor current.
7. Write down your analysis in the table below.
8. Discus your results with the instructor.
9. Put up all the motors.

SUMMARY

Voltage	Speeds	RPM	Nameplate FLA	Running Amps	Condition (Good, open, shorted, grounded, bad bearings)

LAB 37.14 OHMING THREE PHASE MOTORS

LABORATORY OBJECTIVE

The purpose of this lab is to learn how to check three phase motors for faults using an ohm meter.

LABORATORY NOTES

You will check the three phase motors assigned by the instructor.

***FUNDAMENTALS OF HVACR* TEXT REFERENCE**

Unit 37 Motors

REQUIRED MATERIALS PROVIDED BY STUDENT

Volt-Ohm meter
Amp meter
Electrical hand tools

REQUIRED MATERIALS PROVIDED BY SCHOOL

Three phase motors

PROCEDURE

1. Check to see that the shaft turns easily by hand. If it does not turn easily by hand, the motor has a mechanical problem and should not be wired up.
2. Ohm the motor windings and record your readings on the data sheet.
   a. A reading of infinity (OL) indicates an open motor winding.
   b. A reading less than 1.0 ohm indicates a shorted winding.
   c. A reading from the motor winding to the motor frame indicates a grounded winding.

   Single voltage three phase motors should have three equal readings.

   Dual voltage motors will have two groups of readings. The reading in each group should be the same. See text for diagram of dual voltage motors.

Single Voltage Three Phase Motors						
Motor	Lead 1 to Lead 2	Lead 2 to Lead 3	Lead 3 to Lead 1	Lead 1 to Ground	Lead 2 to Ground	Lead 3 to Ground

Dual Voltage Three Phase Motors – Wye Wound							
Motor	T1 to T4	T2 to T5	T3 to T6	T7 to T8	T8 to T9	T9 to T7	All to Ground

Dual Voltage Three Phase Motors – Delta Wound										
Motor	T1 to T4	T1 to T9	T4 to T9	T2 to T7	T2 to T5	T5 to T7	T3 to T6	T3 to T8	T6 to T8	All to Ground

LAB 37.15 WIRE AND OPERATE THREE PHASE MOTORS

LABORATORY OBJECTIVE

You will learn how to wire and operate three phase motors used in HVACR.

LABORATORY NOTES

You will wire three phase motors, operate them, and measure their amp draw.

FUNDAMENTALS OF HVACR **TEXT REFERENCE**

Unit 37 Motors

REQUIRED MATERIALS PROVIDED BY STUDENT

Volt-Ohm meter

Amp meter

Electrical hand tools

REQUIRED MATERIALS PROVIDED BY SCHOOL

Three phase motors

PROCEDURE

1. Use the wiring diagram on the motor and/or the diagrams in the book to wire the three phase motors assigned by the instructor.
2. Be sure to check the motor nameplate voltage and compare it to the voltage you are connecting to the motor.
3. Operate the motor.
4. Use a clamp on amp meter to read the motor current.
5. Write down your reading in the chart below.
6. Discus your results with the instructor.
7. Put up all the motors.

Voltage	RPM	Nameplate FLA	Running Amps (for each speed)

LAB 37.16 EXAMINE ECM MOTOR

LABORATORY OBJECTIVE
You will identify the parts of an ECM motor and explain its operation.

LABORATORY NOTES
You will disassemble the ECM motor assigned by the instructor, identify its parts, discuss its operation, and reassemble the motor.

FUNDAMENTALS OF HVACR TEXT REFERENCE
Unit 37 Motors

REQUIRED MATERIALS PROVIDED BY STUDENT
Volt-Ohm meter
Amp meter
Electrical hand tools

REQUIRED MATERIALS PROVIDED BY SCHOOL
ECM motor for disassembly

PROCEDURE
1. Use a marker to mark where the end bells align with the motor frame.
2. Loosen the nuts on the bolts that hold the motor together.
3. Disassemble the motor.
4. Identify the rotor, stator, and module.
5. Notice the difference between the rotor and the rotors of induction motors you examined earlier.
6. Ohm the windings.
7. Discuss your findings with the instructor.
8. Reassemble the motor.
9. Be careful to get the end bells aligned and check that the shaft turns freely after assembly.

LAB 37.17 OPERATE ECM MOTOR

LABORATORY OBJECTIVE

You will describe the operation of an ECM blower motor and compare its operation to a traditional PSC blower motor.

LABORATORY NOTES

You will operate an ECM blower motor, measure its operating characteristics, and compare them to a traditional PSC blower motor.

***FUNDAMENTALS OF HVACR* TEXT REFERENCE**

Unit 37 Motors

REQUIRED MATERIALS PROVIDED BY STUDENT

Electrical hand tools

REQUIRED MATERIALS PROVIDED BY SCHOOL

ECM motor trainer

PROCEDURE

1. Connect the controls to the ECM trainer.
2. Have the instructor check your work.
3. Operate the blower and notice the blower RPM, amp draw, and Watts.
4. Restrict the airflow and observe the changes in the RPM, amp draw, and Watts.
5. Adjust the speed control and observe the changes in the RPM, amp draw, and Watts.
6. Notice the difference between the rotor and the rotors of induction motors you examined earlier.
7. Remove the airflow restriction and observe the changes in the RPM, amp draw, and Watts.

LAB 37.18 CHECK ECM MOTOR USING TECMATE

LABORATORY OBJECTIVE

You will learn how to check the operation of an ECM blower motor using the TECMATE tool.

LABORATORY NOTES

You will connect a TECMATE to an ECM blower motor. You will then operate and test the motor using the TECMATE to control the motor.

***FUNDAMENTALS OF HVACR* TEXT REFERENCE**

Unit 37 Motors

REQUIRED MATERIALS PROVIDED BY STUDENT

Electrical hand tools

REQUIRED MATERIALS PROVIDED BY SCHOOL

ECM motor trainer
TECMATE

PROCEDURE

1. Connect the TECMATE to the ECM trainer.
2. Have the instructor check your work.
3. Operate the blower with the TECMATE.

LAB 37.19 SET BLOWER CFM ON ECM BLOWER BOARD

LABORATORY OBJECTIVE
You will learn how to adjust the airflow setting on blower control board for an ECM blower motor.

LABORATORY NOTES
You will connect measure the airflow of an ECM equipped blower. Next, you will adjust the airflow using the blower control board settings. Then you will read the new CFM after the adjustment. Finally, you will return the adjustment to its original setting.

***FUNDAMENTALS OF HVACR* TEXT REFERENCE**
Unit 37 Motors

REQUIRED MATERIALS PROVIDED BY STUDENT
Electrical hand tools

REQUIRED MATERIALS PROVIDED BY SCHOOL
Unit with ECM blower and ECM blower control board

PROCEDURE
1. Note the setting on the blower control board.
2. Operate the blower and measure the CFM.
3. Shut off the blower and adjust the CFM on the blower control.
4. Operate the blower and measure the CFM.
5. Return the blower control to its original setting.

LAB 37.20 TROUBLESHOOTING CSR MOTORS

LABORATORY OBJECTIVE
The purpose of this lab is to learn how to check hermetic CSR motors for faults.

LABORATORY NOTES
We will use an amp meter, volt meter, and an ohm meter to test CSR compressors and their starting components. You will determine if the starting components are good and if the motor is good, open, shorted, or grounded. If the motor is good, you will use the ohm readings to identify the Common, Start, and Run terminals.

FUNDAMENTALS OF HVACR TEXT REFERENCE
Unit 37 Motors

REQUIRED MATERIALS PROVIDED BY STUDENT
Volt-Ohm meter
Amp meter
Electrical hand tools

REQUIRED MATERIALS PROVIDED BY SCHOOL
Hermetic compressors with CSR motors

PROCEDURE

1. First check to see that voltage is available to the unit.
2. Check the amp draw through the start capacitor when the compressor starts. The current should spike when the compressor starts and drop to 0 after it gets started.
3. If there is no amp draw thought the start capacitor when the compressor tries to start the problem is most likely with an open start capacitor or start relay.
4. If there is no current through the common wire with voltage available, you need to disconnect the power and ohm the compressor motor.
5. Take ohm readings between each combination of two terminals – there should be three readings
6. Write the readings down.
7. If any of the three readings is infinite, either the motor winding or internal overload is open.
8. If the reading from C to S and C to R are both infinite, but the reading from S to R is not, the internal overload is open. If the motor is hot, the overload may close and reset when the motor cools.
9. If any of the three readings is less than 0.3 ohms, the motor is shorted.
10. Read the resistance between each terminal and one of the copper lines on the compressor.
11. Any reading other than infinite indicates a grounded motor.
12. If the motor is not open, shorted, or grounded you can determine the C, S, and R terminals.
13. The highest reading will be between the start and ruin terminals. Therefore, the terminal that is NOT involved in the highest reading is common.
14. The lowest reading is between Common and run. The terminal that is involved in the lowest reading with the Common terminal is Run.
15. The start terminal will be the terminal left over after identifying both the common and run terminals.
16. Complete the table and be prepared to explain your results to the instructor.

The three most common terminal arrangements are shown in the boxes below.

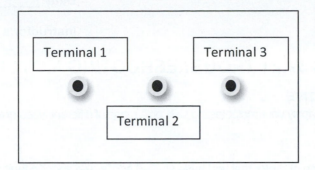

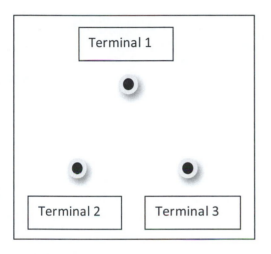

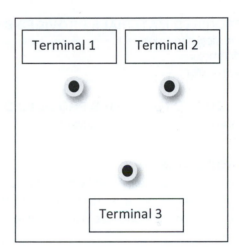

	Term 1 to Term 2	Term 1 to Term 3	Term 2 to Term 3	Terminal 1 to Ground	Terminal 2 to Ground	Terminal 3 to Ground	Condition Good Shorted Open Grounded
			Ohming Compressor Motors Resistance Readings				
Comp 1							
Comp 2							
Comp 3							

	Terminal 1	Terminal 2	Terminal 3
		Determining Common Start and Run Terminal Designations (Common, Start, and Run)	
Comp 1			
Comp 2			
Comp 3			

LAB 37.21 TROUBLESHOOTING MOTORS

LABORATORY OBJECTIVE
The purpose of this lab is to apply your motor testing skills on a range of different types of electric motors.

LABORATORY NOTES
You will test two of each type of motor that you have studied and list the specific fault of each motor. If the motor checks out, you will wire it, operate it, and record the amp reading.

FUNDAMENTALS OF HVACR TEXT REFERENCE
Unit 37 Motors

REQUIRED MATERIALS PROVIDED BY STUDENT
Electrical hand tools
Multimeter
Amp meter

REQUIRED MATERIALS PROVIDED BY SCHOOL
Assortment of electric motors

PROCEDURE
Check the motors assigned by the instructor and record the information requested on the chart. State the fault as bearing failure, winding open or shorted, starting component failure, etc. If the motor checks out, wire and operate it and record the amp reading.

Motor	Voltage	Nameplate FLA	Running Amps	Condition (Good, open, shorted, grounded, bad bearings)
Shaded Pole				
Shaded Pole				
Split phase (open)				
Capacitor Start (open)				
PSC (open)				
PSC (open)				
Three phase				
Three phase				
Capacitor Start (hermetic)				
Capacitor Start Run (hermetic)				

LAB 38.1 WIRING DIAGRAMS

LABORATORY OBJECTIVE
The student will demonstrate how to properly draw a wiring schematic from a connection diagram.

LABORATORY NOTES
The student may draw the wiring schematic from the connection diagram below or from an alternate connection diagram supplied by the Lab Instructor.

FUNDAMENTALS OF HVACR TEXT REFERENCE
Unit 38 Electrical Diagrams

REQUIRED TOOLS AND EQUIPMENT
None

SAFETY REQUIREMENTS
None

The connection diagram shown in Figure 38-1-1 is for an air cooled condensing unit. The control system consists of a wiring panel enclosing the compressor starter, the start relay, the run capacitor, a thermostat and switch combination, a start capacitor, and a junction box.

External to the wiring panel are the fan motor, power supply, compressor motor, junction box, and high/low pressure control.

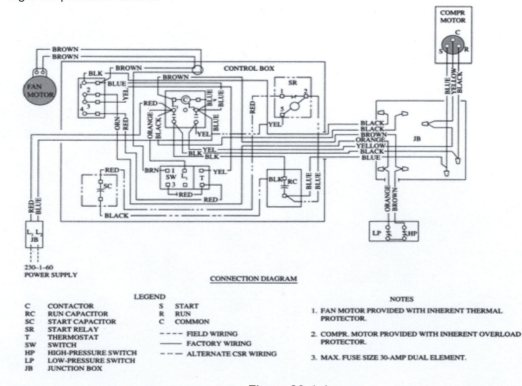

LEGEND

C	CONTACTOR	S	START
RC	RUN CAPACITOR	R	RUN
SC	START CAPACITOR	C	COMMON
SR	START RELAY		
T	THERMOSTAT	- - - -	FIELD WIRING
SW	SWITCH	———	FACTORY WIRING
HP	HIGH-PRESSURE SWITCH	- - -	ALTERNATE CSR WIRING
LP	LOW-PRESSURE SWITCH		
JB	JUNCTION BOX		

NOTES

1. FAN MOTOR PROVIDED WITH INHERENT THERMAL PROTECTOR.

2. COMPR. MOTOR PROVIDED WITH INHERENT OVERLOAD PROTECTOR.

3. MAX. FUSE SIZE 30-AMP DUAL ELEMENT.

Figure 38-1-1

PROCEDURE
Step 1
The first step in preparing the schematic is to locate the loads and determine the number of circuits. There are five loads and therefore five circuits.

A. The five loads are:

a)

b)

c)

d)

e)

Step 2
The second step in preparing the schematic is to locate the switches in each circuit.

A. The switches in each circuit are:

a)

b)

c)

d)

e)

Step 3
The third step in preparing the schematic is to draw each circuit.

A. Draw the compressor <u>motor</u> circuit and start relay coil circuit in the space provided below.

433

B. Draw the fan <u>motor</u> circuit, the compressor contactor circuit, and the green test lamp circuit in the space provided below.

C. Combine Part A and B together into one complete schematic wiring diagram in the space provided below.

QUESTIONS

(Circle the letter that indicates the correct answer.)

1. The three types of wiring diagrams are:
   - A. internal, external, and connection.
   - B. external, connection, and schematic.
   - C. internal, connection, and schematic.
   - D. external, connection, and plan.

2. The electric diagram symbol shown below is a:

   - A. thermal relay.
   - B. thermal cutout.
   - C. flow switch.
   - D. closed contact.

3. The electric diagram symbol shown below is a:

   - A. thermal relay.
   - B. transformer.
   - C. flow switch.
   - D. closed contact.

4. The electric diagram symbol shown below is a:

   - A. thermal relay.
   - B. transformer.
   - C. flow switch.
   - D. closed contact.

5. The electric diagram symbol shown below is a:

   - A. motor winding.
   - B. transformer.
   - C. flow switch.
   - D. closed contact.

6. The electric diagram symbol shown below will:

   - A. close on rising temperature.
   - B. open on rising temperature.
   - C. close on rising pressure.
   - D. open on rising pressure.

7. The electric diagram symbol shown below will open on rising temperature.

   - A. True.
   - B. False.

LAB 38.2 WIRING TRANSFORMERS

LABORATORY OBJECTIVE
You will demonstrate your understanding of transformer operation by safely wiring a transformer to control a contactor with a 24-volt coil.

LABORATORY NOTES
In this lab exercise you will construct a simple circuit with the SPST switch controlling the coil of a 24-volt contactor. A 110-volt primary to 24-volt secondary transformer will be the low voltage power supply.

FUNDAMENTALS OF HVACR TEXT REFERENCE
Unit 36 Electrical Components / Unit 38 Electrical Diagrams / Unit 39 Control Systems

REQUIRED MATERIALS PROVIDED BY STUDENT
Safety glasses
Multimeter
Wire stripper
Crimp tool
Wire cutter
6-in-1 screwdriver

REQUIRED MATERIALS PROVIDED BY SCHOOL
120-volt power cord
Wire
Electrical terminals
Wire nuts
SPST switch
120-volt primary 24-volt secondary transformer
Contactor with 24-volt coil

PROCEDURE
POWER CORD SHOULD BE UNPLUGGED WHILE CONSTRUCTION THE CIRCUIT!
1. The two wires from the power supply cord should feed the transformer 120-volt primary.
2. One side of the transformer secondary should connect to the switch.
3. The other side of the switch should connect to one side of the contactor coil.
4. The other side of the contactor coil should connect to the remaining wire on the transformer secondary side.
5. Have an instructor check your work before energizing it.
6. With the circuit energized and the switch off, measure the voltage across the switch.
7. With the circuit energized and the switch off, measure the voltage at the contactor coil.
8. With the circuit energized and the switch off, measure the resistance across one set of contactor contacts.
9. With the circuit energized and the switch on, measure the voltage across the switch.
10. With the circuit energized and the switch on, measure the voltage at the contactor coil.
11. With the circuit energized and the switch on, measure the voltage across one set of contactor contacts.

QUESTIONS

1. When the switch is off, where did you read voltage?

2. When the switch is on, where did you read volts?

3. When did you read continuity across the contactor contacts?

LAB 38.3 WIRING CONTACTORS

LABORATORY OBJECTIVE
You will demonstrate your understanding of contactor operation by safely wiring a 24-volt coil contactor to control a 120-volt light.

LABORATORY NOTES
In this lab exercise you will construct a simple circuit with the SPST switch controlling the coil of a 24-volt contactor. The contactor contacts will control the 120-volt light. A 110-volt primary to 24-volt secondary transformer will be the low voltage power supply.

FUNDAMENTALS OF HVACR TEXT REFERENCE
Unit 36 Electrical Components
Unit 38 Electrical Diagrams
Unit 39 Control Systems

REQUIRED MATERIALS PROVIDED BY STUDENT
Safety glasses
Multimeter
Wire stripper
Crimp tool
Wire cutter
6-in-1 screwdriver

REQUIRED MATERIALS PROVIDED BY SCHOOL
120-volt power cord
Wire
Electrical terminals
Wire nuts
SPST switch
120-volt primary 24-volt secondary transformer
Contactor with 24-volt coil
120-volt light

PROCEDURE
POWER CORD SHOULD BE UNPLUGGED WHILE CONSTRUCTION THE CIRCUIT!

1. The two wires from the power supply cord should wire to L1 and L2 of the contactor contacts.
2. The primary side of the transformer should also wire to L1 and L2 on the contactor.
3. The light should be wired to T1 and T2 of the contactor contacts.
4. One side of the transformer secondary should connect to the switch.
5. The other side of the switch should connect to one side of the contactor coil.
6. The other side of the contactor coil should connect to the remaining wire on the transformer secondary side.
7. Have an instructor check your work before energizing it.
8. With the circuit energized and the switch off, measure the voltage across the switch.
9. With the circuit energized and the switch off, measure the voltage at the contactor coil.

10. With the circuit energized and the switch off, measure the voltage between L1 and L2 on the contactor.
11. With the circuit energized and the switch off, measure the voltage between T1 and T2 on the contactor.
12. With the circuit energized and the switch off, measure the voltage between L1 and T1 on the contactor.
13. With the circuit energized and the switch on, measure the voltage across the switch.
14. With the circuit energized and the switch on, measure the voltage at the contactor coil.
15. With the circuit energized and the switch on, measure the voltage between L1 and L2 on the contactor.
16. With the circuit energized and the switch on, measure the voltage between T1 and T2 on the contactor.
17. With the circuit energized and the switch on, measure the voltage between L1 and T1 on the contactor.

QUESTIONS

1. When the switch is off, where did you read voltage?

2. When the switch is on, where did you read voltage?

LAB 38.4 WIRING RELAYS

LABORATORY OBJECTIVE

You will demonstrate your understanding of relay operation by safely wiring a 24-volt coil relay to control two 120-volt lights; one light to operate only when the relay coil is de-energized and the other light to operate only when the relay coil is energized.

LABORATORY NOTES

In this lab exercise you will construct a simple circuit with the SPST switch controlling the coil of a 24-volt relay. The NC relay contacts will control one 120-volt light; the NO contacts will control the other 120-volt light. A 110-volt primary to 24-volt secondary transformer will be the low voltage power supply.

FUNDAMENTALS OF HVACR TEXT REFERENCE

Unit 36 Electrical Components
Unit 38 Electrical Diagrams
Unit 39 Control Systems

REQUIRED MATERIALS PROVIDED BY STUDENT

Safety glasses
Multimeter
Wire stripper/crimper/cutter
6-in-1 screwdriver

REQUIRED MATERIALS PROVIDED BY SCHOOL

120-volt power cord
Wire
Electrical terminals
Wire nuts
SPST switch
120-volt primary 24-volt secondary transformer
Relay with 24-volt coil, 1 NO and 1 NC
Two 120-volt lights

PROCEDURE

Use a 110-volt to 24-volt transformer and a SPST switch to control a 24-volt relay coil.

The relay contacts should be wired to control two 110-volt lights.

One light should come on when the relay is energized and the other should come on when the relay is de-energized.

The switch should be wired in series with the transformer secondary and the relay coil.

The relay normally open contacts should be wired in series with one light.

The relay normally closed contacts should be wired in series with the other light.

Have an instructor check your work before energizing it.

LAB 38.5 IDENTIFY ELECTRIC COMPONENT SYMBOLS

LABORATORY OBJECTIVE
You will identify ten common HVACR electrical components and draw the symbol that represents each of them.

LABORATORY NOTES
In this lab exercise you will write down the name of the component and draw the symbol used to represent that component in electrical diagrams.

***FUNDAMENTALS OF HVACR* TEXT REFERENCE**
Unit 38 Diagrams

REQUIRED MATERIALS PROVIDED BY STUDENT
Pencil

REQUIRED MATERIALS PROVIDED BY SCHOOL
Ten common HVACR electrical components.

PROCEDURE
Name and draw the symbols for 10 electrical components provided by the instructor.

Component 1

Component 2

Component 3

Component 4

Component 6

Component 7

Component 8

Component 10

LAB 38.6 IDENTIFY COMPONENTS ON AIR CONDITIONING DIAGRAM

LABORATORY OBJECTIVE
You will identify common HVACR electrical components on an air conditioning unit and locate their symbols on the unit electrical diagram.

LABORATORY NOTES
In this lab exercise you will identify common electrical components on an air conditioner. You will then identify the symbols in the unit diagram that represent the components.

FUNDAMENTALS OF HVACR TEXT REFERENCE
Unit 36 Electrical Components / Unit 38 Electrical Diagrams

REQUIRED MATERIALS PROVIDED BY STUDENT
Safety glasses
Non-contact voltage detector
6-in-1 screwdriver

REQUIRED MATERIALS PROVIDED BY SCHOOL
Air Conditioner with electrical components and an electrical diagram

PROCEDURE
Turn off the power to the packaged air conditioner assigned by the instructor.
Remove the electrical panel to the unit.
Check with the non-contact voltage detector to insure that the panel is not energized.
DO NOT TOUCH ANYTHING YOU DON'T HAVE TO!
You may use the unit component location diagram or pictorial diagram to locate the components.
You should be able to locate the component on the diagram and on the unit.
You must point out the components and their symbols to the instructor.

> Condenser fan motor
> (outdoor fan motor)
> Evaporator fan motor
> (indoor blower motor)
> Contactor
> Transformer
> Compressor
> Capacitor
> Indoor fan relay
> Thermostat

LAB 38.7 IDENTIFY COMPONENTS ON HEAT PUMP DIAGRAM

LABORATORY OBJECTIVE
You will identify common HVACR electrical components on a heat pump and locate their symbols on the unit electrical diagram.

LABORATORY NOTES
In this lab exercise you will identify common electrical components on a heat pump. You will then identify the symbols in the unit diagram that represent the components.

FUNDAMENTALS OF HVACR TEXT REFERENCE
Unit 36 Electrical Components
Unit 38 Electrical Diagrams

REQUIRED MATERIALS PROVIDED BY STUDENT
Safety glasses
Non-contact voltage detector
6-in-1 screwdriver

REQUIRED MATERIALS PROVIDED BY SCHOOL
Heat pump with electrical components and an electrical diagram

Procedure
Turn off the power to the packaged heat pump assigned by the instructor.
Remove the electrical panel to the unit.
Check with the non-contact voltage detector to insure that the panel is not energized.
DO NOT TOUCH ANYTHING YOU DON'T HAVE TO!
You may use the unit component location diagram or pictorial diagram to locate the components.
You should be able to locate the component on the diagram and on the unit.
You must point out the components and their symbols to the instructor.

Compressor	High pressure switch
Supplementary heater	Thermal overload for electric heater
Indoor fan motor	Indoor fan relay
Transformer	Contactor
Defrost control	Defrost thermostat
Reversing Valve	Outdoor fan motor
High pressure switch	

LAB 39.1 DRAW WINDOW UNIT DIAGRAM

(This lab is done at home, NOT during lab time.)

LABORATORY OBJECTIVE
You will draw a schematic diagram for a window air conditioning unit.

Note: You must have completed this lab PRIOR to starting Lab 39.2, which is wiring this diagram.

LABORATORY NOTES
You will draw a schematic diagram of a window air conditioning unit. This diagram will be used to wire the circuit in Lab 39.2.

FUNDAMENTALS OF HVACR TEXT REFERENCE
Unit 38 Electrical Diagrams
Unit 39 Control Systems

REQUIRED MATERIALS PROVIDED BY STUDENT
Pencil
Paper

PROCEDURE
On a separate sheet of paper, draw a schematic diagram to meet the following specifications:
1. UNIT TYPE: Window Air Conditioner
2. COMPRESSOR TYPE: PSC Motor
3. FAN MOTOR TYPE: PSC Single Speed Motor
4. CONTROLS: Line voltage
5. SWITCH: Line voltage rotary selector switch to control fan motor and compressor
6. THERMOSTAT: Line Voltage

LAB 39.2 WIRE WINDOW UNIT DIAGRAM

LABORATORY OBJECTIVE
You will wire the electrical components for a window air conditioning unit using the diagram you drew in Lab 39.1.
Note: You must have completed Lab 39.1 PRIOR to starting this Lab.

LABORATORY NOTES
You will wire the electrical components of a window air conditioning unit using the schematic diagram you drew in Lab 39.1.

FUNDAMENTALS OF HVACR TEXT REFERENCE
Unit 38 Electrical Diagrams
Unit 39 Control Systems

REQUIRED MATERIALS PROVIDED BY STUDENT
Safety glasses
Multimeter
Ammeter
Wire cutter
Wire crimper/stripper
6-in-1 screwdriver

REQUIRED MATERIALS PROVIDED BY SCHOOL
120-volt power cord.
Wire
Wire connectors
Wire nuts
Line voltage thermostat
Line voltage rotary control switch
Light to represent compressor.
Light to represent fan motor.

PROCEDURE
ALL CIRCUIT BUILDING SHOULD BE DONE WITH THE POWER CORD UNPLUGGED!
Build the diagram you drew in Lab 39.1.
Have instructor check you completed diagram BEFORE energizing it.
Energize and operate your diagram.

LAB 39.3 DRAW PACKAGED AC UNIT DIAGRAM

(This lab is done at home, NOT during lab time.)

LABORATORY OBJECTIVE
You will draw a schematic diagram for a packaged central air conditioning unit.

Note: You must have completed this lab PRIOR to starting Lab 39.4, which is wiring this diagram.

LABORATORY NOTES
You will draw a schematic diagram of a packaged air conditioning unit. This diagram will be used to wire the circuit in Lab 39.4.

FUNDAMENTALS OF HVACR TEXT REFERENCE
Unit 38 Electrical Diagrams
Unit 39 Control Systems

REQUIRED MATERIALS PROVIDED BY STUDENT
Pencil
Paper

PROCEDURE
On a separate sheet of paper, draw a schematic diagram to meet the following specifications:
On a separate sheet of paper, draw a schematic diagram to meet the following specifications:
1. UNIT TYPE: Packaged Unit
2. COMP. TYPE: PSC Motor with Internal Overload
3. CFM TYPE: PSC Motor
4. IFM TYPE: Shaded Pole Motor
5. CONTROL SYSTEM: 24-volt Control System
6. THERMOSTAT: 24-volt thermostat to control both indoor fan and cooling
7. PROTECTION: LP & HP Switch in low voltage circuit to Protect Compressor

LAB 39.4 WIRE PACKAGED AC DIAGRAM

LABORATORY OBJECTIVE
You will wire the electrical components for a window air conditioning unit using the diagram you drew in Lab 39.3.
Note: You must have completed Lab 39.3 PRIOR to starting this Lab.

LABORATORY NOTES
You will wire the electrical components of the packaged air conditioning unit using the schematic diagram you drew in Lab 39.3.

***FUNDAMENTALS OF HVACR* TEXT REFERENCE**
Unit 38 Electrical Diagrams
Unit 39 Control Systems

REQUIRED MATERIALS PROVIDED BY STUDENT
Safety glasses
Multimeter
Ammeter
Wire cutter
Wire crimper/stripper
6-in-1 screwdriver

REQUIRED MATERIALS PROVIDED BY SCHOOL
120-volt power cord.
Wire
Wire connectors
Wire nuts
Transformer
Low voltage heat-cool thermostat
24-volt contactor
24-volt relay
Three Lights to represent compressor, indoor fan motor, and outdoor fan motor.

PROCEDURE
ALL CIRCUIT BUILDING SHOULD BE DONE WITH THE POWER CORD UNPLUGGED!
Build the diagram you drew in Lab 39.3.
Have instructor check you completed diagram BEFORE energizing it.
Energize and operate your diagram.

LAB 39.5 DRAW 2 STAGE HEAT, 2 STAGE COOL DIAGRAM

(This lab is done at home, NOT during lab time.)

LABORATORY OBJECTIVE
You will draw a schematic diagram for a two stage heating, two stage cooling air conditioning unit.

Note: You must have completed this lab PRIOR to starting Lab 39.6, which is wiring this diagram.

LABORATORY NOTES
You will draw a schematic diagram of a two stage heating, two stage cooling air conditioning unit. This diagram will be used to wire the circuit in Lab 39.6.

FUNDAMENTALS OF HVACR TEXT REFERENCE
Unit 38 Electrical Diagrams / Unit 39 Control Systems

REQUIRED MATERIALS PROVIDED BY STUDENT
Pencil

Paper

PROCEDURE
On a separate sheet of paper, draw a schematic diagram to meet the following specifications:

1. UNIT TYPE: 2 stage heat, 2 stage cool
2. COMP. TYPE: Two PSC compressor Motors with Internal Overload (one for each stage)
3. CFM TYPE: PSC Motor
4. IFM TYPE: PSC Motor
5. COOL: One compressor for each stage
6. HEAT: Electric strip heat – 1 heater for each stage.
7. CONTROL SYSTEM: 24-volt Control System, 2 stage heat, 2 stage cool
8. THERMOSTAT: 24-volt thermostat to control both indoor fan and cooling
9. PROTECTION: LP & HP Switch in low voltage circuit to Protect each Compressor
10. PROTECTION: Limit switch in series with each heat strip to protect from overheating

LAB 39.6 WIRE 2 STAGE HEAT 2 STAGE COOL DIAGRAM

LABORATORY OBJECTIVE
You will wire the electrical components for a 2 stage heating, 2 stage cooling air conditioning unit using the diagram you drew in Lab 39.5.

Note: You must have completed Lab 39.5 PRIOR to starting this Lab.

LABORATORY NOTES
You will wire the electrical components of the packaged air conditioning unit using the schematic diagram you drew in Lab 39.5.

FUNDAMENTALS OF HVACR TEXT REFERENCE
Unit 38 Electrical Diagrams / Unit 39 Control Systems

REQUIRED MATERIALS PROVIDED BY STUDENT
Safety glasses
Multimeter
Ammeter
Wire cutter
Wire crimper/stripper
6-in-1 screwdriver

REQUIRED MATERIALS PROVIDED BY SCHOOL
120-volt power cord.
Wire
Wire connectors
Wire nuts
Transformer
Low voltage heat-cool thermostat
Four 24-volt contactors
24-volt relay
Eight lights to represent 2 compressors, 2 condenser fan motors, 2 outdoor fan motors, and an indoor fan motor.

PROCEDURE
ALL CIRCUIT BUILDING SHOULD BE DONE WITH THE POWER CORD UNPLUGGED!
Build the diagram you drew in Lab 39.5.
Have instructor check you completed diagram BEFORE energizing it.
Energize and operate your diagram.

LAB 39.7 WIRE A HEATING AND COOLING THERMOSTAT

LABORATORY OBJECTIVE

The purpose of this lab is to learn how to wire a low voltage thermostat to a gas furnace and split system air conditioner.

LABORATORY NOTES

You will wire a single stage heating-cooling thermostat to control a gas furnace with a split system air conditioner. You will need to wire the control wiring at the thermostat, furnace control panel, and condensing unit.

FUNDAMENTALS OF HVACR TEXT REFERENCE

Unit 38 Electrical Diagrams
Unit 39 Control Systems

REQUIRED TOOLS AND MATERIALS

Low voltage thermostat
Gas furnace
Split system air conditioner
Multimeter
Electrical hand tools

PROCEDURE

1. Read the installation instructions for the thermostat, furnace, and air conditioner.
2. Turn off the power to the furnace and air conditioning unit.
3. Check the voltage to the unit with a multimeter to be sure the power is off.
4. Wire the 5-wire at the thermostat and furnace.
5. Wire the 2 wire at the furnace and condensing unit.
6. Have an instructor check your wiring.
7. Turn the power on and operate the unit in both heating and cooling.

Name _____

Date _____

Instructor's OK ☐

LAB 39.8 WIRE A PACKAGED HEAT PUMP THERMOSTAT

LABORATORY OBJECTIVE

The purpose of this lab is to learn how to wire a low voltage thermostat to a packaged heat pump.

LABORATORY NOTES

You will wire a two stage heat pump thermostat to control a packaged heat pump.

FUNDAMENTALS OF HVACR TEXT REFERENCE

Unit 38 Electrical Diagrams
Unit 39 Control Systems

REQUIRED TOOLS AND MATERIALS

Low voltage heat pump thermostat
Packaged heat pump
Multimeter
Electrical hand tools

PROCEDURE

1. Read the installation instructions for the thermostat and heat pump.
2. Turn off the power to the heat pump.
3. Check the voltage to the unit with a multimeter to be sure the power is off.
4. Wire the thermostat to control the unit.
5. Have an instructor check your wiring.
6. Turn the power on and operate the unit.

LAB 39.9 DESIGN AND OPERATE A SIMPLE 120-VOLT RELAY CIRCUIT

LABORATORY OBJECTIVE
The purpose of this lab is to learn how to design and wire a simple relay circuit.

LABORATORY NOTES
You will design, wire, and operate a simple relay circuit. A toggle switch will control the coil on a 120-volt relay. And the normally open relay contacts will control a 120-volt light.

FUNDAMENTALS OF HVACR TEXT REFERENCE
Unit 38 Electrical Diagrams
Unit 39 Control Systems

REQUIRED TOOLS AND MATERIALS
Relay with 120-volt coil
Toggle switch
120-volt light
120-volt power cord
Electrical hand tools

PROCEDURE
1. Draw a diagram with a toggle switch controlling a 120-volt relay and the normally open relay contacts operating a 120-volt light. There will be two circuits – one with the relay coil as the load, and one with the light as the load. The toggle switch will be in series with the relay coil to control it. The relay normally open contacts will be in series with the light to control it.
2. Wire the circuit you have drawn and have an instructor check it.
3. Operate your circuit.
4. Draw a diagram, wire and operate a relay control 110-volt fan motor with a single pole switch. (Relay has 24-volt coil)
5. Draw a diagram, wire and operate a contactor to control a 230-volt motor with a single pole switch. Contactor has 230-volt coil)

LAB 39.10 DESIGN AND OPERATE A SIMPLE 24-VOLT RELAY CIRCUIT

LABORATORY OBJECTIVE
The purpose of this lab is to learn how to design and wire a simple contactor circuit.

LABORATORY NOTES
You will design, wire, and operate a simple contactor circuit. A toggle switch will control the coil on a 24-volt contactor. The normally open contactor contacts will control a 120-volt light.

FUNDAMENTALS OF HVACR TEXT REFERENCE
Unit 38 Electrical Diagrams
Unit 39 Control Systems

REQUIRED TOOLS AND MATERIALS
Contactor with 24-volt coil
Transformer
Toggle switch
120-volt light
120-volt power cord
Electrical hand tools

PROCEDURE
1. Draw a diagram with a toggle switch controlling a 24-volt contactor coil. The normally open contactor contacts will operate a 120-volt light. There will be three circuits – one for the 120-volt primary transformer winding, one with the 24-volt transformer secondary winding as the source and the contactor coil as the load, and one with the 120-volt light as the load. The toggle switch will be in series with the contactor coil to control it. The contactor normally open contacts will be in series with the light to control

2. Wire the circuit you have drawn and have an instructor check it.

3. Operate your circuit.

LAB 39.11 DESIGN AND OPERATE A SIMPLE 24-VOLT CONTACTOR CIRCUIT

LABORATORY OBJECTIVE
The purpose of this lab is to learn how to design and wire a simple contactor circuit.

LABORATORY NOTES
You will design, wire, and operate a simple contactor circuit. A toggle switch will control the coil on a 24-volt contactor. The normally open contactor contacts will control a 120-volt light.

FUNDAMENTALS OF HVACR TEXT REFERENCE
Unit 38 Electrical Diagrams
Unit 39 Control Systems

REQUIRED TOOLS AND MATERIALS
Contactor with 24-volt coil
Transformer
Toggle switch
120-volt light
120-volt power cord
Electrical hand tools

PROCEDURE
1. Draw a diagram with a toggle switch controlling a 24-volt contactor coil. The normally open contactor contacts will operate a 120-volt light. There will be three circuits – one for the 120-volt primary transformer winding, one with the 24-volt transformer secondary winding as the source and the contactor coil as the load, and one with the 120-volt light as the load. The toggle switch will be in series with the contactor coil to control it. The contactor normally open contacts will be in series with the light to control it.

2. Wire the circuit you have drawn and have an instructor check it.

3. Operate your circuit.

LAB 39.12 WIRING A RELAY CONTROL SYSTEM

LABORATORY OBJECTIVE
The purpose of this lab is to learn how to design and wire a control system using relays and contactors.

LABORATORY NOTES
You will Design a control system to control three lights. Each light will be controlled by a relay. Each relay will be controlled by its own toggle switch. All the lights will be 120-volt lights. The relays will each have a different coil voltage. One will be 24-volts, another 120-volts, and the other 230-volts.

FUNDAMENTALS OF HVACR TEXT REFERENCE
Unit 38 Electrical Diagrams
Unit 39 Control Systems

REQUIRED TOOLS AND MATERIALS
24-volt relay
120-volt relay
230-volt relay
Transformer
3 toggle switches
3 120-volt lights
Multimeter
Electrical hand tools

PROCEDURE
1. Draw a ladder type schematic diagram for the control system.
2. Have the instructor check your design.
3. Gather and check the components. Note: we have MANY BAD components – check them BEFORE you wire the circuit.
4. Wire the control system.
5. Have the instructor check the wiring before energizing the circuit.
6. Operate the control system and troubleshoot any problems.

Name _____

Date _____

Instructor's OK ☐

LAB 39.13 WIRING ANTI-SHORT CYCLE TIME DELAYS

LABORATORY OBJECTIVE
The purpose of this lab is to learn how to wire an anti-short cycle time delay to control a contactor.

LABORATORY NOTES
You will wire an anti-short cycle time delay to control a 24-volt contactor coil. The contactor should energize when a thermostat calls for cooling. If the system is de-energized, the anti short cycle timer should prevent the contactor from coming back on until after a delay of 3 to 5 minutes.

FUNDAMENTALS OF HVACR TEXT REFERENCE
Unit 38 Electrical Diagrams / Unit 39 Control Systems

REQUIRED TOOLS AND MATERIALS
Low voltage thermostat
Transformer
24-volt contactor
Time delay
Multimeter
Electrical hand tools

PROCEDURE
1. Draw a ladder type schematic diagram of the time delay controlling the contactor.
2. Have the instructor check your design.
3. Wire the control system.
4. Have the instructor check the wiring before energizing the circuit.
5. Operate the control system and troubleshoot any problems.

LAB 39.14 WIRING AIR CONDITIONING CONTROL SYSTEM

LABORATORY OBJECTIVE
You will draw a ladder diagram for an air conditioning system using the specifications on the next page. You will then select the components necessary to wire the air conditioning control system you drew. You will then wire and operate the system.

LABORATORY NOTES
You will select the components necessary to construct the air conditioning control system that you drew. After building the system, you will operate it to verify that the controls work properly.

FUNDAMENTALS OF HVACR TEXT REFERENCE
Unit 38 Electrical Diagrams / Unit 39 Control Systems

REQUIRED MATERIALS PROVIDED BY STUDENT
Safety glasses
Gloves
Electrical hand tools
Multimeter
Amp Meter

REQUIRED MATERIALS PROVIDED BY SCHOOL
Transformer
Low voltage thermostat
Relays
Contactors
Motors
Lights to simulate loads
Pressure Switches
Power Wire
Control Wire

PROCEDURE
Check your components before wiring. It will save you time in the long run.
Arranging your components neatly and running your wires neatly will make life easier for the instructor that has to check it. Easy to check = good grade. Hard to check = bad grade.
Make all connections secure. Wires pulling loose add confusion and difficulty.
Do NOT energize the circuit until the instructor has checked it.

SPECIFICATIONS

Unit Type: Air Cooled Packaged Unit

Power Voltage: 230/1/60

Control Voltage: 24-volts

Condenser Fan Motor: PSC motor [230/1/60]

Evaporator Fan Motor: Shaded Pole Motor {230/1/60}

Safety Controls:

    Low-pressure switch

    High Pressure Switch

    Internal Overload for Compressor

CFM will be cycled by a thermostat connected in series with the motor, thus controlling the head pressure.

Power feeding the CFM will be from the load side of the contactor.

Thermostat will be a three-wire low voltage cooling thermostat.

LAB 39.15 WIRING COMMERCIAL PACKAGED UNIT

LABORATORY OBJECTIVE
You will draw a ladder diagram for a commercial air conditioning system using the specifications on the next page. You will then select the components necessary to wire the commercial packaged unit control system you drew. You will then wire and operate the system.

LABORATORY NOTES
You will select the components necessary to construct the commercial packaged unit control system that you drew. After building the system, you will operate it to verify that the controls work properly.

FUNDAMENTALS OF HVACR TEXT REFERENCE
Unit 38 Electrical Diagrams / Unit 39 Control Systems

REQUIRED MATERIALS PROVIDED BY STUDENT
Safety glasses
Gloves
Electrical hand tools
Multimeter
Amp Meter

REQUIRED MATERIALS PROVIDED BY SCHOOL
Transformer
Low voltage thermostat
Relays
Contactors
Motors
Lights to simulate loads
Pressure Switches
Power Wire
Control Wire

PROCEDURE
Check your components before wiring. It will save you time in the long run.
Arranging your components neatly and running your wires neatly will make life easier for the instructor that has to check it. Easy to check = good grade. Hard to check = bad grade.
Make all connections secure. Wires pulling loose add confusion and difficulty.
Do NOT energize the circuit until the instructor has checked it.

SPECIFICATIONS

1.   Unit Type: 10 Ton Condensing Unit with Two Stage Cooling

2.   Power Voltage: 208/3/60

3.   Control Voltage: 24-volts

4.   Each Condenser fan and compressor to have its own contactor

5.   Control Relay to Operate Compressor Contactor and condenser fan contactor.

6.   Coil for the Compressor contactor and condenser fan contactor to be 208 Volts.

7.   First Stage Cool: Controls Compressor, condenser fans, and indoor fan

8.   Second Stage Cool: Controls Solenoid Valve for #2 Refrigerant Circuit.

9.   Indoor Fan Motor Controlled by Thermostat.

10.   Magnetic Starter Controlling each Compressor.

11.   Compressor Voltage: 208/3/60

12.   Condenser Fan Motors{2}: 203/1/60 with overload protection

13.   Second CFM controlled by HPS to maintain constant head pressure.

14.   Compressor protected by current type pilot duty overload.

15.   Crankcase heater controlled thorough a control relay to energize heater on off cycle.

16.   CFM motors controlled through control relay to operate when CR energized and CFM thermostat.

17.   LP & HP switch for safety.

LAB 39.16 WIRING TWO STAGE AIR CONDITIONING SYSTEM

LABORATORY OBJECTIVE

You will draw a ladder diagram for a two stage air conditioning system using the specifications on the next page. You will then select the components necessary to wire the two stage air conditioning control system you drew. You will then wire and operate the system.

LABORATORY NOTES

You will select the components necessary to construct the two stage commercial air conditioning control system that you drew. After building the system, you will operate it to verify that the controls work properly.

FUNDAMENTALS OF HVACR TEXT REFERENCE

Unit 38 Electrical Diagrams / Unit 39 Control Systems

REQUIRED MATERIALS PROVIDED BY STUDENT

Safety glasses
Gloves
Electrical hand tools
Multimeter
Amp Meter

REQUIRED MATERIALS PROVIDED BY SCHOOL

Transformer
Low voltage thermostat
Relays
Contactors
Motors
Lights to simulate loads
Pressure Switches
Power Wire
Control Wire

PROCEDURE

Check your components before wiring. It will save you time in the long run.

Arranging your components neatly and running your wires neatly will make life easier for the instructor that has to check it. Easy to check = good grade. Hard to check = bad grade.

Make all connections secure. Wires pulling loose add confusion and difficulty.

Do NOT energize the circuit until the instructor has checked it.

SPECIFICATIONS

1. Unit Type: Roof Top Two-Stage Cool and Two-Stage Heating.
2. Power Voltage: 230/1/60
3. Control Voltage: 24-volts
4. Compressors{2} Type: 3 phase Motors
5. Condenser Fan Motors{2} Type: PSC Motors
6. Indoor Fan Motor: PSC Motor
7. Two Stage Heating & Cooling Low Voltage Thermostat
8. Electric Heaters 230/1/60

9. 1st Compressor: 1st Stage Cool
10. 2nd Compressor: 2ng Stage Cool
11. 1st Heater: 1st Stage Heat
12. 2nd Heater: 2nd Stage Heat
13. Pressure switch in series with #2 CFM to control head pressure.
14. Safety Controls:
   a. Low-pressure switch
   b. High Pressure Switch
   c. Pilot Duty Current Overload Relays
   d. Limit Switch to interrupt power if temperature rises above 180°F

Note: rating of transformer must be considered in this diagram because of the
Number of 24 loads. You may use any type pilot duty system that you desire.

LAB 41.1 USING SLING PSYCHROMETER

LABORATORY OBJECTIVE

You will demonstrate your ability to use a sling psychrometer to take wet bulb readings.

LABORATORY NOTES

You will take wet bulb and dry bulb temperature readings in the lab, outside, in an air conditioning return air stream, and in an air conditioning supply air stream. You will use these readings and the scale on the psychrometer to determine the relative humidity in each of these locations.

***FUNDAMENTALS OF HVACR* TEXT REFERENCE**

Unit 41 Psychrometrics and Airflow
Unit 73 Testing and Balancing Air Systems

REQUIRED MATERIALS PROVIDED BY SCHOOL

Sling psychrometer
Psychrometric chart
Operating air conditioner

PROCEDURE

1. Wet the sock on the wet bulb of the psychrometer with a few drops of room temperature water.
2. Spin the psychrometer in the air or hold in an airstream for approximately one minute.
3. Note: it is not necessary to spin the psychrometer when reading a moving air stream.
4. Read the wet bulb temperature first, and then the dry bulb temperature.
5. Use a psychrometric chart to determine the relative humidity and dew point.
6. Repeat to complete table.

Sling Psychrometer Readings				
	DRY BULB	**WET BULB**	**%HUMID**	**DEW POINT**
Lab				
Outside				
AC Return Air				
AC Supply Air				

LAB 41.2 USING ELECTRONIC HYGROMETER (DIGITAL PSYCHROMETER)

LABORATORY OBJECTIVE
You will demonstrate your ability to use a digital psychrometer to take wet bulb readings.

LABORATORY NOTES
You will take wet bulb and dry bulb temperature readings in the lab, outside, in an air conditioning return air stream, and in an air conditioning supply air stream. You will use the digital psychrometer to read the dry bulb, wet bulb, relative humidity, and dew point in each of these locations.

FUNDAMENTALS OF HVACR TEXT REFERENCE
Unit 41 Psychrometrics and Airflow / Unit 73 Testing and Balancing Air Systems

REQUIRED MATERIALS PROVIDED BY SCHOOL
Digital psychrometer
Psychrometric chart
Operating air conditioner

PROCEDURE
1. Turn on the electronic hygrometer and note the location of the dry bulb, wet bulb, and humidity readouts.
2. Open the sensing probe area with a gentle twist.
3. Read the dry bulb temperature, wet bulb temperature, and relative humidity.
4. Push the select button to read the dew point.
5. Repeat to complete table.

Digital Psychrometer Readings				
	DRY BULB	**WET BULB**	**%HUMID**	**DEW POINT**
Lab				
Outside				
AC Return Air				
AC Supply Air				

LAB 42.1 CHANGING PANEL FILTERS

LABORATORY OBJECTIVE

You will demonstrate your ability to change the disposable panel filters in an air conditioning system.

LABORATORY NOTES

You will remove the existing disposable panel filter and replace it with a clean filter, making sure that the filter is oriented properly.

***FUNDAMENTALS OF HVACR* TEXT REFERENCE**

Unit 42 Air Filters

REQUIRED MATERIAL PROVIDED BY STUDENT

6-in-1 screwdriver
5/16" nut driver
Respirator (if allergic to dust)

REQUIRED MATERIAL PROVIDED BY SCHOOL

Air conditioner with disposable panel filter
Clean disposable panel filter

PROCEDURE

Turn the power off to the unit.

Open the access door or filter access panel.

Remove the existing filter.

Replace the filter with a new, clean filter of the same dimensions.

Pay attention to the arrows on the filter indicating the proper airflow direction.

The arrows should point toward the unit.

Replace the access panel and turn the unit back on.

LAB 42.2 SERVICING MEDIA FILTERS

LABORATORY OBJECTIVE

You will demonstrate your ability to change the media in a high efficiency media filter.

LABORATORY NOTES

You will remove the existing filter media replace it with clean filter media, making sure that the filter is oriented properly..

FUNDAMENTALS OF HVACR **TEXT REFERENCE**

Unit 42 Air Filters

REQUIRED MATERIAL PROVIDED BY STUDENT

6-in-1 screwdriver

5/16" nut driver

Respirator (if allergic to dust)

REQUIRED MATERIAL PROVIDED BY SCHOOL

Air conditioner with high efficiency media filter

New replacement filter media material

PROCEDURE

Turn the power off to the unit.

Open the filter access panel.

Remove the filter box.

Study the way the media is in the box.

Carefully unsnap the combs one at a time by tilting them at an angle until they unsnap and then lifting them out.

Unsnap the edge holding the filter media in.

Remove the old media and place it in a plastic bag.

Lay the new media in place.

Snap the edges down to hold it in place.

Install the first comb being careful to match up the comb teeth with the pleats in the media.

Continue to install the combs until they are all in place.

Reinstall the filter into the cabinet – paying attention to the airflow directional arrows.

The arrows should point toward the unit.

Replace the access panel and turn the unit back on.

LAB 42.3 ELECTRONIC AIR CLEANERS

LABORATORY OBJECTIVE
You will demonstrate your ability to use a high voltage probe to measure the voltage output of an electronic air cleaner power supply.

LABORATORY NOTES
You will use a high voltage probe to measure the voltage output of the charging section of the air cleaner.

SAFETY NOTE: The power supply produces 6,000 volts! Do NOT try to measure the voltage with your standard volt meter! Your meter will be fried and you along with it. A high voltage adapter is required!

FUNDAMENTALS OF HVACR TEXT REFERENCE
Unit 42 Air Filters

REQUIRED MATERIAL PROVIDED BY STUDENT
6-in-1 screwdriver
5/16" nut driver
Respirator (if allergic to dust)

REQUIRED MATERIAL PROVIDED BY SCHOOL
Air conditioner with electronic air cleaner
High voltage probe
Digital multimeter with correct impedance to work with high voltage probe
Source of dry, compressed air

PROCEDURE
Turn on the power supply with the fan off. Does the indicator light come on?

Now, turn on the fan. What reaction do you get from the indicator light?

Turn off the power supply.

With the power off, remove the electronic cells.

Examine the cell and identify the ionizer wires and plates and the collector plates.

You will require instructor help for the next step.

SAFETY NOTE: The cell produces 6,000 volts! Do NOT touch ANYTHING on the power cell!

Take the power supply to the bench and connect 120 volts to the incoming voltage section with an extension cord.

Check the voltage output using a high-voltage probe.

SAFETY NOTE: The cell produces 6,000 volts! Do NOT try to measure the voltage with your standard volt meter! Your meter will be fried and you along with it. A high voltage adapter is required!

What voltage is read on the high voltage probe?

Use the air hose to blow air across the airflow sensor.

Read the voltage on the high voltage probe while blowing air across the air flow sensor.

What voltage is read on the high voltage probe when air is blowing across the sensor?

472

LAB 46.1 DUCT COMPONENTS

LABORATORY OBJECTIVE
You will demonstrate your ability to identify common duct system components.

LABORATORY NOTES
You will identify common duct system components and explain their purpose in the system.
SAFETY NOTE: Sheet metal duct can cause serious cuts and lacerations. Always handle carefully and wear gloves!

***FUNDAMENTALS OF HVACR* TEXT REFERENCE**
Unit 46 Duct Installation

REQUIRED MATERIAL PROVIDED BY STUDENT
Gloves

REQUIRED MATERIAL PROVIDED BY SCHOOL
An assortment of duct components and an assembled duct system.

PROCEDURE
Examine the duct system components assigned by the instructor.
Be prepared to name each of the components and discuss their function.

Specific components should include:
Round metal duct
Rectangular metal duct
S-lock
Drive cleat
Flex duct
transition
elbow
plenum
endcap
boot
wye
takeoff

LAB 47.1 BASIC REFRIGERATION SYSTEM STARTUP

LABORATORY OBJECTIVE
The student will demonstrate the correct procedure for installing a gauge manifold to read system operating pressures when conducting routine preventative maintenance.

LABORATORY NOTES
This basic startup assumes the system had been working in the past and is still expected to be operable but has not been used for several months. This procedure is similar to the maintenance you would perform on an ice cream parlor that has been closed for the winter and will open for the summer, or a school that has been closed for the summer. It is your job to perform the pre-start visual inspection, install the gauges, check system idle pressures, and to operate the system in a normal mode, making a judgment as to current system operating condition.

***FUNDAMENTALS OF HVACR* TEXT REFERENCE**
Unit 47 Troubleshooting Split Air-Conditioning Systems

REQUIRED TOOLS AND EQUIPMENT
Gloves and Goggles
Service Valve Wrench
Gauge Manifold
Operating Refrigeration System

SAFETY REQUIREMENTS
A. Wear safety goggles and gloves when working with refrigerants. Liquid refrigerant can cause frostbite when in contact with eyes and skin.

B. Use low loss hose fittings, or wrap cloth around hose fittings before removing the fittings from a pressurized system or cylinder. Inspect all fittings before attaching hoses.

PROCEDURE
Step 1
Familiarize yourself with the major components in the refrigeration system including the condenser, compressor, evaporator, and metering device. Determine where the high side and low side connections for the system are located.

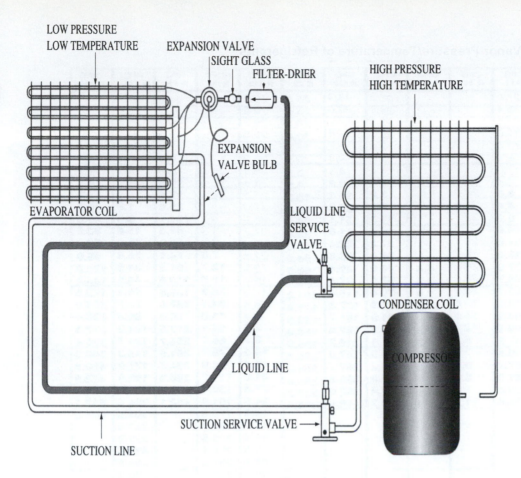

LOW PRESSURE
LOW TEMPERATURE
EXPANSION VALVE
SIGHT GLASS
FILTER-DRIER
HIGH PRESSURE
HIGH TEMPERATURE
EXPANSION VALVE BULB
EVAPORATOR COIL
LIQUID LINE SERVICE VALVE
CONDENSER COIL
COMPRESSOR
LIQUID LINE
SUCTION SERVICE VALVE
SUCTION LINE

Figure 47-1-1

Step 2
The refrigerant type and required amount of charge can be found on the refrigeration system nameplate.

F.L.A.		F.L.A.	
L.R.A.		L.R.A.	
H.P.		H.P.	
VOLTS		VOLTS	
HERTZ PHASE		HERTZ PHASE	
REFRIGERANT 22	5.0. LB.		KG.

Figure 47-1-2

Step 3
Determine the minimum operating head pressure. Use a pressure –temperature (P-T) relationship chart such as shown in Table 47-1-1.

TABLE 47-1-1 Vapor Pressure/Temperature of Refrigerants

Temp (Deg F)	CFC R 11	CFC R12	CFC R 113	CFC R 114	CFC R 500	CFC R 502	HCFC R 22	HCFC R 123	HCFC R124	HFC R 125	HFC R 134a	HFC R 410a
-100	*29.8*	*27.0*			*26.4*	*23.3*	*25.0*	*29.9*	*29.2*	*24.4*	*27.8*	
-90	*29.7*	*25.7*			*24.9*	*20.6*	*23.0*	*29.8*	*28.8*	*21.7*	*26.9*	
-80	*29.6*	*24.1*			*22.9*	*17.2*	*20.2*	*29.7*	*28.2*	*18.1*	*25.6*	
-70	*29.4*	*21.8*			*20.3*	*12.8*	*16.6*	*29.6*	*27.4*	*13.3*	*23.8*	
-60	*29.2*	*19.0*			*17.0*	*7.2*	*12.0*	*29.5*	*26.3*	*7.1*	*21.5*	
-50	*28.9*	*15.4*			*12.8*	*0.2*	*6.2*	*29.2*	*24.8*	**0.3**	*18.5*	**5.8**
-40	*28.4*	*11.0*			*7.6*	**4.1**	**0.5**	*28.9*	*22.8*	**4.9**	*14.7*	**11.7**
-30	*27.8*	*5.4*			*1.2*	**9.2**	**4.9**	*28.5*	*20.2*	**10.6**	*9.8*	**18.9**
-20	*27.0*	**0.6**	*29.0*	*22.8*	**3.2**	**15.3**	**10.2**	*27.8*	*16.9*	**17.4**	*3.8*	**27.5**
-10	*26.0*	**4.4**	*28.6*	*20.5*	**7.8**	**22.6**	**16.4**	*27.0*	*12.7*	**25.6**	**1.8**	**38.8**
0	*24.7*	**9.2**	*28.1*	*17.7*	**13.3**	**31.1**	**24.0**	*26.0*	*7.6*	**35.1**	**6.3**	**49.8**
10	*23.1*	**14.6**	*27.5*	*14.3*	**19.7**	**41.0**	**32.8**	*24.7*	*1.4*	**46.3**	**11.6**	**63.9**
20	*21.1*	**21.0**	*26.7*	*10.1*	**27.2**	**52.4**	**43.0**	*23.0*	**3.0**	**59.2**	**18.0**	**80.2**
30	*18.6*	**28.4**	*25.7*	*5.1*	**36.0**	**65.6**	**54.9**	*20.8*	**7.5**	**74.1**	**25.6**	**99.0**
40	*15.6*	**37.0**	*24.4*	**0.4**	**46.0**	**80.5**	**68.5**	*18.2*	**12.7**	**91.2**	**34.5**	**120.0**
50	*12.0*	**46.7**	*22.9*	**3.9**	**57.5**	**97.4**	**84.0**	*15.0*	**18.8**	**110.6**	**44.9**	**144.9**
60	*7.8*	**57.7**	*20.9*	**7.9**	**70.6**	**116.4**	**101.6**	*11.2*	**25.9**	**132.8**	**56.9**	**172.5**
70	*2.8*	**70.2**	*18.6*	**12.6**	**85.3**	**137.6**	**121.4**	*6.6*	**34.1**	**157.8**	**70.7**	**203.6**
80	**1.5**	**84.2**	*15.8*	**18.0**	**101.9**	**161.2**	**143.6**	*1.1*	**43.5**	**186.0**	**86.4**	**238.4**
90	**4.9**	**99.8**	*12.4*	**24.2**	**120.4**	**187.4**	**168.4**	**2.6**	**54.1**	**217.5**	**104.2**	**277.3**
100	**8.8**	**117.2**	*8.5*	**31.2**	**141.1**	**216.2**	**195.9**	**6.3**	**66.2**	**252.7**	**124.3**	**320.4**
110	**13.1**	**136.4**	*3.8*	**39.1**	**164.0**	**247.9**	**226.4**	**10.5**	**79.7**	**291.6**	**146.8**	**368.2**
120	**18.3**	**157.7**	**0.8**	**48.0**	**189.2**	**282.7**	**259.9**	**15.4**	**94.9**	**334.3**	**171.9**	**420.9**
130	**24.0**	**181.0**	**3.8**	**58.0**	**217.0**	**320.8**	**296.8**	**21.0**	**111.7**	**380.3**	**199.8**	**478.9**
140	**30.4**	**206.6**	**7.3**	**69.1**	**247.4**	**362.6**	**337.2**	**27.3**	**130.4**	**430.2**	**230.5**	**542.5**
150	**37.7**	**234.6**	**11.2**	**81.4**	**280.7**	**408.4**	**381.5**	**34.5**	**151.0**	**482.1**	**264.4**	**612.1**
160								**42.5**	**173.6**		**301.5**	**684.0**
170								**51.5**	**198.4**		**342.0**	
180								**61.4**	**225.6**		**385.9**	
190								**72.5**	**255.1**		**433.6**	
200								**84.7**	**287.3**		**485.0**	
210								**98.1**	**322.1**		**540.3**	
220								**112.8**	**359.9**			
230								**128.9**	**400.6**			
240								**146.3**	**444.5**			
250								**165.3**	**491.8**			

italics are in inches of mercury vacuum (in/Hg or " Hg)

bold are in pounds per squair inch pressure (psig)

A. A rule of thumb for checking the high side for proper operation is to take a temperature reading of the ambient air temperature. In order to maintain a proper condensing temperature at the condenser, there must be approximately a 20 to 35°F difference in ambient temperature and condensing temperature.

B. EXAMPLE: For an 80°F ambient temperature and a 20°F temperature difference using refrigerant R-410a, the condensing temperature is 100°F. Looking at a P-T chart, this means that the high side gauge should read at least 320.4 psig pressure.

C. In analyzing the low side (compound gauge) pressure, suppose the unit is designed for an evaporator temperature of 30°F at 65°F ambient and 40°F at 85°F ambient. Using R-410A, the pressure of boiling refrigerant in the evaporator at 30°F would be 99.0 psig and at 40°F it would be 120.0 psig.

Step 4
Connect the gauge manifold using the procedure provided below.

476

A. Remove the valve stem caps as shown in Figure 47-1-3 from the equipment service valves and check to be sure that both service valves are back seated (turned all the way out – counterclockwise).

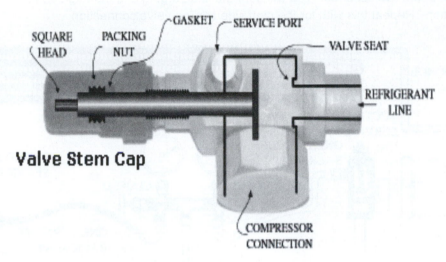

Figure 47-1-3

B. Remove the service (gauge) port caps from both service valves

C. Connect the center hose from the gauge manifold to a refrigerant cylinder, using the same type of refrigerant that is in the system and open both valves on the gauge manifold.

D. Open the valve on the refrigerant cylinder for about two seconds and then close it. This will purge any contaminants from the gauge manifold and hoses as shown in Figure 47-1-4.

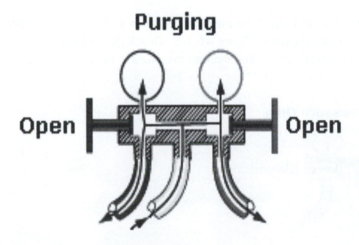

Figure 47-1-4

E. Connect the gauge manifold hoses to the gauge ports – the low pressure compound gauge to the suction service valve and the high pressure gauge to the high side service valve as shown in Figure 47-1-5.

F. Purge the air from each hose one at a time. This is done by first opening the valve on the refrigerant cylinder. Then slightly loosen the hose connection at the suction line service valve for about two seconds. You will hear the line purging and then tighten the connection. Repeat this with for the high side service valve connection.

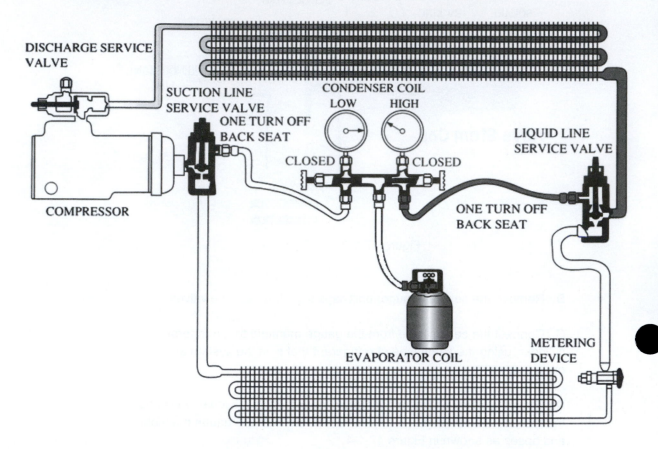

Figure 47-1-5

G. After the lines are purged of air then close the cylinder valve and front seat (close) both valves on the gauge manifold as shown in Figure 47-1-6. Crack (turn clockwise) both service valves one turn off the back seat. The system pressure is now allowed to register on each gauge.

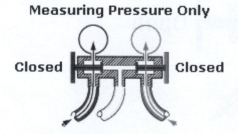

Figure 47-1-6

478

Step 5

Record the idle pressures and then turn the system on. Observe the high side pressure go up and the low side pressure go down. Allow five minutes of operation for the system to obtain normal pressures and then record these.

IDLE

High Side Pressure _____   Low Side Pressure _____

RUNNING

High Side Pressure _____   Low Side Pressure _____

Step 6

Once the readings are recorded, prepare to remove the gauge manifold from the system using the following procedure.

      A. Back seat the liquid line first (counterclockwise).

      B. Mid seat the valves on the gauge manifold to allow the remaining liquid in the high pressure hose to be pulled into the suction side of the system.

      C. After the pressures equalize, back seat the suction service valve on the compressor.

      D. Remove the hoses from the gauge ports and seal the ends of the hoses by reattaching them to the back of the gauge manifold set.

QUESTIONS

1. What are the expected suction and discharge pressures for this system?

2. How did you determine what these pressures should be?

3. How do the pressures you recorded compare with the expected pressures?

(Circle the letter that indicates the correct answer.)

4. If the discharge pressure is higher than expected, this could be due to:
    A. a dirty condenser.
    B. bad compressor bearings.
    C. excessive compressor lubrication.
    D. an inadequate supply of refrigerant.

5. If the discharge pressure is lower than expected, this could be due to:
    A. moisture in the refrigerant.
    B. air trapped in the system.
    C. a low charge of refrigerant.
    D. insufficient condenser cooling.

6. If the suction pressure is lower than expected, this could be due to:
    A. excessive condenser cooling.
    B. the expansion valve not opening.
    C. an excessive charge of refrigerant.
    D. air trapped in the evaporator.

7. When service valves are back seated:
    A. the gauge port is closed from system pressure.
    B. the gauge port is open to system pressure.
    C. there will be no refrigerant flow through the valve.
    D. The entire valve is closed.

8. Whenever purging air from gauge manifold hoses:
    A. be sure to minimize the amount of refrigerant release.
    B. never use the same refrigerant as that found in the system.
    C. always use a refrigerant with a higher boiling point to purge.
    D. always use a refrigerant with a lower boiling point to purge.

9. Low loss gauge manifold hose fittings:
    A. reduce the amount of refrigerant lost.
    B. prevent the refrigerant from being released and causing skin burns.
    C. Both A and B are correct.
    D. Neither A or B is correct.

10. An overcharge of refrigerant would be indicated by:
    A. a lower than expected discharge pressure.
    B. a higher than expected discharge pressure.
    C. a lower than expected suction pressure.
    D. a suction pressure of exactly 0 psig.

11. An worn compressor could be indicated by:
    A. a very cold evaporator.
    B. a very hot condenser.
    C. a chattering expansion valve.
    D. a high suction pressure and low discharge pressure.

480

12. Precautions to take when using a gauge manifold include:
- A. never dropping or abusing the gauge manifold.
- B. keeping the ports or charging lines capped when not in use.
- C. never using any fluid other than clean oil and refrigerant.
- D. All of the above are correct.

13. If the high and low pressure readings on a gauge manifold are identical:
- A. both valves on the gauge manifold are open.
- B. both valves on the gauge manifold are shut.
- C. the compressor short cycling.
- D. the condenser is dirty.

14. Under EPA regulations, air conditioning technicians are allowed to purge the gauge hoses provided it is a de minimis release.
- A. True
- B. False

15. A gauge manifold can measure:
- A. pressure and volume only.
- B. pressure only, along with corresponding saturation temperature.
- C. both pressure and temperature.
- D. temperature only.

LAB 48.1 PRINCIPLES OF COMBUSTION

LABORATORY OBJECTIVE
You will demonstrate the effect of the correct fuel-air mixture for combustion.

LABORATORY NOTES
You will operate an atmospheric gas burner, observe the flames and measure the CO concentration in the flue gas. You will then close the primary air shutters, observe the flames, and measure the CO concentration. Finally, you will summarize the effect of primary air on proper gas combustion.

FUNDAMENTALS OF HVACR TEXT REFERENCE
Unit 48 Principles of Combustion and Safety

REQUIRED MATERIALS PROVIDED BY STUDENT
Safety glasses

REQUIRED MATERIALS PROVIDED BY SCHOOL
Operating gas furnace with primary air shutters
Meter with CO measurement capability

PROCEDURE

1. Examine the furnace burners.

    a. Where does the primary air enter?

    b. Where does the secondary air enter?

2. Operate a furnace that has atmospheric burners with a primary air adjustment.

3. Observe the flames.

4. Measure the CO content of the flue gas.

5. Close the primary air adjustment all the way and observe the flames.

6. Measure the CO content of the flue gas.

    a. How did reducing the primary air affect the CO content of the flue gas?

7. Open the primary air shutter until all the yellow flame tips disappear.

8. Measure the CO content of the flue gas.

LAB 48.2 GAS SAFETY INSPECTION

LABORATORY OBJECTIVE
You will describe typical gas safety hazards.

LABORATORY NOTES
You will describe common gas safety hazards to the instructor using a gas furnace to show where to look for the hazard.

FUNDAMENTALS OF HVACR TEXT REFERENCE
Unit 48 Principles of Combustion and Safety

REQUIRED MATERIALS PROVIDED BY STUDENT
Safety glasses

REQUIRED MATERIALS PROVIDED BY SCHOOL
Gas furnace

PROCEDURE
Describe at least five gas-related safety hazards, recommend corrective action for each safety hazard and ways to avoid these hazards. If the hazard does not actually exist in the shop, you can describe the hazard and how it would affect gas safety. Hazards should include one from each of the following categories:

- Combustion Air

- Venting

- Gas Leaks

- Clearance

- Flammable Material Storage

LAB 48.3 SERVICING GAS BURNERS

LABORATORY OBJECTIVE
You will adjust the primary air on an atmospheric gas burner to achieve a clean –burning, efficient flame.

FUNDAMENTALS OF HVACR TEXT REFERENCE
Unit 48 Principles of Combustion and Safety

REQUIRED MATERIALS PROVIDED BY STUDENT
Safety glasses
Hand tools

REQUIRED MATERIALS PROVIDED BY SCHOOL
Operating gas furnace with primary air adjustment

PROCEDURE
1. Identify the type of burners in the furnaces assigned by the instructor. They could be
   - atmospheric - slotted port
   - atmospheric - drilled port
   - atmospheric - ribbon
   - atmospheric – in shot
   - power burner

2. Adjust the primary air on the burners assigned by an instructor.

3. Close off the primary air shutter on one burner. What happens? Why?

4. Open the primary air shutters all the way open on all the burners. What happens? Why?

5. Turn off the call for heat. Call for heat and watch the lighting of the burners. DO NOT GET YOUR FACE RIGHT IN FRONT OF THE FURNACE!!! What happens? Why?

6. Readjust the burners to the proper primary air setting. Now, turn the furnace off and then on again. How do the burners light now?

LAB 49.1 GAS FURNACE CHARACTERISTICS

LABORATORY OBJECTIVE
You will identify gas furnaces by their characteristics; including blower configuration, fuel type, efficiency, and type of draft.

LABORATORY NOTES
You will examine the gas furnaces assigned by the instructor, identify the different furnace characteristics, and record them in the data table below.

FUNDAMENTALS OF HVACR TEXT REFERENCE
Unit 49 Gas Furnaces

REQUIRED MATERIALS PROVIDED BY STUDENT
Safety glasses

REQUIRED MATERIALS PROVIDED BY SCHOOL
Gas furnaces

PROCEDURE
Examine each gas furnace for the blower configuration, type of fuel, efficiency, and type of draft.

Fuel
The fuel for a gas furnace will be either natural gas or propane, indicated as LP. The furnace data-plate will indicate the type of fuel. Most furnace data-plates are on the inside of the furnace.

Blower configuration
Locate the blower. The air goes across the blower first, so the blower compartment will be opposite where the air leaves.

Upflow
A vertically oriented furnace with the blower on the bottom is an upflow because it is blowing the air up.

Downflow or Counterflow
A vertically oriented furnace with the blower on top is a downflow because it is blowing the air down.

Horizontal
Air moves horizontally, or sideways, through a horizontal furnace.

Multipoise
A multiposition, or multipoise furnace can be installed in more than one position. Multipoise furnaces typically indicate their ability to be installed in multiple positions somewhere on the furnace itself.

Draft
Natural
Natural draft furnaces do not have a draft blower. Instead, they have either a draft hood or a draft diverter. They use metal vent pipe.

Induced
Induced draft furnaces have a draft blower that sucks air through the combustion chamber.

Forced
Forced draft furnaces have a draft blower that blow, or forces air through the combustion chamber.

Efficiency
Gas furnaces fall into three large efficiency categories: 60%, 80%, and 90%.

60% Efficiency
Natural draft, standing pilot furnaces are no longer manufactured, but many still exist. They typically operate at 50%-70% efficiency.

80% Efficiency
Furnaces that do not have a standing pilot, have an induced draft blower, but still use metal venting are 80% efficient.

90% Efficiency
Furnaces that have plastic PVC vents are 90%-98% efficient.

	Furnace 1	Furnace 2	Furnace 3	Furnace 4
Fuel *Natural* *LP*				
Configuration *Upflow* *Horizontal* *Downflow* *Multipoise*				
Draft *Natural* *Induced* *Forced*				
Efficiency *60%* *80%* *90%*				

LAB 49.2 NATURAL DRAFT FURNACE COMPONENTS

LABORATORY OBJECTIVE
You will identify the components on a natural gas furnace.

FUNDAMENTALS OF HVACR **TEXT REFERENCE**
Unit 49 Gas Furnaces

REQUIRED MATERIALS PROVIDED BY STUDENT
Safety glasses

REQUIRED MATERIALS PROVIDED BY SCHOOL
Natural draft gas furnace

PROCEDURE
You will examine the natural draft gas furnace assigned by the instructor and identify the listed components.

- Gas line
- Gas valve
- Gas manifold
- Pilot light
- Burners
- Heat Exchanger
- Indoor blower
- Draft Diverter
- Vent

LAB 49.3 NATURAL DRAFT FURNACE SEQUENCE OF OPERATION

LABORATORY OBJECTIVE
You will describe the sequence of operation for a standing pilot, natural draft furnace.

FUNDAMENTALS OF HVACR TEXT REFERENCE
Unit 49 Gas Furnaces

REQUIRED MATERIALS PROVIDED BY STUDENT
Safety glasses

REQUIRED MATERIALS PROVIDED BY SCHOOL
Natural draft gas furnace

PROCEDURE
Set the thermostat to call for heat and observe the furnace operating sequence.

List the order that things happen in the data table.

Natural Draft Furnace Operating Sequence	
Order	Action
1	Thermostat set to heat
2	
3	
4	Thermostat set to OFF
5	
6	

Once the indoor fan is running, set the thermostat to "Off".

Observe the order that things shut off and record in the table.

QUESTIONS

1. What controls the gas pressure?

2. How are the gas and combustion air mixed?

3. How is the gas-air mixture ignited?

4. How is the heat from combustion transferred into the house?

5. What happens to the combustion gasses?

LAB 49.4 INDUCED DRAFT FURNACE COMPONENTS

LABORATORY OBJECTIVE
You will identify the components on an induced draft natural gas furnace.

FUNDAMENTALS OF HVACR TEXT REFERENCE
Unit 49 Gas Furnaces

REQUIRED MATERIALS PROVIDED BY STUDENT
Safety glasses

REQUIRED MATERIALS PROVIDED BY SCHOOL
Induced draft gas furnace

PROCEDURE
You will examine the natural draft gas furnace assigned by the instructor and identify the listed components.

- Gas line
- Gas valve
- Gas manifold
- Igniter (do not touch the igniter)
- Burners
- Heat Exchanger
- Inducer Motor
- Indoor blower

LAB 50.1 INSTALL REPLACEMENT THERMOSTAT

LABORATORY OBJECTIVE
The purpose of this lab is to demonstrate your ability to install a replacement thermostat.

LABORATORY NOTES
Many thermostats are multiday programmable, and may even be multi-zone programmable. Older thermostats may have a mercury switch and the proper disposal for this type is at a toxic waste collection site. Check with your instructor regarding the necessary procedures for handling mercury switches.

***FUNDAMENTALS OF HVACR* TEXT REFERENCE**
Unit 50 Gas Furnace Controls

REQUIRED TOOLS AND EQUIPMENT
Operating HVAC unit
Tool kit
Multimeter
Clamp-on Ammeter
Temperature sensor
Thermostat

SAFETY REQUIREMENTS
A. Check all circuits for voltage before doing any service work.

B. Stand on dry nonconductive surfaces when working on live circuits.

C. Never bypass any electrical protective devices.

PROCEDURE

Collect the HVAC unit data and fill in the chart.

HVAC Unit Data

Unit Description	
Fuel Type	
Ignition System	
Control Transformer	VA rating = _____ Location _____
Fan Relay Type and Number	
Heating Circuit Device and Rated Amperage Draw	
Cooling Capacity	
Number of Stages of Heat	
Number of Stages of Cooling	
List Other Functions to Control or Supervise	

Step 2

Collect the thermostat data and fill in the chart.

Thermostat Data

Make:	Model Number:

Stages of Heat:	Stages of Cooling:

Heat Anticipator: (circle one)

    Adjustable   Nonadjustable

Cool Anticipator: (circle one)

    Adjustable   Nonadjustable

Subbase Model Number:

Subbase Switching (circle all that apply)

    HEAT   OFF   AUTO   COOL   FAN ON   FAN AUTO

List special thermostat features if any:

Step 3

Complete the Thermostat Installation Check List and make sure to check each step off in the appropriate box as you finish it. This will help you to keep track of your progress.

Thermostat Installation Check List

STEP	PROCEDURE	CHECK
1	Make sure all power to the unit is off. Lock and tag the power panel before removing any parts.	
2	Check existing thermostat wire and size.	
3	Thermostat wire size = _____ (18 gauge is typical)	
4	Number of conductors = _____	
5	Color of wires (circle all that apply). White Red Green Blue Yellow Other	
6	Install and level new thermostat subbase in a stable location.	

7	Match and record thermostat wire color with subbase terminals. W1_____ W2_____ O_____ B_____ R or RH_____ RY_____ G_____ Y1_____ Y2_____	
8	Install a jumper from RH to W1.	
9	Turn the power to the unit on and observe the heat operation.	
10	Measure the amperage draw of the heating circuit. Amperage = _____	
11	Turn the power to the unit off. Lock and tag the power panel before removing any parts.	
12	Install the thermostat on the subbase.	
13	Set the heat anticipator to the measured amperage draw.	

Step 4

After installing the thermostat you are now ready to check for proper operation. You will follow the same procedure of working through a check list. This will help you to keep track of your progress. Complete the Operational Check List and make sure to check each step off in the appropriate box as you finish it.

Operational Check List

STEP	PROCEDURE	CHECK
1	Turn the thermostat to the off position.	
2	Obtain an accurate room temperature at the thermostat location.	
3	Turn the set-point setting until the contacts open.	
4	Compare the set-point with the room temperature. Set-point_____ Room temperature_____	
5	Calibrate set-point to equal room temperature.	
6	Turn the power on.	
7	Start and set clock if required.	
8	Use the thermostat instructions to program if programmable.	

9	Turn heat-off-cool switch to heat position and turn set-point to above room temperature.	
10	Observe heat operation.	
11	Turn fan switch to "ON" position. Observe fan turn on.	
12	Turn fan switch to "AUTO" position. Observe fan turn off.	
13	Turn heat-off-cool switch to "COOL" position.	
14	Turn set-point to below room temperature.	
15	Observe cooling operation.	
16	Turn set-point to above room temperature.	
17	Observe cooling off.	
18	Install thermostat cover.	

LAB 50.2 PROGRAM THERMOSTAT

LABORATORY OBJECTIVE

The purpose of this lab is to demonstrate your ability to program a thermostat.

LABORATORY NOTES

Energy conservation is important for heating and air conditioning systems. The electronic programmable clock thermostat is a popular energy management tool for the residential and light commercial market. This device allows the customer to automatically change the set-point temperatures for occupied times and set back or set up the unoccupied temperature depending on the heating or cooling mode.

Since each programmable thermostat operates differently, keeping the instruction booklet is essential. Certain features have common functions but the function name may vary from brand to brand. Some are full seven day programmable while others are 5-day /2-day clocks with all week days treated the same and both weekend days treated the same. A business would usually require a full 7-day programmable thermostat. Typically there are a maximum of four daily times (wake, leave, come home, and sleep). The four times are designed for two set back periods; one during the night and one during the day when people are at work. This type of thermostat can be used for business applications by setting the midday setback at one temperature all day or stacking the middle time points on one point in time.

***FUNDAMENTALS OF HVACR* TEXT REFERENCE**

Unit 50 Gas Furnace Controls

REQUIRED TOOLS AND EQUIPMENT

Programmable Thermostat

SAFETY REQUIREMENTS

A. Check all circuits for voltage before doing any service work.

PROCEDURE

Step 1

Obtain a 5-day / 2-day programmable thermostat with four daily setback periods. Complete the Thermostat Programming Check List and make sure to check each step off in the appropriate box as you finish it. This will help you to keep track of your progress.

Thermostat Programming Check List (5-day / 2-day)

STEP	PROCEDURE	CHECK
1	Read and follow the manufacturer's instructions.	
2	Set the clock to the correct time of day.	
3	Set the 5-day (weekday) setting for: 6:00 AM (wake) – set to 70°F 7:30 AM (leave) - set to 62°F 4:30 PM (home) - set to 70°F 10:00 PM (sleep) - set to 62°F	
4	Set the 2-day (weekend) setting for: 8:00 AM (wake) – set to 70°F 11:00 AM (leave) - set to 65°F 5:00 PM (home) - set to 70°F 11:30 PM (sleep) - set to 60°F	

Step 2

Obtain a 7-day programmable thermostat. Complete the Thermostat Programming Check List and make sure to check each step off in the appropriate box as you finish it.

Thermostat Programming Check List (7-day)

STEP	PROCEDURE	CHECK
1	Read and follow the manufacturer's instructions.	
2	Set the clock to the correct time of day.	
3	Set the 7-day setting for a typical retail store: 9:00 AM to 8:00 PM Monday through Thursday – set to 70ºF 9:00 AM to 11 PM Friday - set to 70ºF 8:00 Am to 11:30 PM Saturday - set to 70ºF 9:00 AM to 11 PM Sunday - set to 70ºF Provide a 60ºF temperature for the unoccupied time.	
4	Allow lead time for morning warm up based on a 5ºF / hr pick-up. This means that at 8:00 AM every day except for Saturday which would be 7:00 AM, the temperature would be set for 65ºF. The thermostat could also be set to allow for the unoccupied setting temperature to begin 1/2 hour before store closing and this would still allow for comfortable temperature and reduce energy usage.	

LAB 50.3 PILOT TURNDOWN TEST – THERMOCOUPLE TYPE

LABORATORY OBJECTIVE
The purpose of this lab is to test the proper operation of a thermocouple type pilot for a gas furnace.

LABORATORY NOTES
A pilot turndown test is a test to determine the smallest possible pilot capable of proving flame presence to the control circuit and lighting the main burner safely. It is performed by lighting the pilot, measuring the pilot flame signal, and observing a series of main burner ignitions using various pilot sizes. Remember a cold burner is more difficult to ignite and in the performance of this test the burner becomes warmed up; you must allow sufficient cool-down time to perform the final test. A pilot turndown test can also point to other pilot problems.

***FUNDAMENTALS OF HVACR* TEXT REFERENCE**
Unit 50 Gas Furnace Controls

REQUIRED TOOLS AND EQUIPMENT
Operating gas furnace
Tool kit
Millivolt-meter

SAFETY REQUIREMENTS
A. Never allow gas flow without a flame. If gas is allowed to build up and then suddenly ignite, this can create a serious hazard.

B. If the furnace contains a pilot, always ensure it is properly lit prior to starting the unit.

C. Check all circuits for voltage before doing any service work.

PROCEDURE
Step 1
Obtain any standard thermocouple installed or not installed.

   A. Complete the Thermocouple Open Circuit Test (30 MV) Check List and make sure to check each step off in the appropriate box as you finish it. This will help you to keep track of your progress.

Thermocouple Open Circuit Test (30 MV) Check List

STEP	PROCEDURE	CHECK
1	Remove the threaded connection from the gas valve or pilot safety switch if necessary.	
2	Connect a millivolt meter from the outer copper line to the inner core lead at the end of the thermocouple.	
3	Heat the enclosed end of the thermocouple with a torch or a normal pilot assembly if installed.	
4	A voltage of up to 30 millivolts will be read on the meter. If the meter goes down, then reverse the leads.	
5	Record the highest millivoltage. Voltage = _____mV	

Step 2

Obtain a thermocouple installed on a standard combination gas valve or pilot safety switch.

    A. Complete the Thermocouple Closed Circuit Test (30 MV) Check List.

Thermocouple Closed Circuit Test (30 MV) Check List

STEP	PROCEDURE	CHECK
1	Obtain the thermocouple adaptor for the valve end.	
2	Light the pilot for a normal pilot flame.	
3	Read the millivolts at the adaptor. Voltage = _____mV	
4	Blow out the pilot light and then relight it within 30 seconds of going out.	
5	Why does gas still come out of the pilot burner after the pilot is out?	
6	Blow out the pilot again and observe the voltage output.	
7	Record the voltage in millivolts when the pilot valve closes stopping the pilot gas so that the pilot will not relight. This value should be between 9 and 12 mV. Voltage = _____mV	
8	Adjust the size of the pilot flame for the smallest pilot flame capable of proving the pilot and lighting the main burner smoothly and safely.	

LAB 50.4 PILOT TURNDOWN TEST – FLAME ROD TYPE

LABORATORY OBJECTIVE
The purpose of this lab is to test the proper operation of a flame rod type pilot for a gas furnace. Perform this test on any furnace equipped with a pilot, a separate spark igniter, and a flame rod.

LABORATORY NOTES
A pilot turndown test is a test to determine the smallest possible pilot capable of proving flame presence to the control circuit and lighting the main burner safely. It is performed by lighting the pilot, measuring the pilot flame signal, and observing a series of main burner ignitions using various pilot sizes. Remember a cold burner is more difficult to ignite and in the performance of this test the burner becomes warmed up; you must allow sufficient cool-down time to perform the final test. A pilot turndown test can also point to other pilot problems.

FUNDAMENTALS OF HVACR TEXT REFERENCE
Unit 50 Gas Furnace Controls

REQUIRED TOOLS AND EQUIPMENT
Operating gas furnace
Tool kit
Micro-ammeter

SAFETY REQUIREMENTS
A. Never allow gas flow without a flame. If gas is allowed to build up and then suddenly ignite, this can create a serious hazard.

B. If the furnace contains a pilot, always ensure it is properly lit prior to starting the unit.

C. Check all circuits for voltage before doing any service work.

PROCEDURE
Step 1

Perform this test on any furnace equipped with a pilot, a separate spark igniter, and a flame rod.

   B.   Complete the Flame Rod Pilot System Test Check List and make sure to check each step off in the appropriate box as you finish it. This will help you to keep track of your progress.

Flame Rod Pilot System Test Check List

STEP	PROCEDURE	CHECK
1	Locate the pilot adjustment screw. This is usually a needle valve screw located under a screw cap labeled *pilot adj*.	
2	Turn off main power.	
3	Install the micro-ammeter in series with the flame rod sensor wire to sensor wire terminal of the control box.	
4	Turn on power, and obtain a pilot only flame by turning off manual main burner valve or pull wire labeled main at redundant gas valve. Consult wiring diagram as required to locate main gas wire.	
5	Temporarily wire a small toggle switch in series with the main burner wire and the main gas valve. Turn switch off.	
6	Obtain operation of pilot only flame.	
7	Turn on main burner toggle and observe main burner on.	
8	Read flame signal of pilot only in micro-amps. Current = _____ micro Amps	
9	Turn system off.	
10	Remove and clean flame rod with steel wool.	

11	Read flame signal of pilot only in micro-amps. Current = _____ micro Amps	
12	Has the signal changed from step #8?	
13	Reposition the flame rod to obtain a better contact with the clean blue flame to gain a stronger signal if possible.	
14	Turn the pilot adjustment screw and observe the pilot flame and corresponding micro-amp signal get smaller.	
15	Obtain the smallest flame possible that is capable of proving the pilot flame. Current = _____ micro Amps	
16	Turn on the main burner toggle and observe the main burner light.	
17	Cycle the burner on and off several times and observe the ignition	
18	Enlarge the pilot as required to provide for smooth main burner ignition. Current = _____ micro Amps	

LAB 50.5 PILOT TURNDOWN TEST – FLAME IGNITER TYPE

LABORATORY OBJECTIVE
The purpose of this lab is to test the proper operation of a flame igniter type pilot for a gas furnace. Perform this test on any furnace equipped with a pilot and a combination pilot igniter/pilot proving device.

LABORATORY NOTES
A pilot turndown test is a test to determine the smallest possible pilot capable of proving flame presence to the control circuit and lighting the main burner safely. It is performed by lighting the pilot, measuring the pilot flame signal, and observing a series of main burner ignitions using various pilot sizes. Remember a cold burner is more difficult to ignite and in the performance of this test the burner becomes warmed up; you must allow sufficient cool-down time to perform the final test. A pilot turndown test can also point to other pilot problems.

FUNDAMENTALS OF HVACR TEXT REFERENCE
Unit 50 Gas Furnace Controls

REQUIRED TOOLS AND EQUIPMENT
Operating gas furnace
Tool kit

SAFETY REQUIREMENTS
A. Never allow gas flow without a flame. If gas is allowed to build up and then suddenly ignite, this can create a serious hazard.

B. If the furnace contains a pilot, always ensure it is properly lit prior to starting the unit.

C. Check all circuits for voltage before doing any service work.

PROCEDURE
Step 1
Perform this test on any furnace equipped with a pilot and a combination pilot igniter/pilot proving device.

    A. Complete the Flame Igniter Pilot System Test Check List and make sure to check each step off in the appropriate box as you finish it. This will help you to keep track of your progress.

Flame Igniter Pilot System Test Check List

STEP	PROCEDURE	CHECK
1	Locate the pilot adjustment screw. This is usually a needle valve screw located under a screw cap labeled **pilot adj**.	
2	Turn off main power.	
3	Install an appropriate toggle switch in series with main gas terminal of redundant gas valve.	
4	Turn on power and obtain a pilot only flame.	
5	Turn pilot adjustment valve in (clockwise, cw) and observe the pilot flame get smaller.	
6	Continue turning in until the pilot goes out and pilot igniter goes out.	
7	Turn pilot adjustment out (counter-clockwise, ccw) and observe the pilot relight.	
8	With the pilot on, turn on main gas toggle and observe main gas burner turn on.	
9	With the main burner on, adjust the pilot flame smaller.	
10	Observe the main burner go off when the pilot flame goes out.	
11	Adjust the pilot flame to a size that will light easily and ignite the main burner assembly.	

LAB 50.6 SEPARATE PILOT GAS PRESSURE REGULATOR

LABORATORY OBJECTIVE
The purpose of this lab is to test the proper operation of a separate pilot gas pressure regulator for a gas furnace. This Laboratory Worksheet deals with any commercial burner controller utilizing a separate pilot gas pressure regulator, such as a Honeywell commercial RA89F series controller. This procedure will use a plug in flame monitor jack or wire meter in series with the flame rod.

LABORATORY NOTES
A pilot turndown test is a test to determine the smallest possible pilot capable of proving flame presence to the control circuit and lighting the main burner safely. It is performed by lighting the pilot, measuring the pilot flame signal, and observing a series of main burner ignitions using various pilot sizes.

FUNDAMENTALS OF HVACR TEXT REFERENCE
Unit 50 Gas Furnace Controls

REQUIRED TOOLS AND EQUIPMENT
Operating gas furnace
Tool kit
Micro-ammeter
Plug-in flame monitor

SAFETY REQUIREMENTS
A. Never allow gas flow without a flame. If gas is allowed to build up and then suddenly ignite, this can create a serious hazard.

B. If the furnace contains a pilot, always ensure it is properly lit prior to starting the unit.

C. Check all circuits for voltage before doing any service work.

PROCEURE
Step 1
Perform this test on commercial gas furnace power burners.

    A.   Complete the Separate Pilot Gas Pressure Regulator Test Check List and make sure to check each step off in the appropriate box as you finish it. This will help you to keep track of your progress.

Separate Pilot Gas Pressure Regulator Test Check List

STEP	PROCEDURE	CHECK
1	Turn off power, main gas valve, and pilot gas valve.	
2	Obtain a micro-ammeter and a plug-in flame monitor jack or wire the micro-ammeter in series with the flame rod.	
3	Remove the primary control cover.	
4	Open the pilot gas valve and turn on the power.	
5	Observe the pilot flame only.	
6	Record the pilot micro-amp reading. Current = _____ micro Amps	
7	Decrease he gas pressure by turning the gas pressure regulator out (counter-clockwise, ccw). Observe the pilot flame get smaller and the flame signal reduce in current (micro-amps)	
8	Turn the pilot down until the controller relay opens. Current = _____ micro Amps	

9	Increase the pilot flame size until the controller relay opens. Current = _____ micro Amps	
10	Observe the main burner go off when the pilot flame goes out.	
11	Turn on the main burner gas and power.	
12	Observe the main burner ignition.	
13	Adjust the pilot flame as required to obtain smooth burner ignition during cold start. Remember a cold burner is more difficult to ignite and in the performance of this test the burner becomes warmed up; you must allow sufficient cool-down time to perform the final test.	

LAB 50.7 CHECK/TEST/REPLACE HOT SURFACE IGNITER

LABORATORY OBJECTIVE

The purpose of this lab is to check, test, and replace a hot surface igniter for a gas furnace

LABORATORY NOTES

Hot surface ignition and flame proving is the most common method of burner control in modern gas furnaces. The major replacement item in this type of system is the hot surface igniter.

Hot surface igniters are sometimes called glow coils because they glow when energized. An igniter that does not glow is the first indication that it has failed. A typical startup sequence will occur as follows: the thermostat calls for heat, the draft fan comes on, the hot surface igniter glows, the gas valve opens (you can hear the click), the gas ignites, the gas flame reaches the flame rod and proves, and the normal heat mode is in progress. When the surface igniter fails to glow, you hear the click of the gas valve opening but no flame will appear and since the flame does not prove, the gas valve will close.

***FUNDAMENTALS OF HVACR* TEXT REFERENCE**

Unit 50 Gas Furnace Controls

REQUIRED TOOLS AND EQUIPMENT

Operating gas furnace
Tool kit
Multimeter

SAFETY REQUIREMENTS

A. Check all circuits for voltage before doing any service work.

B. Never allow gas flow without a flame.

PROCEDURE

Step 1

Familiarize yourself with the gas furnace components.

      A. Complete Gas Furnace Component Identification Check List and make sure to check each step off in the appropriate box as you finish it. This will help you to keep track of your progress.

Gas Furnace Component Identification Check List

STEP	PROCEDURE	CHECK
1	Locate the inducer fan and induced draft vent system.	
2	Locate the hot surface igniter.	
3	Locate the flame rod for the flame proving system.	
4	Locate the electronic module or the flame system control.	
5	Locate and write down the terminal on the module that the flame rod wire connects to.	
6	Locate the plug or wire connection from the control system to the hot surface igniter.	

Step 2

Observe a normal trial for an ignition sequence.

A. Complete the Normal Ignition Sequence Trial Check List and make sure to check each step off in the appropriate box as you finish it.

Normal Ignition Sequence Trial Check List

STEP	PROCEDURE	CHECK
1	Adjust the thermostat to call for heat.	
2	Observe the draft inducer fan come on.	
3	Does the hot surface igniter? (this takes about a minute after the fan comes on) (circle one) YES NO	
4	If the surface igniter glows, the gas valve should open and the burner flame should light and the furnace should operate normally and the trial is complete. If the surface igniter does not glow and the burner fails to light, proceed to the next step.	
5	If the surface igniter does not glow, did you hear the gas valve click as it opened? (circle one) YES NO	

6	If the surface igniter did not glow, the gas valve clicked open, and the burner did not light, then the surface igniter must be checked.	
7	Make sure all power to the furnace is off. Lock and tag the power panel before removing any parts.	
8	Disconnect the wire nuts from the surface igniter leads or unplug it and test the surface igniter for continuity with a multimeter.	
9	If there is continuity, the surface igniter may not be the problem and the voltage to the igniter will need to be verified. This will involve troubleshooting and possibly replacing the module for the flame system control.	
10	If there is no continuity (infinite resistance), then the surface igniter has an open and it will need to be replaced.	
11	Obtain a replacement surface igniter of the same configuration for both the shape of the coil and the connection and replace the defective surface igniter (make sure to supply any shield supplied to protect the coil).	
12	After installing the new surface igniter, turn the power on for the furnace and cycle it through a normal stat mode with the thermostat calling for heat. Observe the main burner ignition and proper operation.	

LAB 50.8 NATURAL DRAFT FURNACE ELECTRICAL COMPONENTS

LABORATORY OBJECTIVE
You will identify common electrical components on a standing pilot, natural draft furnace and explain their function.

FUNDAMENTALS OF HVACR TEXT REFERENCE
Unit 50 Gas Furnace Controls

REQUIRED MATERIALS PROVIDED BY STUDENT
Safety glasses

REQUIRED MATERIALS PROVIDED BY SCHOOL
Operating natural draft gas furnace

PROCEDURE
You will examine the natural draft gas furnace assigned by the instructor and identify the listed components.

- Gas valve
- Thermocouple
- Limit Switch
- Auxiliary Limit Switch
- Fan Switch
- Indoor blower
- Door Switch
- Transformer

LAB 50.9 INDUCED DRAFT ELECTRICAL COMPONENTS

LABORATORY OBJECTIVE

You will identify common electrical components on an induced draft furnace and explain their function.

FUNDAMENTALS OF HVACR TEXT REFERENCE

Unit 50 Gas Furnace Controls

REQUIRED MATERIALS PROVIDED BY STUDENT

Safety glasses

REQUIRED MATERIALS PROVIDED BY SCHOOL

Operating induced draft gas furnace

PROCEDURE

You will examine the natural draft gas furnace assigned by the instructor and identify the listed components.

- Ignition control
- Transformer
- Inducer motor
- Draft Pressure Switch
- Gas valve
- Igniter
- Flame Rod
- Limit Switch
- Roll out switch
- Door Switch
- Indoor blower

LAB 50.10 INDUCED DRAFT FURNACE SEQUENCE OF OPERATION

LABORATORY OBJECTIVE
You will describe the sequence of operation for an induced draft furnace.

FUNDAMENTALS OF HVACR TEXT REFERENCE
Unit 50 Gas Furnace Controls

REQUIRED MATERIALS PROVIDED BY STUDENT
Safety glasses

REQUIRED MATERIALS PROVIDED BY SCHOOL
Induced draft gas furnace

PROCEDURE
Set the thermostat to call for heat and observe the furnace operating sequence.

List the order that things happen in the data table.

Induced Draft Furnace Operating Sequence	
Order	Action
1	Thermostat set to heat
2	
3	
4	
5	
6	Thermostat set to OFF
7	
8	
9	

Once the indoor fan is running, set the thermostat to "Off".

Observe the order that things shut off and record in the table.

Pull a wire off one side of the draft switch and set the thermostat to call for heat again.

Observe the furnace operation. What happens?

Turn the thermostat to the off position, replace the draft switch wire and remove the wire from the flame sensor.

Set the thermostat to call for heat.

Observe the operation. What happens?

Show the instructor how to adjust the length of the fan running cycle.

Pull one wire off the high limit and set the thermostat to call for heat.

Observe the operation. What happens?

QUESTIONS

1. What comes on first?

2. What controls the gas pressure?

3. How are the gas and combustion air mixed?

4. How is the gas-air mixture ignited?

5. How is the heat from combustion transferred into the house?

6. What happens to the combustion gasses?

LAB 50.11 WIRE A NATURAL DRAFT FURNACE

LABORATORY OBJECTIVE
You will completely wire a natural draft furnace, including all internal wiring.

FUNDAMENTALS OF HVACR TEXT REFERENCE
Unit 50 Gas Furnace Controls

REQUIRED MATERIALS PROVIDED BY STUDENT
Safety glasses
Non-contact voltage probe
Multimeter
Wire cutter/stripper/crimper

REQUIRED MATERIALS PROVIDED BY SCHOOL
Natural draft gas furnace with all internal wiring removed
Wire
Electrical terminals

PROCEDURE
The instructor will remove all the electrical wiring from a standing pilot, natural draft furnace.

Turn the furnace disconnect off .

Check to make sure the power is off.

You will wire the furnace according to the furnace electrical wiring diagram.

Have instructor check your work BEFORE energizing the furnace.

Operate the furnace and verify correct operation.

LAB 50.12 WIRE AN INDUCED DRAFT GAS FURNACE

LABORATORY OBJECTIVE
You will completely wire an induced draft furnace, including all internal wiring.

FUNDAMENTALS OF HVACR TEXT REFERENCE
Unit 50 Gas Furnace Controls

REQUIRED MATERIALS PROVIDED BY STUDENT
Safety glasses
Non-contact voltage probe
Multimeter
Wire cutter/stripper/crimper

REQUIRED MATERIALS PROVIDED BY SCHOOL
Induced draft gas furnace with all internal wiring removed
Wire
Electrical terminals

PROCEDURE
The instructor will remove all the electrical wiring from an induced draft furnace.

Turn the furnace disconnect off.

Check to make sure the power is off.

You will wire the furnace according to the furnace electrical wiring diagram.

Have instructor check your work BEFORE energizing the furnace.

Operate the furnace and verify correct operation.

LAB 50.13 GAS VALVE INSPECTION

LABORATORY OBJECTIVE
You will dismantle and inspect a diaphragm gas valve to learn how it works.

FUNDAMENTALS OF HVACR **TEXT REFERENCE**
Unit 50 Gas Furnace Controls

REQUIRED MATERIALS PROVIDED BY STUDENT
Safety glasses
Allen wrench set
Screw drivers

REQUIRED MATERIALS PROVIDED BY SCHOOL
Gas Valve for Dismantling

PROCEDURE
Disassemble and examine the gas valves assigned by the instructor.

Be prepared to identify each valve type and describe its operation. Reassemble the valves after explaining them to the instructor. Clean up your work area.

Examine the gas valves on equipment assigned by your instructor. Be prepared to identify the types of the valves and their operation. After discussing the valves with the instructor, put all the covers back on the units and clean up your work area.

LAB 50.14 ADJUSTING GAS REGULATORS

LABORATORY OBJECTIVE
You will adjust the gas valve regulator to maintain the manufacturer's specified manifold pressure on a gas furnace.

FUNDAMENTALS OF HVACR TEXT REFERENCE
Unit 50 Gas Furnace Controls

REQUIRED MATERIALS PROVIDED BY STUDENT
Safety glasses
Allen wrench set
Screw drivers
Manometer

REQUIRED MATERIALS PROVIDED BY SCHOOL
Operating Gas Furnace

PROCEDURE
Find the specified manifold pressure on the unit data plate.

With the unit off, connect your manometer to the manifold pressure tap on the gas valve.

Operate the furnace and check the operating manifold pressure.

Turn the adjustment in clockwise until you see a change in the manometer reading.

Turn the adjustment out counterclockwise back to the starting point.

Adjust the manifold pressure to meet the pressure specified on the data plate.

Show your results to the instructor and cleanup your work area.

LAB 50.15 TESTING STANDING PILOT SAFETY DEVICES

LABORATORY OBJECTIVE
You will measure the milli-voltage output of a thermocouple.

FUNDAMENTALS OF HVACR TEXT REFERENCE
Unit 50 Gas Furnace Controls

REQUIRED MATERIALS PROVIDED BY STUDENT
Safety glasses
Multimeter that can read DC millivolts

REQUIRED MATERIALS PROVIDED BY SCHOOL
Operating Gas Furnace with a standing pilot light and thermocouple
Thermocouple milli-voltage adapter

PROCEDURE
Remove the thermocouple from the gas valve connection.

Install the thermocouple millivoltage adapter into the threaded thermocouple connector on the gas valve.

Install the thermocouple into the thermocouple adapter.

Turn the gas valve know to "Pilot", press in the knob, and light the pilot light.

Hold the knob in for at least a minute after lighting the pilot light.

Release the knob, the pilot light should remain burning.

Measure the millivolt output of the thermocouple on the adapter.

LAB 50.16 CHECKING INTERMITTENT PILOT IGNITION SYSTEMS

LABORATORY OBJECTIVE
You will observe the operating sequence of an intermittent pilot ignition system and simulate common failure symptoms.

FUNDAMENTALS OF HVACR TEXT REFERENCE
Unit 50 Gas Furnace Controls

REQUIRED MATERIALS PROVIDED BY STUDENT
Safety glasses
Electrical tools

REQUIRED MATERIALS PROVIDED BY SCHOOL
Operating Gas Furnace with intermittent pilot ignition system

PROCEDURE

1. Study the wiring diagram of a furnace using an intermittent spark pilot system.

2. Explain the sequence of operation to an instructor.

3. Run the furnace and observe the operating sequence.

4. Turn the furnace off and pull a wire loose from the draft switch.

5. Turn the furnace on and set the thermostat to call for heat.

6. What happens?

7. Turn the furnace off, replace the wire on the draft switch, and turn the furnace back

    on.

8. With the main burners going, turn the gas valve to the OFF position. What happens?

9. Does the system try to relight?

10. Run the furnace. When all burners are lit and the spark is no longer present, pull the

    wire off of the flame sensor. What happens? Why? Turn the unit off and replace the

    wire.

LAB 50.17 DIRECT SPARK IGNITION CONTROLS

LABORATORY OBJECTIVE

You will observe the operating sequence of a direct spark ignition system and simulate common failure symptoms.

FUNDAMENTALS OF HVACR **TEXT REFERENCE**

Unit 50 Gas Furnace Controls

REQUIRED MATERIALS PROVIDED BY STUDENT

Safety glasses
Electrical tools

REQUIRED MATERIALS PROVIDED BY SCHOOL

Operating Gas Furnace with direct spark ignition system

PROCEDURE

1. Study the wiring diagram of a furnace using a direct ignition system
2. Operate the DSI control system assigned by the instructor.
3. Does this system use a pre-purge cycle? If so, how long is it?
4. How long did the spark igniter operate?
5. Cycle the system off.
6. Does this system have a post purge cycle? If so, how long is it?
7. Turn the gas valve to the "OFF" position and call for heat once more.
8. How long does an ignition trial take?
9. How many ignition trials does the system have before lockout?
10. Turn the gas valve back to the "ON" position.
11. Reset the control by killing the call for heat and re-establishing it.
12. After flame is proved, turn the gas valve back to the "OFF" position.
13. Does the system attempt to re-establish combustion?
14. How many trial periods does it go through before it locks out?
15. Does it use a purge cycle in between the trials?
16. How long are they?
17. Turn the unit off and return the gas valve to the "ON" position.
18. Give a complete description of the operation cycle of this control.

LAB 50.18 TROUBLESHOOTING DSI CONTROL SYSTEMS

LABORATORY OBJECTIVE
Given a DSI gas furnace with an ignition problem, you will identify the problem, its root cause, and recommend corrective action.

FUNDAMENTALS OF HVACR TEXT REFERENCE
Unit 50 Gas Furnace Controls

REQUIRED MATERIALS PROVIDED BY STUDENT
Safety glasses
Hand tools
Manometer
Multimeter

REQUIRED MATERIALS PROVIDED BY SCHOOL
DSI Gas furnace with a problem

PROCEDURE
Troubleshoot the DSI gas furnace assigned by the instructor. Be sure to be complete in your description of the problem. You should include:

What is the furnace is doing wrong?

What component or condition is causing this?

What tests did you perform that told you this?

How would you correct this?

LAB 51.1 GAS FURNACE STARTUP

LABORATORY OBJECTIVE

The purpose of this lab is to demonstrate your ability to go through the necessary sequence for a typical gas furnace startup.

LABORATORY NOTES

You will need to familiarize yourself with gas furnace arrangements and types. Always remember to make the visual check of all system components prior to starting a gas furnace.

FUNDAMENTALS OF HVACR TEXT REFERENCE

Unit 51 Gas Furnace Installation

REQUIRED TOOLS AND EQUIPMENT

Operating gas furnace
Tool kit
Multimeters
Clamp-on Ammeter

SAFETY REQUIREMENTS

A. Never allow gas flow without a flame. If gas is allowed to build up and then suddenly ignite, this can create a serious hazard.

B. If the furnace contains a pilot, always ensure it is properly lit prior to starting the unit.

C. Check all circuits for voltage before doing any service work.

D. Stand on dry nonconductive surfaces when working on live circuits.

E. Never bypass any electrical protective devices.

PROCEDURE
Step 1

Collect the gas furnace data and fill in the chart.

Gas Furnace Unit Data

Furnace Make:		Model Number:	Blower Motor Amperage Rating:	
Furnace Type (circle one)	Upflow	Counterflow	Basement	Horizontal
Blower Type (circle one)	Direct drive	Belt drive		
Blower speed (circle one)	Single speed	Two speed	Three speed	Four speed
System type (circle one)	Heat only	Heat and humidify	Heat and cool	
Burner type (circle one)	Atmospheric	Induced draft	Power	

Step 2

Complete the Prestart Check List and make sure to check each step off in the appropriate box as you finish it. This will help you to keep track of your progress.

Prestart Check List

STEP	PROCEDURE	CHECK
1	Turn thermostat down below the room temperature.	
2	Make sure all power to the furnace is off. Lock and tag the power panel before removing any parts.	
3	Vent connector connected with three screws per joint.	
4	Fuel line installed properly, with no apparent leaks.	
5	Spin all fans to be sure they are loose and turn freely.	
6	Combustion area free from debris.	
7	Electrical connections complete – main power.	
8	Electrical connections complete – thermostat.	
9	All doors and panels available and in place.	
10	Thermostat installed and operating correctly.	

Step 3

After finishing the Prestart Check List you are now ready to start the furnace and check its operation. You will follow the same procedure of working through a check list. This will help you to keep track of your progress. Complete the Start-Up Check List and make sure to check each step off in the appropriate box as you finish it.

Start-Up Check List

STEP	PROCEDURE	CHECK
1	Check the main fuse or breaker and the amperage rating and make sure it is the correct rating for the furnace.	
2	Measure the incoming supply voltage.	
3	If the Steps 1 & 2 have been completed and are satisfactory, then turn on the supply power for the furnace.	
4	Set the thermostat to – Fan On - to obtain the fan only operation and observe that the blower is operating. If the thermostat is not so equipped then consult with your instructor.	
5	Light the pilot if so equipped, following the manufacturer's instructions provided with the furnace.	
6	Turn the fan off and the thermostat to heat.	
7	Adjust the thermostat setting to 10 °F above the room temperature.	

8	Observe the flame sequence begin and the flame comes on.	
9	Observe that the blower begins running after the heat exchanger comes up to temperature.	
10	Use a clamp-on ammeter to measure the blower motor amperage and compare this value to its rating.	
11	Turn the thermostat setting down below the room temperature and observe the flame go off.	
12	Observe the blower stop running approximately three minutes after the burner turns off.	

LAB 51.2 MEASURE GAS USAGE

LABORATORY OBJECTIVE
The purpose of this lab is to demonstrate your ability to determine the gas usage for a furnace.

LABORATORY NOTES
You will need to familiarize yourself with gas furnace arrangements and types. Always remember to make the visual check of all system components prior to starting a gas furnace.

FUNDAMENTALS OF HVACR TEXT REFERENCE
Unit 51 Gas Furnace Installation

REQUIRED TOOLS AND EQUIPMENT
Operating gas furnace
Tool kit
Multimeters
Clamp-on Ammeter

SAFETY REQUIREMENTS
A. Never allow gas flow without a flame. If gas is allowed to build up and then suddenly ignite, this can create a serious hazard.

B. If the furnace contains a pilot, always ensure it is properly lit prior to starting the unit.

C. Check all circuits for voltage before doing any service work.

D. Stand on dry nonconductive surfaces when working on live circuits.

E. Never bypass any electrical protective devices.

PROCEDURE
Step 1

Collect the gas furnace data and fill in the chart.

Gas Furnace And Fuel Meter Data

Furnace Make:		Model Number:	Blower Motor Amperage Rating:	
Burner Make:				
Number of Main Burner Orifices:				
Input Rating in BTUH:				
Burner type (circle one)	Atmospheric	Induced draft	Power	
Meter type (circle one)	½ ft^3	2 ft^3		

Step 2

Complete the Prestart Check List and prepare to start the furnace as outlined in the procedure provided in Lab 51.1.

Step 3

After finishing the Prestart Check List and then starting the furnace to check its operation, you can prepare to clock the main burner. You will follow the same procedure of working through a check list. This will help you to keep track of your progress. Complete the Clock Main Burner Check List and make sure to check each step off in the appropriate box as you finish it.

Clock Main Burner Check List

STEP	PROCEDURE	CHECK
1	Isolate the main burner.	
2	Leave the pilot on for normal operation.	
3	Operate the main burner to be tested.	
4	Observe the rotating dial of the gas meter. This is typically either a ½ ft^3 or 2 ft^3 dial	
5	Measure the time in seconds for one revolution of the dial. Time in seconds =	
6	Calculate cubic feet per hour (CFH). CFH = 3600 / seconds x 1/2 ft^3 or 2 ft^3 dial) =	
7	Calculate actual burner input. The approximate heating value for natural gas is 1000 BTU per ft^3. You may contact the gas supplier to obtain a more accurate value if desired.	

	Actual burner input = CFH x 1000 BTU/ft^3 = BTUH	
8	Compare actual measured input with the nameplate input rating.	
9	If there is a difference of greater than 5% between the actual measured input as compared to the nameplate input rating, then the gas manifold pressure should be checked.	

LAB 51.3 GAS FURNACE COMBUSTION TESTING

LABORATORY OBJECTIVE
The purpose of this lab is to demonstrate your ability to test a gas furnace to determine proper combustion.

LABORATORY NOTES
Combustion testing is primarily used when setting power burners on which there is an adjustment that will allow total over-combustion to the burners. The traditional standard setting is at 50% excess air to ensure complete combustion. Consult the manufacturer recommendation for the burner, furnace, or boiler type for more accurate settings. When using one of the newer electronic combustion analyzers, the appearance of CO will show when complete combustion is taking place.

FUNDAMENTALS OF HVACR TEXT REFERENCE
Units 51 Gas Furnace Installation

REQUIRED TOOLS AND EQUIPMENT
Operating gas furnace
Tool kit
Combustion analyzer
Temperature sensor

SAFETY REQUIREMENTS
A. Never allow gas flow without a flame. If gas is allowed to build up and then suddenly ignite, this can create a serious hazard.

B. If the furnace contains a pilot, always ensure it is properly lit prior to starting the unit.

C. Vent pipes will be hot with the furnace operating. Be careful not to touch hot surfaces when inserting, removing, or reading instruments.

PROCEDURE
Step 1

Collect the gas furnace data and fill in the chart.

Gas Furnace Unit Data

Furnace Make:		Model Number:	Blower Motor Amperage Rating:	
Furnace Type (circle one)	Upflow	Counterflow	Basement	Horizontal
Blower Type (circle one)	Direct drive	Belt drive		
Blower speed (circle one)	Single speed	Two speed	Three speed	Four speed
System type (circle one)	Heat only	Heat and humidify	Heat and cool	
Burner type (circle one)	Atmospheric	Induced draft	Power	

Step 2

Complete the Prestart Check List and prepare to start the furnace as outlined in the procedure provided in Lab 51.1. Before starting the furnace complete the Preparation for Combustion Test Check List.

Preparation For Combustion Test Check List

STEP	PROCEDURE	CHECK
1	Prior to starting the furnace locate CO_2 test openings and the temperature probe location for an undiluted vent gas sample.	
2	Insert the thermometer probe and check the CO_2 for position.	
3	Start the furnace and allow it to run until the vent gas temperature is at its normal maximum.	

Step 3

With the furnace now operating, begin taking your initial readings and record them in the Combustion Test Chart.

   A.   Initial readings are taken prior to making any adjustments.
   B.   For the yellow flame test, close the primary air shutter until a lazy flame is present.
   C.   For the excess air test, open the primary air shutter to a maximum amount.
   D.   For the 8% CO_2 test, close the air shutter to obtain 8%.
   E.   Calculate the combustion efficiency using a slide rule calculator with the natural gas slide.

Combustion Test Chart

Combustion Test Readings	Test 1 (initial)	Test 2 (yellow)	Test 3 (excess)	Test 4 (8% CO_2)
CO_2				
Actual Stack Temperature				
Net Stack (gross–room)				
Combustion Efficiency				

   F.   Which flame is the most efficient?

LAB 51.4 MEASURE GAS FURNACE THERMAL EFFICIENCY

LABORATORY OBJECTIVE
The purpose of this lab is to demonstrate your ability to measure thermal efficiency of a gas furnace and to draw a correlation, if there is any, between the airflow through the furnace and the thermal efficiency of the furnace.

LABORATORY NOTES
Thermal efficiency is BTU output divided by BTU input. Input is the measure of gas consumed while output is measured by the sensible heat formula which is (temperature in ºF) x (1.08) x (airflow in CFM). The thermal efficiency will be calculated at three different airflows.

FUNDAMENTALS OF HVACR TEXT REFERENCE
Unit 51 Gas Furnace Installation

REQUIRED TOOLS AND EQUIPMENT
Operating gas furnace with variable speed blower
Tool kit
Temperature sensor
Airflow hood

SAFETY REQUIREMENTS
A. Never allow gas flow without a flame. If gas is allowed to build up and then suddenly ignite, this can create a serious hazard.

B. If the furnace contains a pilot, always ensure it is properly lit prior to starting the unit.

C. Vent pipes will be hot with the furnace operating. Be careful not to touch hot surfaces when inserting, removing, or reading instruments.

PROCEDURE
Step 1

Collect the gas furnace data and fill in the chart.

Gas Furnace Unit Data

Furnace Make:		Model Number:	Blower Motor Amperage Rating:	
Furnace Type (circle one)	Upflow	Counterflow	Basement	Horizontal
Blower Type (circle one)	Direct drive	Belt drive		
Blower speed (circle one)	Single speed	Two speed	Three speed	Four speed
System type (circle one)	Heat only	Heat and humidify	Heat and cool	
Burner type (circle one)	Atmospheric	Induced draft	Power	

Step 2

Complete the Prestart Check List and prepare to start the furnace as outlined in the procedure provided in Lab 51.1. Before starting the furnace complete the Preparation for Efficiency Test Check List.

Preparation For Efficiency Test Check List

STEP	PROCEDURE	CHECK
1	Prepare to measure the airflow on the inlet and/or the outlet of the furnace using an air flow hood. It may be necessary to build a bracket to install the airflow hood.	
2	If the airflow hood is located on the outlet, the airflow measurement must be taken prior to starting the furnace as the airflow hood will be damaged if exposed to temperatures above 140°F (Fan only operation).	
3	Operate the furnace in the normal heating mode. The airflow hood may be left on the return air during this test. Airflow will typically go down as the air is heated due to the air restriction of the furnace and the increased air volume of the heated air.	

Step 3

With the furnace now operating, begin taking your initial readings and record them in the Efficiency Test Chart.

    A.   Initial readings are taken prior to making any adjustments.
    B.   For the reduced airflow test, reduce the blower speed or throttle the air damper.
    C.   For the increased airflow test, increase the blower speed or open the air damper.
    D.   For the 8% CO_2 test, close the air shutter to obtain 8%.
    E.   Calculate the combustion efficiency using a slide rule calculator with the natural gas slide.

Efficiency Test Chart

Efficiency Test Readings	Test 1 (initial)	Test 2 (reduced airflow)	Test 3 (increased airflow)
Air Inlet Temperature			
Air outlet Temperature			
Airflow (CFM)			

Step 4

Calculate the BTU output for the three test conditions using the formula:

(outlet temperature – inlet temperature) x (1.08) x (airflow in CFM)

BTU Output

	Test 1 (initial)	Test 2 (reduced airflow)	Test 3 (increased airflow)
BTU Output			

Step 5

Calculate the efficiency for the three test conditions using the formula:

(BTU output from Step 4) / (Burner rated BTU input)

Calculated Efficiency

	Test 1 (initial)	Test 2 (reduced airflow)	Test 3 (increased airflow)
Calculated Efficiency			

    A.   What correlation can be drawn between the thermal efficiency and the airflow of the furnace?

LAB 51.5 CHECKING GAS PRESSURE

LABORATORY OBJECTIVE
You will demonstrate your ability to use a manometer to measure gas pressure.

***FUNDAMENTALS OF HVACR* TEXT REFERENCE**
Unit 51 Gas Furnace Installation

REQUIRED MATERIALS PROVIDED BY STUDENT

Safety glasses	Manometer
Non-contact voltage probe	Allen wrench set

REQUIRED MATERIALS PROVIDED BY SCHOOL
Operational gas furnace

PROCEDURE
1. Turn the power off to the furnace and verify that it is off.
2. Close the manual gas valve in the gas line.
3. Remove the plug on the gas valve inlet test port .
4. Install the manometer in the gas valve inlet test port.
5. Open the gas line valve.
6. The manometer should read the incoming gas line pressure.
7. Turn the power on and set the thermostat to heat.
8. Read the gas line pressure with the furnace operating.
9. Turn the thermostat to "OFF".
10. Turn the power off to the furnace and verify that it is off.
11. Close the manual gas valve in the gas line.
12. Replace the plug on the gas valve inlet test port .
13. Remove the plug on the gas valve outlet test port .
14. Install the manometer in the gas inlet test port on the gas valve.
15. Open the gas line valve.
16. The manometer should read a manifold gas pressure of 0.
17. Turn the power on and set the thermostat to heat.
18. Read the manifold gas pressure after the furnace lights.
19. Turn the thermostat to "OFF".
20. Close the manual gas valve in the gas line.
21. Replace the plug on the gas valve outlet test port.

Gas Pressure Measurements		
	Furnace Off	Furnace Operating
Gas Inlet Pressure		
Manifold Pressure		

LAB 51.6 SIZING GAS LINES

(This lab is done at home.)

LABORATORY OBJECTIVE
You will correctly size gas lines to meet manufacturer's specifications.

FUNDAMENTALS OF HVACR TEXT REFERENCE
Unit 51 Gas Furnace Installation

REQUIRED MATERIALS PROVIDED BY SCHOOL
Manufacturer's installation instructions for gas furnace.

PROCEDURE
The National Fire Protection Association, otherwise known as NFPA, is the most widely recognized authority on fire prevention. The sizing procedures shown here are not from any specific government code, but from the NFPA Bulletin 54 on the "Installation of Gas Appliances and Gas Piping."

The five factors used to determine gas-piping size are:

1. allowable pressure loss from the gas source to the unit
2. maximum gas consumption to be provided
3. piping length + equivalent length of all fittings and valves
4. specific gravity of the gas
5. diversity factor (more on this later)

The **maximum allowable pressure drop** is the difference in the gas pressure between the source and the appliance. This is limited by the difference between the minimum source pressure available and the minimum pressure needed at the appliance. For instance, if the gas pressure at the source is 3.5" W.C. and the unit requires 3.0" W.C. to operate, the maximum allowable pressure drop could be no more than 1/2" W.C. Generally, gas piping is sized for a maximum pressure loss of between 0.35" W.C. and 0.5" W.C.

The **maximum consumption** of the unit is the most cubic feet per hour that the unit would use. This is easily found by dividing the unit's rated input in BTU's per hour by the BTU content of the gas it is using. This is approximately 1000 BTU's per cubic foot for natural gas, 2500 BTU's per cubic foot for propane. These figures are close enough for most calculations; however, the local gas distributor can give you exact figures for the product they sell.

The **piping length** is the actual length in feet between the gas source and the gas appliance plus the equivalent length in feet of all fittings, valves, or obstructions. The tables listed here have allowance for a "normal" amount of fittings so equivalent length determination is needed only for systems with an unusual number of fittings, valves, or accessories. For our purposes, we'll assume that all systems in this section are "normal" systems.

The **diversity factor** is the ratio of maximum PROBABLE demand to maximum POSSIBLE demand. In other words, how much gas is LIKELY to be used versus how much gas COULD POSSIBLY be used. Take the case of a house with a 100,000 BTU input gas furnace and a 50,000 BTU input gas air conditioner. Since both would logically never be running at the same time, the maximum PROBABLE demand would be the higher demand of the two, the gas furnace at 100,000 BTU's. The maximum POSSIBLE demand would of course be the sum of the two, or 150,000 BTU's. In this case, the diversity factor would be 100,000 / 150,000 or 2/3. This means that instead

546

of sizing the main line supplying the system for 150,000 BTU's it would be sized for 150,000 BTU's x 2/3 or 100,000 BTU's. We will not be considering diversity factor in our examples.

To use the pipe sizing chart:

1. Determine the length of piping to the appliance that is the greatest distance from the gas source.
2. This distance will be used to size ALL piping.
3. Find a row with a length equal to or greater than that length.
4. Use that row to find a cubic foot capacity equal to or greater than that required for each section of gas pipe.
5. The column at the top gives the nominal pipe size.
6. This row is used for sizing ALL appliances and ALL sections of pipe.

Capacity in Cubic Feet of Gas for 0.6 SG with 0.5" wc pressure drop					
Nominal Pipe Size Rigid Schedule 40 Iron Pipe					
Length	1/2	3/4	1	1 1/4	11/2
10	201	403	726	1260	1900
20	138	277	499	865	1310
30	111	222	401	695	1050
40	95	190	343	594	898
50	84	169	304	527	796
60	76	153	276	477	721
70	70	140	254	439	663
80	65	131	236	409	617
90	61	123	221	383	579
100	58	116	209	362	547
125	51	103	185	321	485
150	46	93	168	291	439

Size the following systems using the specifications given.

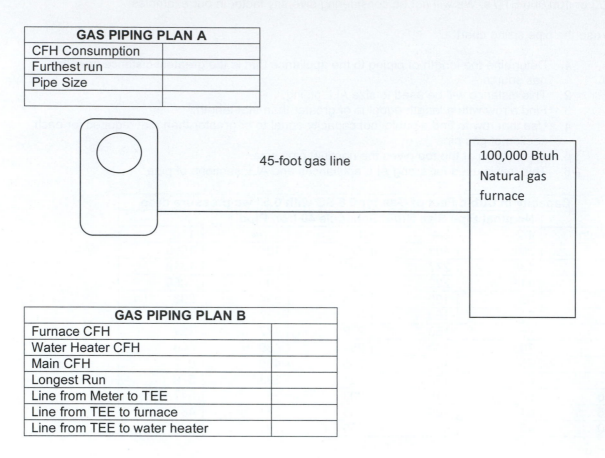

GAS PIPING PLAN A	
CFH Consumption	
Furthest run	
Pipe Size	

45-foot gas line

100,000 Btuh Natural gas furnace

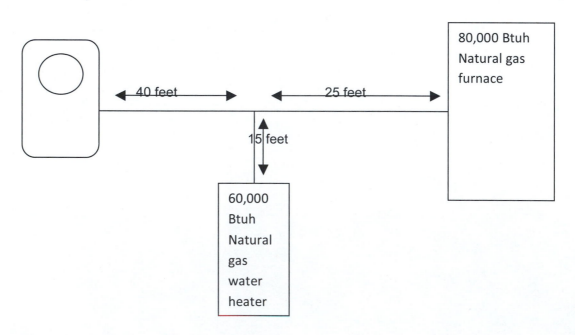

GAS PIPING PLAN B	
Furnace CFH	
Water Heater CFH	
Main CFH	
Longest Run	
Line from Meter to TEE	
Line from TEE to furnace	
Line from TEE to water heater	

40 feet

25 feet

15 feet

80,000 Btuh Natural gas furnace

60,000 Btuh Natural gas water heater

GAS PIPING PLAN C

Longest Run Section	CFH	Pipe Size
A 20 ft		
B 15 ft		
C 10 ft		
D 20 ft		
E 10 ft		
F 10 ft		
G 15 ft		
H 5 ft		
I 30 ft		
J 5 ft		
K 30 ft		

A

B   30,000 Btuh
Natural gas
stove

C

D   25,000 Btuh
Natural gas
dryer

40,000 Btuh
Natural gas
grill

K   I   E

G

J   H   F   40,000
Btuh
Natural
gas water
heater

50,000 Btuh
Natural gas
fireplace

80,000 Btuh
Natural gas
furnace

LAB 51.7 INSTALLING GAS PIPING

LABORATORY OBJECTIVE
You will size, install, and test gas piping.

FUNDAMENTALS OF HVACR TEXT REFERENCE
Unit 51 Gas Furnace Installation

REQUIRED MATERIALS PROVIDED BY STUDENT
Safety Glasses
Non-contact voltage probe
Manometer
Allen wrench set

REQUIRED MATERIALS PROVIDED BY SCHOOL
Operational gas furnace
Gas pipe
Pipe fittings
Pipe threading tools
Pipe wrenches
Pipe thread sealing compound
Soap bubbles

PROCEDURE

Use the information on the furnace data plate to size the gas line.

The gas line should include the following:

       Manual shutoff valve

       Drip leg

       Union near the furnace or flexible gas connector

Have the instructor check the piping before turning on the gas.

Remove the plug on the gas valve outlet test port .

Open the gas line valve.

Check all joints for leaks using soap bubbles or an electronic combustible gas leak detector.

Check both the gas line inlet pressure and manifold pressure with the furnace off and with the furnace operating.

Gas Pressure Measurements		
	Furnace Off	Furnace Operating
Gas Inlet Pressure		
Manifold Pressure		

LAB 51.8 INSPECTING GAS REGULATORS

LABORATORY OBJECTIVE
You will dismantle and inspect a gas regular to learn how it works.

FUNDAMENTALS OF HVACR TEXT REFERENCE
Unit 51 Gas Furnace Installation

REQUIRED MATERIALS PROVIDED BY STUDENT
Safety Glasses
Allen wrench set
Screw drivers

REQUIRED MATERIALS PROVIDED BY SCHOOL
Gas Regulator for Dismantling

PROCEDURE
Remove the plug on the bottom of the regulator.

Manually push the regulator plunger up until it is firmly seated.

Listen as you do this.

You should hear the sound of air leaving the regulator vent through the orifice.

Now release it and listen.

You should hear the sound of air entering the orifice in the regulator vent.

Disassemble the gas regulator.

Identify the diaphragm, seat, and plunger.

Note where the gas enters the regulator and where it leaves the regulator.

LAB 51.9 MEASURING DRAFT

LABORATORY OBJECTIVE
You will measure the draft pressure in a Category I furnace vent.

FUNDAMENTALS OF HVACR TEXT REFERENCE
Unit 51 Gas Furnace Installation

REQUIRED MATERIALS PROVIDED BY STUDENT
Safety Glasses
Hand tools

REQUIRED MATERIALS PROVIDED BY SCHOOL
Operating Category I Gas Furnace connected vent
Magnehelic, inclined manometer, or draft gauge

PROCEDURE
If the vent stack does not have a hole for testing vent pressure, drill a 5/16" hole in the vent.

Operate the furnace assigned by the instructor.

Insert the test probe into the hole.

The vent should have a negative pressure of at least 0.02" wc.

LAB 51.10 MEASURING TEMPERATURE RISE

LABORATORY OBJECTIVE
You will measure the temperature rise on a gas furnace.

FUNDAMENTALS OF HVACR **TEXT REFERENCE**
Unit 51 Gas Furnace Installation

REQUIRED MATERIALS PROVIDED BY STUDENT
Safety Glasses
Hand tools

REQUIRED MATERIALS PROVIDED BY SCHOOL
Operating gas furnace

PROCEDURE
Measure the return air temperature.

Measure the supply air temperature at a point that will not "see" the heat exchanger and pick up radiant heat.

Calculate the temperature rise Temp Rise = Supply air temp – return air temp

Turn the fan switch on the thermostat to "On"

This will cause an increase in fan speed on most furnaces.

Furnace Temperature Rise	
Manufacturer's Temperature Rise Specification	
Return Air Temperature	
Supply Air Temperature	
Furnace Operating Temperature rise (Supply Air Temp – Return Air Temp)	

LAB 52.1 GAS FURNACE PREVENTATIVE MAINTENANCE (PM)

LABORATORY OBJECTIVE

The purpose of this lab is to demonstrate your ability to conduct the proper preventative maintenance procedure for a gas furnace.

LABORATORY NOTES

You will need to familiarize yourself with gas furnace arrangements and types. Always remember to make the visual check of all system components prior to starting a gas furnace.

***FUNDAMENTALS OF HVACR* TEXT REFERENCE**

Unit 52 Troubleshooting Gas Furnaces

REQUIRED TOOLS AND EQUIPMENT

Operating gas furnace
Tool kit
Multimeters
Clamp-on Ammeter

SAFETY REQUIREMENTS

A. Never allow gas flow without a flame. If gas is allowed to build up and then suddenly ignite, this can create a serious hazard.

B. If the furnace contains a pilot, always ensure it is properly lit prior to starting the unit.

C. Check all circuits for voltage before doing any service work.

D. Stand on dry nonconductive surfaces when working on live circuits.

E. Never bypass any electrical protective devices.

PROCEDURE
Step 1

Collect the gas furnace data and fill in the chart.

Gas Furnace Unit Data

Furnace Make:		Model Number:	Blower Motor Amperage Rating:	
Furnace Type (circle one)	Upflow	Counterflow	Basement	Horizontal
Blower Type (circle one)	Direct drive	Belt drive		
Blower speed (circle one)	Single speed	Two speed	Three speed	Four speed
System type (circle one)	Heat only	Heat and humidify	Heat and cool	
Burner type (circle one)	Atmospheric	Induced draft	Power	

Step 2

Complete the Prestart Check List and prepare to start the furnace as outlined in the procedure provided in Lab 51.1. With the furnace operating complete the Running Check List.

Running Check List

STEP	PROCEDURE	CHECK
1	Observe a normal sequence of operation (burner on, fan on, burner off, fan off). Check for any unusual noise – bearing noise.	
2	Inspect operating pilot and note the color, size, shape, and position.	
3	Inspect burner light off for flame lifting, floating, noise and smooth ignition.	
4	Measure gas manifold pressure.	

Step 3

After finishing the Running Check List, shut down the furnace. Make sure all power to the furnace is off. Lock and tag the power panel before removing any parts. Complete the Blower Maintenance Check List.

Blower Maintenance Check List

STEP	PROCEDURE	CHECK
1	Make sure that the power to the blower motor is secured. Verify this with a multimeter voltage test.	
2	Remove the wires from the blower motor at an accessible location.	
3	Remove screws or bolts securing the blower assembly.	

4	Inspect belt for cracks and signs of wear.	
5	Inspect pulleys for wear, grooving, and alignment.	
6	Spin blower by hand and observe pulleys turn.	
7	Inspect for pulley wobble and alignment.	
8	Listen for bearing noise, drag, or movement.	
9	Inspect blower shaft for signs of wear.	
10	Clean blower and motor with air pressure, brushes, scrapers and cleaning solution as required.	
11	Reassemble blower, taking care to check pulley alignment and correct belt tension.	
12	Check the condition of the heat exchanger before re-installing the blower.	

Step 4

After finishing the Blower Maintenance Check List you can perform the necessary burner maintenance. Again, make sure all power to the furnace is off. Lock and tag the power panel before removing any parts. Complete the Gas Burner Maintenance Check List.

Gas Burner Maintenance Check List

STEP	PROCEDURE	CHECK
1	Remove and clean pilot assembly.	
2	Clean and inspect flame sensor and igniter.	
3	Remove and clean main burners, inspect and mark burners for original location. Note: All burners are not interchangeable.	
4	Remove vent connector, draft diverter, and any flue baffles.	
5	Vacuum and brush all soot, rust, and solid particles from the fire side of the heat exchanger.	
6	Insert a light into the combustion area.	
7	If possible turn off the lights in the furnace room.	
8	With the light in each burner, inspect the heat exchanger from both the fan side and the plenum side for holes.	
9	Reinstall the blower assembly after the heat exchanger check.	
10	Reinstall main and pilot burners.	

11	Reconnect vent components.	
12	Prepare to start the furnace for a running check following the start up procedure from Lab 51.1.	
13	Light pilot and adjust for proper size, configuration, and location.	
14	Turn on main burner and adjust gas pressure as required.	
15	Adjust air shutter to obtain correct flame color and CO_2.	

LAB 52.2 HOT SURFACE IGNITION OPERATION

LABORATORY OBJECTIVE

You will observe the operating sequence of a hot surface ignition system and simulate common failure symptoms.

FUNDAMENTALS OF HVACR TEXT REFERENCE

Unit 52 Troubleshooting Gas Furnaces

REQUIRED MATERIALS PROVIDED BY STUDENT

Safety Glasses

Electrical tools

REQUIRED MATERIALS PROVIDED BY SCHOOL

Operating Gas Furnace with a hot surface ignition system

PROCEDURE

1. Study the wiring diagram of a furnace using a hot surface ignition system
2. Operate the furnace assigned by the instructor.
3. Does this system use a pre-purge cycle? If so, how long is it?
4. How long did the igniter operate?
5. Cycle the system off.
6. Does this system have a post purge cycle? If so, how long is it?
7. Turn the gas valve to the "OFF" position and call for heat once more.
8. How long does an ignition trial take?
9. How many ignition trials does the system have before lockout?
10. Turn the gas valve back to the "ON" position.
11. Reset the control by killing the call for heat and re-establishing it.
12. After flame is proved, turn the gas valve back to the "OFF" position.
13. Does the system attempt to re-establish combustion?
14. How many trial periods does it go through before it locks out?
15. Does it use a purge cycle in between the trials?
16. How long are they?
17. Turn the unit off and return the gas valve to the "ON" position.
18. Give a complete description of the operation cycle of this control.

LAB 52.3 TROUBLESHOOTING HOT SURFACE IGNITION SYSTEMS

LABORATORY OBJECTIVE

Given a hot surface ignition gas furnace with an ignition problem, you will identify the problem, its root cause, and recommend corrective action.

FUNDAMENTALS OF HVACR TEXT REFERENCE

Unit 52 Troubleshooting Gas Furnaces

REQUIRED MATERIALS PROVIDED BY STUDENT

Safety Glasses
Hand tools
Manometer
Multimeter

REQUIRED MATERIALS PROVIDED BY SCHOOL

Hot surface ignition gas furnace with a problem

PROCEDURE

Troubleshoot the hot surface ignition gas furnace assigned by the instructor. Be sure to be complete in your description of the problem. You should include:

What is the furnace is doing wrong?

What component or condition is causing this?

What tests did you perform that told you this?

How would you correct this?

LAB 52.4 VENT PROBLEMS

LABORATORY OBJECTIVE
Given a gas furnace with a venting problem, you will identify the problem, its root cause, and recommend corrective action.

FUNDAMENTALS OF HVACR TEXT REFERENCE
Unit 51 Gas Furnace Installation / Unit 52 Troubleshooting Gas Furnaces

REQUIRED MATERIALS PROVIDED BY STUDENT
Safety Glasses
Hand tools
Manometer
Multimeter

REQUIRED MATERIALS PROVIDED BY SCHOOL
DSI Gas furnace with a problem

PROCEDURE
Troubleshoot the DSI gas furnace assigned by the instructor. Be sure to be complete in your description of the problem. You should include:

What is the furnace is doing wrong?

What component or condition is causing this?

What tests did you perform that told you this?

How would you correct this?

LAB 52.5 INSPECTING HEAT EXCHANGERS

LABORATORY OBJECTIVE
You will learn to identify defects in furnace heat exchangers.

FUNDAMENTALS OF HVACR TEXT REFERENCE
Unit 52 Troubleshooting Gas Furnaces

REQUIRED MATERIALS PROVIDED BY STUDENT
Safety Glasses
Hand tools

REQUIRED MATERIALS PROVIDED BY SCHOOL
Gas Furnace

PROCEDURE
Inspect the heat exchanger(s) assigned by the instructor. Note any cracks, holes, or other aberrations. Also remember to look for heat exchangers, which are sooted up and clean as necessary. Different inspection techniques are listed below.

Flamedance
Normally, the heat exchanger keeps the combustion gas completely separate from the air passing over the heat exchanger; however, a large enough crack can let air blow into the heat exchanger and blow the flames around. If you see a lot of flame movement when the fan comes on, there could be a hole in the heat exchanger. This is especially true if the flames blow in just one of the sections.

Fans Eye View
Probably the best overall view of the heat exchanger is looking at it from the blower's point of view. To do this, you need to remove the blower. Removing the blower is not as difficult as it sounds because most furnace blowers slide in and out and are secured by a handful of screws. Be sure to turn power off and disconnect the wiring to the blower BEFORE removing it. After the blower has been removed, look up into the heat exchanger for cracks and holes. Often, you will need a flashlight and an inspection mirror to really see up inside the heat exchanger. Remember that the unit is still hot, DON'T TOUCH THE SIDES OF THE HEAT EXCHANGER. If you smell burning flesh, remove your hands from the furnace. Pay particular attention to bends and crimps in the heat exchanger because these receive the greatest thermal stress. Also look for rusty areas.

Borescope
If you are still uncertain, the next step is to remove the burners so that you can get a better look inside the heat exchanger. Remove the burners to gain clear access to the heat exchanger. Use a borescope to go up inside each section or tube looking for cracks , discolored metal, or excessive rust. A large amount of rust on top of the burners and piled up around the bottom of the heat exchanger could mean that it has rusted through. Also, notice

any visible differences between the sections. On natural draft furnaces you can also go in through the draft hood or draft diverter.

Smoke Bomb

Finally, there is one sure-fire check: a smoke bomb. With the unit operating, put a smoke bomb in the bottom of the heat exchanger. If there are cracks or holes, smoke will be released into the ductwork and into the house. This can be somewhat unsettling to the homeowners, so it is usually reserved for last. Having smoke pour out of the heating registers produces high levels of anxiety among finicky customers but it does yield a rather positive diagnosis. After you have filled the customer's house with smoke, inform them that that could be carbon monoxide and they could not wake up morning. Then suggest that they contact the sales office because you are disabling their furnaces so that it won't work. Finally, tell them to have a nice day.

90% Furnaces

Suppose when you pulled the blower out all you saw was a coil? You must be checking a high efficiency furnace which uses a condensing recuperative coil. In this case, the only thing a visual inspection from underneath can possibly show would be possibly a leak in the recuperative coil. If you see water dripping, it could be a leaky recuperative coil. BE SURE that it is not simply a leaky drain. It would be quite embarrassing to condemn a furnace because of a leaky drain.

Draft Pressure Check

On many high efficiency furnaces the heat exchanger is too narrow to see up inside, the underside is hidden by the recuperative coil, and the flames may not be readily visible. What then? Watching for changes in the vent draft pressure of a Category I furnace can also indicate a leak in the heat exchanger. Category I furnaces operate with a negative draft pressure. A leak in the heat exchanger can allow air from the indoor blower to enter the heat exchanger, causing an increase in the pressure inside the heat exchanger and vent. If the pressure in the vent changes from a negative pressure to a positive pressure after the indoor blower starts, there is a leak in the heat exchanger.

LAB 52.6 GAS COMBUSTION EFFICIENCY

LABORATORY OBJECTIVE

You will measure the combustion efficiency of an operating gas furnace.

FUNDAMENTALS OF HVACR TEXT REFERENCE

Unit 52 Troubleshooting Gas Furnaces

REQUIRED MATERIALS PROVIDED BY STUDENT

Safety Glasses

Hand tools

REQUIRED MATERIALS PROVIDED BY SCHOOL

Operating gas furnace

Combustion Efficiency Analyzer

PROCEDURE

If there is not already a hole in the vent, drill a hole large enough to inset the probe of the combustion analyzer.

Turn on the furnace and wait for the fan to come on.

Using Traditional Hourglass Combustion Analyzers…

Measure the temperature of the air around the furnace.

Measure the stack temperature.

Calculate the net stack temperature = stack temperature – ambient temperature

Measure the CO2% or the O2%.

Line up the O2% and net stack temperature on the efficiency calculator to get combustion efficiency.

Using Electronic Analyzers

Insert probe into hole and follow manufacturer's instructions.

	Natural Draft	80% Induced Draft	90% Induced Draft
Fuel			
Stack Temperature			
Ambient Temperature			
Net Stack Temperature			
CO2 %			
O2%			
CO ppm free air			
Efficiency			

LAB 52.7 TROUBLESHOOTING GAS FURNACE SCENARIO 1

LABORATORY OBJECTIVE

Given a gas furnace with a problem, you will identify the problem, its root cause, and recommend corrective action. Labs 52.7 through 52.12 will all follow the same procedure, however each lab will present a different problem scenario based upon the type of furnace and corrective action required.

FUNDAMENTALS OF HVACR TEXT REFERENCE

Unit 52 Troubleshooting Gas Furnaces

REQUIRED MATERIALS PROVIDED BY STUDENT

Safety Glasses
Hand tools
Manometer
Multimeter

REQUIRED MATERIALS PROVIDED BY SCHOOL

Gas furnace with a problem

PROCEDURE

Troubleshoot the gas furnace assigned by the instructor. Be sure to be complete in your description of the problem. You should include:

What is the furnace is doing wrong?

What component or condition is causing this?

What tests did you perform that told you this?

How would you correct this?

LAB 52.8 TROUBLESHOOTING GAS FURNACE SCENARIO 2

LABORATORY OBJECTIVE
Given a gas furnace with a problem, you will identify the problem, its root cause, and recommend corrective action. Labs 52.7 through 52.12 will all follow the same procedure, however each lab will present a different problem scenario based upon the type of furnace and corrective action required.

FUNDAMENTALS OF HVACR TEXT REFERENCE
Unit 52 Troubleshooting Gas Furnaces

REQUIRED MATERIALS PROVIDED BY STUDENT
Safety Glasses
Hand tools
Manometer
Multimeter

REQUIRED MATERIALS PROVIDED BY SCHOOL
Gas furnace with a problem

PROCEDURE
Troubleshoot the gas furnace assigned by the instructor. Be sure to be complete in your description of the problem. You should include

What is the furnace is doing wrong?

What component or condition is causing this?

What tests did you perform that told you this?

How would you correct this?

LAB 52.9 TROUBLESHOOTING GAS FURNACE SCENARIO 3

LABORATORY OBJECTIVE

Given a gas furnace with a problem, you will identify the problem, its root cause, and recommend corrective action. Labs 52.7 through 52.12 will all follow the same procedure, however each lab will present a different problem scenario based upon the type of furnace and corrective action required.

FUNDAMENTALS OF HVACR TEXT REFERENCE

Unit 52 Troubleshooting Gas Furnaces

REQUIRED MATERIALS PROVIDED BY STUDENT

Safety Glasses
Hand tools
Manometer
Multimeter

REQUIRED MATERIALS PROVIDED BY SCHOOL

Gas furnace with a problem

PROCEDURE

Troubleshoot the gas furnace assigned by the instructor. Be sure to be complete in your description of the problem. You should include:

What is the furnace is doing wrong?

What component or condition is causing this?

What tests did you perform that told you this?

How would you correct this?

LAB 52.10 TROUBLESHOOTING GAS FURNACE SCENARIO 4

LABORATORY OBJECTIVE

Given a gas furnace with a problem, you will identify the problem, its root cause, and recommend corrective action. Labs 52.7 through 52.12 will all follow the same procedure, however each lab will present a different problem scenario based upon the type of furnace and corrective action required.

FUNDAMENTALS OF HVACR TEXT REFERENCE

Unit 52 Troubleshooting Gas Furnaces

REQUIRED MATERIALS PROVIDED BY STUDENT

Safety Glasses

Hand tools

Manometer

Multimeter

REQUIRED MATERIALS PROVIDED BY SCHOOL

Gas furnace with a problem

PROCEDURE

Troubleshoot the gas furnace assigned by the instructor. Be sure to be complete in your description of the problem. You should include:

What is the furnace is doing wrong?

What component or condition is causing this?

What tests did you perform that told you this?

How would you correct this?

LAB 52.11 TROUBLESHOOTING GAS FURNACE SCENARIO 5

LABORATORY OBJECTIVE

Given a gas furnace with a problem, you will identify the problem, its root cause, and recommend corrective action. Labs 52.7 through 52.12 will all follow the same procedure, however each lab will present a different problem scenario based upon the type of furnace and corrective action required.

FUNDAMENTALS OF HVACR TEXT REFERENCE

Unit 52 Troubleshooting Gas Furnaces

REQUIRED MATERIALS PROVIDED BY STUDENT

Safety Glasses
Hand tools
Manometer
Multimeter

REQUIRED MATERIALS PROVIDED BY SCHOOL

Gas furnace with a problem

PROCEDURE

Troubleshoot the gas furnace assigned by the instructor. Be sure to be complete in your description of the problem. You should include:

What is the furnace is doing wrong?

What component or condition is causing this?

What tests did you perform that told you this?

How would you correct this?

LAB 52.12 TROUBLESHOOTING GAS FURNACE SCENARIO 6

LABORATORY OBJECTIVE
Given a gas furnace with a problem, you will identify the problem, its root cause, and recommend corrective action. Labs 52.7 through 52.12 will all follow the same procedure, however each lab will present a different problem scenario based upon the type of furnace and corrective action required.

FUNDAMENTALS OF HVACR TEXT REFERENCE
Unit 52 Troubleshooting Gas Furnaces

REQUIRED MATERIALS PROVIDED BY STUDENT
Safety Glasses
Hand tools
Manometer
Multimeter

REQUIRED MATERIALS PROVIDED BY SCHOOL
Gas furnace with a problem

PROCEDURE
Troubleshoot the gas furnace assigned by the instructor. Be sure to be complete in your description of the problem. You should include:

What is the furnace is doing wrong?

What component or condition is causing this?

What tests did you perform that told you this?

How would you correct this?

573

LAB 54.1 OIL BURNER TUNE-UP

LABORATORY OBJECTIVE
The purpose of this lab is to demonstrate your ability to conduct the proper burner tune-up for an oil fired furnace.

LABORATORY NOTES
You will need to familiarize yourself with oil furnace arrangements and types. Always remember to make the visual check of all system components prior to starting an oil furnace.

FUNDAMENTALS OF HVACR TEXT REFERENCE
Unit 54 Oil Furnace and Boiler Service

REQUIRED TOOLS AND EQUIPMENT
Operating oil furnace
Tool kit
Multimeters
Clamp-on Ammeter

SAFETY REQUIREMENTS
A. Never allow fuel oil to flow into the combustion chamber without a flame. If fuel oil is allowed to build up and then suddenly ignite, this can create a serious hazard.

B. Check all circuits for voltage before doing any service work.

C. Stand on dry nonconductive surfaces when working on live circuits.

D. Never bypass any electrical protective devices.

PROCEDURE
Step 1

Collect the oil fired furnace data and fill in the chart.

Oil Fired Furnace Unit Data

Furnace Make:		Model Number:	Blower Motor Amperage Rating:	
Furnace Type **(circle one)**	Upflow	Counterflow	Basement	Horizontal
Burner Make:			**Burner Model Number:**	
Recommended **Nozzle Data**	Gallons per Hour GPH:		Spray Pattern:	Spray Angle:

Step 2

Complete the Prestart Check List and the Start-Up Check List as outlined in the procedure provided in Lab 55.1. After finishing the Start-Up Check List, shut down the furnace. Make sure all power to the furnace is off. Lock and tag the power panel before removing any parts. Complete the Oil Burner Inspection And Tune-Up Check List.

Oil Burner Inspection And Tune-Up Check List

STEP	PROCEDURE	CHECK
1	Make sure all power to the furnace is off. Lock and tag the power panel before removing any parts.	
2	Open or remove transformer.	

3	Remove nozzle assembly.		
4	Remove and clean electrodes.		
5	Remove nozzle and record nozzle data. Make _____ GPH _____ Angle _____ Pattern _____		
6	Replace the nozzle with a new nozzle that matches the data specifications.		
7	Install and adjust electrodes.		
8	Position electrodes to manufacturer recommended setting. (Typically these dimensions are approximately 1/2 inches above, 1/16 inches forward, and 1/8 inches apart.)		
9	Remove and clean cad cell.		
10	Remove and record resistance of cad cell while the face of the cad cell is covered. Resistance = _____ Ohms		

11	Remove and record resistance of cad cell while the face of the cad cell exposed to room light. Resistance = _____ Ohms	
12	Install nozzle assembly.	
13	Slide nozzle assembly to midpoint of forward/backward adjustment.	
14	Swing up or reinstall ignition transformer.	
15	Make sure that the transformer contact springs touch electrodes as the transformer is positioned.	
16	Secure transformer in place with at least one screw for testing.	
17	Turn on power to the furnace and burner and observe ignition. DO not get too close as oil burners can puff back.	
18	Observe a normal continuous burn for one minute.	
19	Adjust the air shutter back and forth slowly. Close to see smoke and then open until smoky flame disappears.	
20	Slide nozzle assembly forward and back until the flame is quiet and no longer jagged and smoky.	
21	Refer to Lab 54.2 for final burner adjustments.	

LAB 54.2 FINAL OIL BURNER ADJUSTMENT

LABORATORY OBJECTIVE

The purpose of this lab is to demonstrate your ability to conduct the proper final burner adjustments after a tune-up for an oil fired furnace.

LABORATORY NOTES

You will need to familiarize yourself with oil furnace arrangements and types. Always remember to make the visual check of all system components prior to starting an oil furnace.

***FUNDAMENTALS OF HVACR* TEXT REFERENCE**

Unit 54 Oil Furnace and Boiler Service

REQUIRED TOOLS AND EQUIPMENT

Operating oil furnace
Tool kit
Draft gauge
Smoke gun
CO Analyzer

SAFETY REQUIREMENTS

A. Never allow fuel oil to flow into the combustion chamber without a flame. If fuel oil is allowed to build up and then suddenly ignite, this can create a serious hazard.

B. Check all circuits for voltage before doing any service work.

C. Stand on dry nonconductive surfaces when working on live circuits.

D. Never bypass any electrical protective devices.

PROCEDURE

Step 1

Complete the Prestart Check List and the Start-Up Check List as outlined in the procedure provided in Lab 55.1. Complete the Oil Burner Inspection and Start-Up Check List as outlined in Lab 54.1. Obtain a draft gauge, smoke gun, and CO analyzer to perform final settings on the oil burner.

Final Burner Adjustment Check List

STEP	PROCEDURE	CHECK
1	With the burner operating, use the draft gauge and adjust the barometric damper to obtain a -0.1 over fire draft.	

578

2	Measure and record the draft at breach of the burner. A restriction of greater than -0.4 through the heat exchanger indicates the heat exchanger may be plugged with soot and needs to be cleaned. Final measured draft at breach _____	
3	If necessary inspect and clean the fire side of the heat exchanger. Make sure all power to the furnace is off. Lock and tag the power panel before removing any parts.	
4	Once cleaning is complete, restart the burner and perform a smoke spot test with the smoke gun. The reading should be a #0.	
5	Install the CO meter into the breach of the furnace.	
6	Readjust the air shutter to obtain 10% CO_2.	
7	Turn the burner on and off, observing several ignitions.	
8	Verify a quick clean quiet ignition.	
9	Remove and put away all test equipment and tools Install all panels and doors.	

LAB 54.3 OIL FURNACE PREVENTATIVE MAINTENANCE

LABORATORY OBJECTIVE

The purpose of this lab is to demonstrate your ability to conduct the proper preventative maintenance for an oil fired furnace.

LABORATORY NOTES

You will need to familiarize yourself with oil furnace arrangements and types. Always remember to make the visual check of all system components prior to starting an oil furnace.

***FUNDAMENTALS OF HVACR* TEXT REFERENCE**

Unit 54 Oil Furnace and Boiler Service

REQUIRED TOOLS AND EQUIPMENT

Operating oil furnace
Tool kit
Multimeters
Clamp-on Ammeter
Oil pressure gauge
Vacuum

SAFETY REQUIREMENTS

A. Never allow fuel oil to flow into the combustion chamber without a flame. If fuel oil is allowed to build up and then suddenly ignite, this can create a serious hazard.

B. Check all circuits for voltage before doing any service work.

C. Stand on dry nonconductive surfaces when working on live circuits.

D. Never bypass any electrical protective devices.

PROCEDURE
Step 1

Collect the oil fired furnace data and fill in the chart.

Oil Fired Furnace Unit Data

Furnace Make:		Model Number:	Blower Motor Amperage Rating:	
Furnace Type **(circle one)**	Upflow	Counterflow	Basement	Horizontal

Burner Make:		Oil Pressure:	Burner Model Number:	
Recommended Nozzle Data	Gallons per Hour GPH:		Spray Pattern:	Spray Angle:

Step 2

Complete the Prestart Check List and the Start-Up Check List as outlined in the procedure provided in Lab 55.1. After finishing the Start-Up Check List, shut down the furnace. Make sure all power to the furnace is off. Lock and tag the power panel before removing any parts. Complete the Blower Maintenance Check List.

Blower Maintenance Check List

STEP	PROCEDURE	CHECK
1	Make sure that the power to the blower motor is secured. Verify this with a multimeter voltage test.	
2	Remove the wires from the blower motor at an accessible location.	
3	Remove screws or bolts securing the blower assembly.	
4	Inspect belt for cracks and signs of wear.	
5	Inspect pulleys for wear, grooving, and alignment.	

6	Spin blower by hand and observe pulleys turn.	
7	Inspect for pulley wobble and alignment.	
8	Listen for bearing noise, drag, or movement.	
9	Inspect blower shaft for signs of wear.	
10	Clean blower and motor with air pressure, brushes, scrapers and cleaning solution as required.	
11	Oil motor and blower bearings as required.	
12	Reassemble blower, taking care to check pulley alignment and correct belt tension.	
13	Check the condition of the heat exchanger before re-installing the blower.	

Step 3

After finishing the Blower Maintenance Check List you can perform the necessary oil burner maintenance. Again, make sure all power to the furnace is off. Lock and tag the power panel before removing any parts. Complete the Oil Burner Maintenance Check List.

Oil Burner Maintenance Check List

STEP	PROCEDURE	CHECK
1	Make sure all power to the furnace is off. Lock and tag the power panel before removing any parts.	
2	Remove and clean nozzle assembly.	
3	Replace nozzle with a manufacturer recommended nozzle.	
4	Place the old nozzle into a small zipper bag with the new nozzle box. Write your name on the bag, date it, and leave it with the furnace. The bag will keep it from smelling and then on the next service visit, it will be evident what was put in and what was taken out.	
5	Position electrodes to manufacturer recommended setting. (Typically these dimensions are approximately 1/2 inches above, 1/16 inches forward, and 1/8 inches apart.)	
6	Remove the vent connector, barometric damper, and flue baffles or cleanout plugs.	
7	Remove or swing out burner assembly.	
8	Vacuum and brush all soot, rust, and solid particles from the fire side of the heat exchanger. Be careful not to damage the combustion chamber refractory.	

9	Insert a light into the combustion chamber and if possible turn off the lights in the furnace room.	
10	With the light in as far back as possible, inspect the heat exchanger from the fan side and the plenum side for holes, light, or leaky gaskets.	
11	After inspecting the heat exchanger reinstall the blower.	
12	Reinstall the nozzle assembly while the burner is still swung away from the furnace.	
13	After the burner assembly is back in place the fuel oil system can be checked. Disconnect the supply line from the pump to the burner assembly and install an oil pressure gauge on the pump supply line.	
14	Remove the lockout tag and turn on the burner and check the initial fuel oil pressure. Fuel Oil Pressure = _____	
15	Adjust to the manufacturer recommended fuel oil pressure (typically 100 psig). Fuel Oil Pressure = _____	

16	Observe the burner turn off due to flame failure.	
17	Observe oil pressure gauge holding 85 psig quick cut-off pressure.	
18	Reinstall the original oil supply line to the nozzle assembly.	
19	Check the fan control settings. The fan should turn on at a supply air temperature of approximately 90°F and off at 135 °F.	
20	The maximum temperature limit setting can be checked by running the burner with the fan off and at approximately 200 °F, the burner should cycle off.	
21	After all maintenance is complete, turn the burner on and off, observing several ignitions. Verify a quick clean quiet ignition. Remove and put away all test equipment and tools Install all panels and doors.	

LAB 54.4 INSTALL A REPLACEMENT FUEL OIL PUMP

LABORATORY OBJECTIVE
The purpose of this lab is to demonstrate your ability to install a replacement fuel oil pump for an oil fired furnace.

LABORATORY NOTES
The pump on an oil burner is one of the most important parts of the burner assembly. Its job is to pull the fuel oil from the tank and deliver it to the burner nozzle at the recommended supply pressure (typically 100 or 140 psig). Always check the burner nameplate to verify the manufacturer recommended fuel oil supply pressure. Minor adjustments can be made, however if the ump is worn, it must be replaced.

There are dozens of different pumps available. This is because there is more than one pump manufacturer, several major burner manufacturers, and different types of burner configurations. Not every wholesale house has every pump. To find a replacement pump, begin by contacting the dealer of the furnace type of burner that you are working on. You will need the furnace and burner model numbers, and serial numbers.

FUNDAMENTALS OF HVACR TEXT REFERENCE
Unit 54 Oil Furnace and Boiler Service

REQUIRED TOOLS AND EQUIPMENT
Oil furnace tank and operating oil furnace
Tool kit
Oil adsorbent pads
Large empty drip pan

SAFETY REQUIREMENTS:
A. It is good practice to have oil absorbent pads and a large drip pan available to collect any minor oil spills that may occur.

PROCEDURE
Step 1

Collect the oil fired furnace data and fill in the chart.

Oil Fired Furnace Unit Data

Furnace Make:	Model Number:	Serial Number:
Burner Make:	Model Number:	Serial Number:
Pump Make:	Model Number:	Serial Number:

Step 2

Complete the System Inspection Check List and make sure to check each step off in the appropriate box as you finish it. This will help you to keep track of your progress.

System Inspection Check List

STEP	PROCEDURE	CHECK
1	Identify the current fuel oil piping system. (circle one) One Pipe Two Pipe	
2	Measure the lift from the lowest possible operating oil level to the burner pump fitting. Lift = _____feet	
3	If the lift is greater than 2 feet on a one pipe system, we will need to convert to a two pipe system (see Lab 55.3).	

Step 3

After completing the System Inspection Check List you can prepare to replace the fuel oil pump. Complete the Fuel Pump Replacement Check List and make sure to check each step off in the appropriate box as you finish it.

Fuel Pump Replacement Check List

STEP	PROCEDURE	CHECK
1	Make sure all power to the furnace is off. Lock and tag the power panel before removing any parts.	
2	Close the tank shut off valve.	
3	Disconnect the existing oil line or lines. Place an adsorbent pad large empty drip pan beneath the connection to catch any dripping fuel oil.	
4	Loosen and remove the bolts holding the fuel oil pump in position.	
5	Change fittings to the new pump (change to a two pipe system if required – see Lab 55.3).	
6	Bolt new pump into position.	
7	After the replacement fuel oil pump has been installed, the tank shut off valve may be opened, and the power restored to the furnace.	
8	On a two pipe system, the burner should start and the fuel line should self-bleed itself of air with the bypass oil returning to the fuel tank.	

9	On a one pipe system, after the replacement pump has been installed, the fuel line from the tank to the oil pump must be purged of air.	
10	Locate the purge valve on the fuel oil pump located on the furnace burner.	
11	Place the drip pan beneath the oil pump purge valve.	
12	Start the burner and slowly open the oil pump purge valve and fuel oil will squirt out to collect into the drip pan. Allow the oil to flow from the purge valve until all of the air has been purged from the line, then close the purge valve.	
13	If air remains in the fuel line, the burner will not light normally and go out.	
14	After purging the air from the oil line, turn the burner on and off, observing several ignitions. Verify a quick clean quiet ignition.	
15	All fuel removed along with the old filter element, adsorbent pads, and any rags, must be removed and disposed of properly. Never leave any of these at the job site.	

LAB 55.1 OIL FIRED FURNACE STARTUP

LABORATORY OBJECTIVE

The purpose of this lab is to demonstrate your ability to go through the necessary sequence for a typical oil furnace startup.

LABORATORY NOTES

You will need to familiarize yourself with oil furnace arrangements and types. Always remember to make the visual check of all system components prior to starting an oil furnace.

FUNDAMENTALS OF HVACR TEXT REFERENCE

Unit 55 Residential Oil Heating Installation

REQUIRED TOOLS AND EQUIPMENT

Operating forced hot air oil furnace
Tool kit
Multimeters
Clamp-on Ammeter

SAFETY REQUIREMENTS

A. Never allow fuel oil to flow into the combustion chamber without a flame. If fuel oil is allowed to build up and then suddenly ignite, this can create a serious hazard.

B. Check all circuits for voltage before doing any service work.

C. Stand on dry nonconductive surfaces when working on live circuits.

D. Never bypass any electrical protective devices.

PROCEDURE
Step 1
Collect the oil furnace data and fill in the chart.

Oil Furnace Unit Data

Furnace Make:		Model Number:	Blower Motor Amperage Rating:	
Furnace Type (circle one)	Upflow	Counterflow	Basement	Horizontal
Blower Type (circle one)	Direct drive	Belt drive		

Blower speed (circle one)	Single speed	Two speed	Three speed	Four speed

Step 2

Complete the Prestart Check List and make sure to check each step off in the appropriate box as you finish it. This will help you to keep track of your progress.

Prestart Check List

STEP	PROCEDURE	CHECK
1	Turn thermostat down below the room temperature.	
2	Make sure all power to the furnace is off. Lock and tag the power panel before removing any parts.	
3	Vent connector connected with three screws per joint.	
4	Fuel line installed properly, with no apparent leaks.	
5	Spin all fans to be sure they are loose and turn freely.	
6	Combustion area free from debris.	
7	Electrical connections complete – main power.	
8	Electrical connections complete – thermostat.	
9	All doors and panels available and in place.	

STEP	PROCEDURE	CHECK
10	Thermostat installed and operating correctly.	
11	Fuel oil present in tank.	
12	Barometric damper installed and swinging freely.	

Step 3

After finishing the Prestart Check List you are now ready to start the furnace and check its operation. You will follow the same procedure of working through a check list. This will help you to keep track of your progress. Complete the Start-Up Check List and make sure to check each step off in the appropriate box as you finish it.

Start-Up Check List

STEP	PROCEDURE	CHECK
1	Check the main fuse or breaker and the amperage rating and make sure it is the correct rating for the furnace.	
2	Measure the incoming supply voltage.	
3	If the Steps 1 & 2 have been completed and are satisfactory, then turn on the supply power for the furnace.	
4	Set the thermostat to – Fan On - to obtain the fan only operation and observe that the blower is operating. If the thermostat is not so equipped then consult with your instructor.	

5	Turn the fan off and the thermostat to heat.	
6	Adjust the thermostat setting to 10 ºF above the room temperature.	
7	Observe the burner come on.	
8	Observe the fan come on after a reasonable warm-up time.	
9	Use a clamp-on ammeter to measure the blower motor amperage and compare this value to its rating.	
10	Turn down the thermostat and observe the burner shut off, and then fan shut off, in that order.	
11	Turn on and off for several ignitions.	
12	The flame should be orange/white in color, uniform in shape and within the combustion chamber, quiet, quick in ignition and extinction, and no drip from the burner tip.	

LAB 55.2 OIL FURNACE STORAGE TANK MAINTENANCE

LABORATORY OBJECTIVE

The purpose of this lab is to demonstrate your ability to conduct the proper preventative maintenance for an oil fired furnace oil storage tank.

LABORATORY NOTES

You will need to familiarize yourself with oil furnace arrangements and types. Always remember to make the visual check of all system components prior to starting an oil furnace.

***FUNDAMENTALS OF HVACR* TEXT REFERENCE**

Unit 55 Residential Oil Heating Installation

REQUIRED TOOLS AND EQUIPMENT

Oil furnace tank and operating oil furnace
Tool kit
Oil adsorbent pads
Large empty drip pan

SAFETY REQUIREMENTS

A. A full oil storage tank will normally hold 250 gallons of fuel oil or more. Always make sure the shutoff valve is closed before changing the tank filter to prevent the possibility of a large oil spill.

B. It is good practice to have oil absorbent pads and a large drip pan available to collect any minor oil spills that may occur when changing oil tank filters.

PROCEDURE

Step 1

Complete the Oil Furnace Storage Tank Maintenance Check List and make sure to check each step off in the appropriate box as you finish it. This will help you to keep track of your progress.

Oil Furnace Storage Tank Maintenance Check List

STEP	PROCEDURE	CHECK
1	Make sure all power to the furnace is off. Lock and tag the power panel before removing any parts.	
2	Visually check the oil storage tank for leaks. Older tanks will rust from the inside out and leaks may develop.	

3	Close the tank shut off valve located between the tank and the filter assembly.	
4	Place an adsorbent pad beneath the filter assembly and place the large empty drip pan beneath the filter assembly to catch any dripping fuel oil.	
5	Carefully remove the filter assembly and check for water. Water is heavier than the fuel oil and will settle to the bottom of the fuel tank.	
6	If water is present, additional fuel may need to be drained from the tank.	
7	All fuel removed along with the old filter element, adsorbent pads, and any rags, must be removed and disposed of properly. Never leave any of these at the job site.	
8	Always replace the filter assembly gaskets with new ones.	
9	After the new filter assembly is in place and tight, the fuel tank shut off valve can be opened and the air bled through the vent screw located on the top of the filter assembly. Vent air from the assembly until fuel comes from the vent screw. Catch any dripping fuel in the drip pan.	
10	Wipe the filter assembly dry and check for any fuel leaks.	
11	After the new filter as been installed, the fuel line from the tank to the oil pump must be purged of air.	

12	Locate the purge valve on the fuel oil pump located on the furnace burner.		
13	Place the drip pan beneath the oil pump purge valve.		
14	Remove the lock on the furnace power and start the burner. Slowly open the oil pump purge valve and fuel oil will squirt out to collect into the drip pan. Allow the oil to flow from the purge valve until all of the air has been purged from the line, then close the purge valve.		
15	If air remains in the fuel line, the burner will not light normally and go out.		
16	After purging the air from the oil line, turn the burner on and off, observing several ignitions. Verify a quick clean quiet ignition.		
17	All fuel removed along with the old filter element, adsorbent pads, and any rags, must be removed and disposed of properly. Never leave any of these at the job site.		

LAB 55.3 OIL FURNACE TWO PIPE CONVERSION

LABORATORY OBJECTIVE
The purpose of this lab is to demonstrate your ability to convert a one pipe oil system to a two pipe oil system for an oil fired furnace.

LABORATORY NOTES
A one pipe oil system refers to the oil line from the tank to the burner being a single pipe, usually 3/8 inch OD copper. This system is recommended for no more than 2 feet of vertical lift from the oil level in the tank to the burner pump. Any time the tank runs out of oil the air must be manually bled from the line at the pump.

A two pipe conversion involves running a second line back to the tank and installing a bypass plug within the fuel oil pump. The second line at the tank needs to go all the way to the bottom of the tank. If oil entered the top of the tank and fell to the bottom, you would hear the oil fall and splash. Frequently a tank duplex fitting is used on two pipe systems. The fitting is installed in the top of the tank and has two fittings in it that will allow a 3/8 inch line to be pushed through to the bottom of the tank and then pulled up 3 to 4 inches. This is the correct position to install both the supply line and the return line. This reduces the chance of pulling sludge or water off the bottom of the tank.

FUNDAMENTALS OF HVACR TEXT REFERENCE
Unit 55 Residential Oil Heating Installation

REQUIRED TOOLS AND EQUIPMENT
Oil furnace tank and operating oil furnace
Tool kit
Oil adsorbent pads
Large empty drip pan
Tank duplex fitting

SAFETY REQUIREMENTS
A. A full oil storage tank will normally hold 250 gallons of fuel oil or more. Always make sure the shutoff valve is closed to prevent the possibility of a large oil spill.

B. It is good practice to have oil absorbent pads and a large drip pan available to collect any minor oil spills that may occur.

PROCEDURE
Step 1

Complete the System Inspection Check List and make sure to check each step off in the appropriate box as you finish it. This will help you to keep track of your progress.

STEP	PROCEDURE	CHECK
1	Identify the current fuel oil piping system. (circle one) One Pipe Two Pipe	
2	Measure the lift from the lowest possible operating oil level to the burner pump fitting Lift = _____feet	
3	Is there a bypass plug at the pump? Every new oil burner comes with a bypass plug in a cloth bag generally attached to the pump with string. Is it still there? (circle one) Yes No	
4	If yes, you have the plug you need. If no, you will need to get one.	
5	Obtain a bypass plug as required.	
6	Obtain a sufficient length of 3/8 inch OD copper tubing and the required brass fittings to make connection at the pump.	

Step 2

After completing the System Inspection Check List you can prepare to convert to a two pipe system. Complete the Two Pipe System Conversion Check List and make sure to check each step off in the appropriate box as you finish it.

Two Pipe System Conversion Check List

STEP	PROCEDURE	CHECK
1	Make sure all power to the furnace is off. Lock and tag the power panel before removing any parts.	
2	Close the tank shut off valve.	
3	Disconnect the existing one line pipe. Place an adsorbent pad large empty drip pan beneath the connection to catch any dripping fuel oil.	
4	Loosen and remove the bolts holding the fuel oil pump in position.	
5	Inspect the pump fitting opening for the return line to the tank (pumps are generally labeled for openings and plug location).	
6	Inspect the pump for the location of the bypass plug. This is generally a 1/8 inch or 1/16 inch female pipe thread inside the return line opening. Refer to the manufacturer data for the pump as required.	
7	Hold the pump at an upward angle. Use an Allen wrench of sufficient length to install and tighten the bypass plug.	

8	Install the copper tubing to the 3/8 inch flare fitting at the pump.	
9	Mount the pump back onto the burner housing.	
10	Run the 3/8 inch line from the pump outlet into the top of the tank and down into the tank at 3 to 4 inches from the bottom.	
11	Snug the fitting to hold the line in place.	
12	Place supports for the line at appropriate locations from the pump to the tank.	
13	After the line has been installed, the tank shut off valve may be opened, and the power restored to the furnace. The burner should start and the fuel line should self-bleed itself of air with the bypass oil returning to the fuel tank.	
14	All fuel removed along adsorbent pads, and any rags, must be removed and disposed of properly. Never leave any of these at the job site.	

LAB 55.4 INSTALL A REPLACEMENT FUEL OIL BURNER

LABORATORY OBJECTIVE

The purpose of this lab is to demonstrate your ability to install a replacement fuel oil burner for an oil fired furnace.

LABORATORY NOTES

Many oil furnaces are constructed of heavy metal and are quite durable. It is not uncommon for the burner to become worn out while the basic furnace is still in good condition. In such cases the entire burner can be replaced. This has the advantage of a matched nozzle assembly and flame cone along with a new pump, motor, transformer, and primary control. This type of replacement is less expensive, faster, and easier than installing an entirely new furnace.

FUNDAMENTALS OF HVACR **TEXT REFERENCE:** Unit 55 Residential Oil Heating Installation

REQUIRED TOOLS AND EQUIPMENT

Oil furnace tank and operating oil furnace
Tool kit
Oil adsorbent pads
Large empty drip pan
vacuum

SAFETY REQUIREMENTS

A. It is good practice to have oil absorbent pads and a large drip pan available to collect any minor oil spills that may occur.

PROCEDURE
Step 1

Collect the oil fired furnace data and fill in the chart.

Oil Fired Furnace Unit Data

Furnace Make:	Model Number:	Serial Number:
Existing Burner Make:	Model Number:	Serial Number:
New Burner Make:	Model Number:	Serial Number:
Length of Blast Tube Required:		

Identify Piping System:
(circle one) One Pipe   Two Pipe
New Combustion Chamber?
(circle one) Yes   No
New Thermostat?
(circle one) Yes   No

Step 2

After recording the Oil Fired Furnace Unit Data List you can prepare to remove the old burner assembly. Complete the Burner Removal Check List and make sure to check each step off in the appropriate box as you finish it.

Burner Removal Check List

STEP	PROCEDURE	CHECK
1	Make sure all power to the furnace is off. Lock and tag the power panel before removing any parts.	
2	Close the fuel tank shut off valve.	
3	Disconnect the existing oil line or lines. Place an adsorbent pad large empty drip pan beneath the connection to catch any dripping fuel oil.	
4	Install 3/8-inch flare plugs in both fuel lines to prevent oil leakage.	
5	Carefully bend the fuel lines out of the way. You will reuse the same lines if possible	

6	Disconnect the main power and thermostat wire from the burner.	
7	Remove the mounting bolts holding the burner assembly in place.	
8	Remove the mount plate from front of the furnace.	
9	Use a vacuum to clean any debris from the combustion chamber area. Do not damage the combustion chamber refractory.	
10	Inspect the combustion chamber for any signs of cracks or deterioration. Replace as required.	

Step 3

After removing the old burner assembly you may prepare to install the new burner assembly. Complete the Burner Installation Check List and make sure to check each step off in the appropriate box as you finish it.

Burner Installation Check List

STEP	PROCEDURE	CHECK
1	Install new combustion chamber and components if required.	
2	Hold the burner mounting plate in position and measure the distance to the combustion chamber. Distance = _____	

3	Measure the length of the blast tube on the new burner. Length = _____	
4	Exchange blast tube on the burner if the length is not a match.	
5	Check/install the nozzle for correct GPH, angle, and pattern.	
6	Check and adjust electrode position (refer to Lab 54.1).	
7	Bolt new mounting plate to furnace.	
8	Bolt new burner with blast tube onto mounting plate.	
9	Connect oil lines to burner. Install bypass plug for two pipe systems.	
10	Replace oil filter in fuel supply line and bleed air from line (refer to Lab 55.2).	
11	Perform final burner adjustments (refer to Lab 54.2)	

Name _____

LAB 55.5 CHECK/TEST A CAD CELL OIL BURNER PRIMARY CONTROL

LABORATORY OBJECTIVE
The purpose of this lab is to demonstrate your ability to test for the proper operation of a cadmium sulfide cell (cad cell) for an oil fired furnace.

LABORATORY NOTES
The cadmium sulfide cell (cad cell) of an oil burner primary control system proves the presence of an oil flame by observing the visible light from the flame. The cad cell's electrical resistance is greatly reduced in the presence of light. The resistance must be high to enable the primary to initiate a trial for ignition, also called a dark start function. Once a flame is established, the light from the flame causes the cad cell's resistance to drop and the flame will continue. During the trial ignition, a safety switch heater is energized that will open the safety switch contacts and lock out the burner if flame is not proved within the trial for ignition time, usually 30, 45, or 60 seconds. This heater must cool off before the burner can be rest manually and started again. When flame is established and proved by the cad cell, the safety switch heater is de-energized and the contacts remain closed.

***FUNDAMENTALS OF HVACR* TEXT REFERENCE**
Unit 55 Residential Oil Heating Installation

REQUIRED TOOLS AND EQUIPMENT
Operating oil furnace
Tool kit
Ohmmeter
Clamp-on Ammeter
1200 Ohm Resistor
Timer for timing seconds

SAFETY REQUIREMENTS
A. Never allow fuel oil to flow into the combustion chamber without a flame. If fuel oil is allowed to build up and then suddenly ignite, this can create a serious hazard.

B. Check all circuits for voltage before doing any service work.

C. Stand on dry nonconductive surfaces when working on live circuits.

D. Never bypass any electrical protective devices.

PROCEDURE
Step 1

Collect the oil fired furnace data and fill in the chart.

Oil Fired Furnace Unit Data

Furnace Make:		Model Number:	Blower Motor Amperage Rating:	
Furnace Type **(circle one)**	Upflow	Counterflow	Basement	Horizontal
Burner Make:			Burner Model Number:	
Recommended **Nozzle Data**	Gallons per Hour GPH:		Spray Pattern:	Spray Angle:

Step 2

Complete the Prestart Check List and the Start-Up Check List as outlined in the procedure provided in Lab 55.1. After finishing the Start-Up Check List, shut down the furnace. Make sure all power to the furnace is off. Lock and tag the power panel before removing any parts. Complete the CAD Cell Test Check List.

CAD Cell Test Check List

STEP	PROCEDURE	CHECK
1	Make sure all power to the furnace is off. Lock and tag the power panel before removing any parts.	
2	Open or remove transformer.	
3	Locate an unplug cad cell from plug mount.	

4	Inspect and wipe clean the lens cover of the cad cell.	
5	Remove and record resistance of cad cell while the face of the cad cell is covered. Resistance = _____ Ohms	
6	Remove and record resistance of cad cell while the face of the cad cell exposed to room light. Resistance = _____ Ohms	
7	Insert the cad cell into the plug assembly.	
8	Locate yellow wires from the cad cell mount at the primary control terminals F and F.	
9	Remove cad cell wires from F and F and connect to the ohmmeter.	
10	Swing transformer slowly closed with the ohmmeter still connected.	
11	Turn on power to the furnace and burner and observe ignition. DO not get too close as oil burners can puff back.	

12	Read and record the ohms of the cad cell exposed to a normal flame. Resistance = _____ ohms	
13	The flame will shut down and the burner will lockout within 30 seconds because the cad cell is not connected to the primary control.	
14	Obtain a 1200 ohm resistor and connect one end of the resistor to one F terminal of the primary control.	
15	Push the reset button. Wait a minimum of 2 minutes cool down time.	
16	Observe the burner start and the flame ignite.	
17	Carefully connect the second wire of the 1200 ohm resistor to the other F terminal of the primary control within the 30 second trial for ignition time.	
18	Observe the flame continue for another five minutes.	
19	You will time how long it takes in seconds for the flame to shutdown and the burner to lockout when you remove one lead from the resistor, which will simulate a flame failure. Running time after removing resistor = _____ seconds	

608

20	After completing this test, make sure all power to the furnace is off. Lock and tag the power panel before removing any parts.	
21	Remove the resistor and ohmmeter and reconnect the cad cell.	
22	After the cad cell has been reconnected turn the burner on and off, observing several ignitions. Verify a quick clean quiet ignition.	

LAB 59.1 WIRING SEQUENCERS

LABORATORY OBJECTIVE
The purpose of this lab is to learn how to wire sequencers to control an electric furnace.

LABORATORY NOTES
You will wire a sequencer to control an electric furnace with a blower and two sets of strip heat. The sequencer should insure that one strip heater always starts before the other. The blower should operate any time a set of strips is energized.

FUNDAMENTALS OF HVACR TEXT REFERENCE
Unit 38 Diagrams / Unit 39 Control Systems / Unit 59 Electric Heat

REQUIRED TOOLS AND MATERIALS
Low voltage thermostat
Electric furnace that contains -
> Transformer
> 24-volt contactor
> Sequencer(s)
> Electric strip heaters
> Blower

Multimeter
Electrical hand tools

PROCEDURE

1. Draw a ladder type schematic diagram of an electric furnace controlled by sequencers.

2. Have the instructor check your design.

3. Wire the control system.

4. Have the instructor check the wiring before energizing the circuit.

5. Operate the control system and troubleshoot any problems.

LAB 60.1 ELECTRIC FURNACE STARTUP

LABORATORY OBJECTIVE
The purpose of this lab is to demonstrate your ability to go through the necessary sequence for a typical electric furnace startup.

LABORATORY NOTES
A typical electric furnace will have multiple stages of electric heat strips, usually approximately 5 KWH each. A 1 KWH heater would produce 3410 BTUH of heat and pull 4.17 amps at 240 V. A 5KWH heater would use five times that or 20.8 amps. Three 5 KW heaters would pull over 60 amps, too much to energize all the heaters at once. Power companies and some codes require electric furnaces be equipped with a time delay device, such as a sequencer, to stagger when individual strips are energized. This control is part of the furnace and not part of the thermostat. Even with a single-stage heat-only furnace, the electric heat elements come on one at a time, spaced apart by a few seconds at least. Supply air temperatures of lower than 120 ºF can feel rather cool and airflow should be reduced to keep the air temperature to a comfortable level.

FUNDAMENTALS OF HVACR TEXT REFERENCE
Unit 59 Electric Heat

REQUIRED TOOLS AND EQUIPMENT:
Operating electric furnace
Tool kit
Multimeter
Clamp-on Ammeter
Temperature sensor

SAFETY REQUIREMENTS
A. Check all circuits for voltage before doing any service work.

B. Stand on dry nonconductive surfaces when working on live circuits.

C. Never bypass any electrical protective devices.

PROCEDURE
Step 1

Collect the electric furnace data and fill in the chart.

Electric Furnace Unit Data

Furnace Make:		Model Number:		
Electrical Data	Voltage:	Phase:	Amperage:	KW:
Blower Type **(circle one)**	Direct drive	Belt drive		
Blower Speed **(circle one)**	Single speed	Two speed	Three speed	Four speed
System Type **(circle one)**	Heat only	Heat pump and electric backup		
Electric Heaters	1st KW:	2nd KW:	3rd KW:	Total KW:

Step 2

Complete the Prestart Check List and make sure to check each step off in the appropriate box as you finish it. This will help you to keep track of your progress.

Prestart Check List

STEP	PROCEDURE	CHECK
1	Make sure all power to the furnace is off. Lock and tag the power panel before removing any parts.	
2	Check all electrical connections for tightness.	
3	Spin all fans to be sure they are loose and turn freely.	
4	Check to make sure all airflow passages are unobstructed.	
5	All doors and panels available and in place.	
6	Thermostat installed and operating correctly.	

Step 3

After finishing the Prestart Check List you are now ready to start the furnace and check its operation. You will follow the same procedure of working through a checklist. This will help you to keep track of your progress. Complete the Start-Up Check List and make sure to check each step off in the appropriate box as you finish it.

STEP	PROCEDURE	CHECK
1	Check the main fuse or breaker and the amperage rating and make sure it is the correct rating for the furnace.	
2	Measure the incoming supply voltage – Refer to Unit 35 in *Fundamentals of HVACR* for the proper procedure for testing the circuit.	
3	If the Steps 1 & 2 have been completed and are satisfactory, then turn on the supply power for the furnace.	
4	Set the thermostat to – Fan On – to obtain the fan only operation and observe that the blower is operating. If the thermostat is not so equipped then consult with your instructor.	
5	Turn the fan off and the thermostat to heat.	
6	Adjust the thermostat setting to 10 ºF above the room temperature.	
7	Observe the fan start and the heater banks come on.	

8	Use the clamp-on ammeter to measure and record the amperage as the sequence brings on electric banks.	
	1st = _____	
	2nd = _____	
	3rd = _____	
9	Obtain normal heating operation.	
10	Measure and record temperatures.	
	Discharge air temperature = _____	
	Room air temperature = _____	
11	Calculate temperature rise.	
	Discharge air temp. – Room air temp. = Temperature rise	
	Temperature rise = _____	

12	Turn the thermostat down and observe the heaters come off.	
13	Observe the blower stop three minutes after the heaters come off.	

LAB 60.2 CALCUALTE AIRFLOW BY TEMPERATURE RISE

LABORATORY OBJECTIVE

The purpose of this lab is to demonstrate your ability to calculate airflow by the temperature rise methods through a typical electric furnace.

LABORATORY NOTES

Electric heaters are nearly 100% efficient and allow the measurement for the amount of heat produced very accurate. The heaters are located within the duct or furnace and there is no heat lost going up the chimney. We can calculate the airflow by the temperature rise method very accurately. The important thing is to measure the voltage, amperage, and temperatures as accurately as possible. The blower amperage must also be kept separate and not added to the heater amperage.

FUNDAMENTALS OF HVACR **TEXT REFERENCE**

Unit 66 Troubleshooting Heat Pump Systems

REQUIRED TOOLS AND EQUIPMENT

Operating electric furnace
Tool kit
Multimeter
Clamp-on Ammeter
Temperature sensor

SAFETY REQUIREMENTS

A. Check all circuits for voltage before doing any service work.

B. Stand on dry nonconductive surfaces when working on live circuits.

C. Never bypass any electrical protective devices.

PROCEDURE

Step 1

Collect the electric furnace data and fill in the chart.

Electric Furnace Unit Data

Furnace Make:		Model Number:		
Electrical Data	Voltage:	Phase:	Amperage:	KW:
Blower Type (circle one)	Direct drive	Belt drive		
Blower Speed (circle one)	Single speed	Two speed	Three speed	Four speed
System Type (circle one)	Heat only	Heat pump and electric backup		
Electric Heaters	1st KW:	2nd KW:	3rd KW:	Total KW:

Step 2

Complete the Prestart Check List and make sure to check each step off in the appropriate box as you finish it. This will help you to keep track of your progress.

Prestart Check List

STEP	PROCEDURE	CHECK
1	Make sure all power to the furnace is off. Lock and tag the power panel before removing any parts.	
2	Check all electrical connections for tightness.	
3	Spin all fans to be sure they are loose and turn freely.	
4	Check to make sure all airflow passages are unobstructed.	
5	All doors and panels available and in place.	
6	Thermostat installed and operating correctly.	
7	Allocate and count each heater element contactor. How many are there and what is the KW rating? _____ of _____ KWH	

Step 3

After finishing the Prestart Check List you are now ready to start the furnace and check its operation. You will follow the same procedure of working through a checklist. This will help you to keep track of your progress. Complete the Start-Up Check List and make sure to check each step off in the appropriate box as you finish it.

Start-Up Check List

STEP	PROCEDURE	CHECK
1	Check the main fuse or breaker and the amperage rating and make sure it is the correct rating for the furnace.	
2	Measure the incoming supply voltage – Refer to Unit 35 in *Fundamentals of HVACR* for the proper procedure for testing the circuit.	
3	If the Steps 1 & 2 have been completed and are satisfactory, then turn on the supply power for the furnace.	
4	Set the thermostat to – Fan On – to obtain the fan only operation and observe that the blower is operating. If the thermostat is not so equipped then consult with your instructor.	
5	Turn the fan off and the thermostat to heat.	
6	Adjust the thermostat setting to 10 °F above the room temperature.	

7	Observe the fan start and the heater banks come on.	
8	Use the clamp-on ammeter to measure and record the amperage as the sequence brings on electric banks. 1st = _____ 2nd = _____ 3rd = _____	
9	Obtain normal heating operation.	
10	Measure and record temperatures. Discharge air temperature = _____ Room air temperature = _____	

11	Calculate temperature rise. Discharge air temp. − Room air temp. = Temperature rise Temperature rise = _____	
12	Read total amperage of electric heaters only. Amperage = _____	
13	Calculate BTUH. Voltage x Amperage x 3.414 = BTUH BTUH = _____	
14	Calculate airflow by the temperature rise method. Airflow (CFM) = $\dfrac{\text{BTUH (from step 13)}}{(1.08) \times \text{Temp. rise (from step 11)}}$ CFM = _____	

15	Measure the blower motor amperage. Blower amperage = _____	
16	Rated fan motor amperage from nameplate on motor. Rated blower amperage = _____	
17	The measured amperage should be lower than the rated blower amperage.	
18	Turn the thermostat down and observe the heaters come off.	
19	Observe the blower stop three minutes after the heaters come off.	

LAB 62.1 TYPES OF HEAT PUMPS

LABORATORY OBJECTIVE
You will examine different heat pumps and identify their characteristics; including whether they are packaged or split, air source or water source, and the type of auxiliary heat.

LABORATORY NOTES
You will examine several heat pumps and determine if they are packaged or split, air source or water source. You will also identify the type of auxiliary heat.

FUNDAMENTALS OF HVACR TEXT REFERENCE
Unit 62 Heat Pump System Fundamentals

REQUIRED MATERIALS PROVIDED BY STUDENT
Safety Glasses

REQUIRED MATERIALS PROVIDED BY SCHOOL
Different Types of Heat Pumps

PROCEDURE

Unit 1

Is this unit a packaged unit or a split system?

Is this unit an air source unit or a water source unit?

What type of auxiliary heat does this system use?

If this is a water source unit, is it an open loop or closed loop system?

Unit 2

Is this unit a packaged unit or a split system?

Is this unit an air source unit or a water source unit?

What type of auxiliary heat does this system use?

If this is a water source unit, is it an open loop or closed loop system?

Unit 3

Is this unit a packaged unit or a split system?

Is this unit an air source unit or a water source unit?

What type of auxiliary heat does this system use?

If this is a water source unit, is it an open loop or closed loop system?

Unit 4

Is this unit a packaged unit or a split system?

Is this unit an air source unit or a water source unit?

What type of auxiliary heat does this system use?

If this is a water source unit, is it an open loop or closed loop system?

LAB 62.2 WINDOW UNIT HEAT PUMP REFRIGERATION CYCLE

LABORATORY OBJECTIVE
You will identify the refrigeration cycle components and describe their function in a window unit heat pump.

LABORATORY NOTES
You will identify the refrigeration components on a window unit heat pump system and describe their function. You should be prepared to show each component to the instructor and describe what happens to the refrigerant going through each component.

FUNDAMENTALS OF HVACR **TEXT REFERENCE**
Unit 62 Heat Pump System Fundamentals

REQUIRED MATERIALS PROVIDED BY STUDENT
Safety Glasses

REQUIRED MATERIALS PROVIDED BY SCHOOL
Window Unit Heat Pump

PROCEDURE

Examine the window unit heat pump.

Locate the compressor.

Locate the reversing valve.

What does the reversing valve actually reverse?

Locate the outdoor coil.

What is the function of this coil in cooling?

What is the function of this coil in heating?

Locate the metering device.

What type of metering device does this unit use?

Locate the indoor coil.

What is the function of this coil in cooling?

What is the function of this coil in heating?

Start at the compressor discharge and show the refrigerant path through the system until it returns to the suction of the compressor in cooling.

Describe the refrigerant changes through the cycle.

Start at the compressor discharge and show the refrigerant path through the system until it returns to the suction of the compressor in heating.

Describe the refrigerant changes through the cycle.

LAB 62.3 PACKAGED UNIT HEAT PUMP REFRIGERATION CYCLE

LABORATORY OBJECTIVE
You will identify the refrigeration cycle components and describe their function in a packaged unit heat pump.

LABORATORY NOTES
You will identify the refrigeration components on a packaged unit heat pump system and describe their function. You should be prepared to show each component to the instructor and describe what happens to the refrigerant going through each component.

FUNDAMENTALS OF HVACR TEXT REFERENCE
Unit 62 Heat Pump System Fundamentals

REQUIRED MATERIALS PROVIDED BY STUDENT
Safety Glasses
6-in-1 Screwdriver

REQUIRED MATERIALS PROVIDED BY SCHOOL
Packaged Unit Heat Pump

PROCEDURE

Examine the packaged unit heat pump.

Locate the compressor.

Locate the reversing valve.

What does the reversing valve actually reverse?

Locate the outdoor coil.

What is the function of this coil in cooling?

What is the function of this coil in heating?

Locate the outdoor metering device.

What type of metering device does this unit use for the outdoor coil?

●

When does this device meter refrigerant?

Locate the indoor coil.

What is the function of this coil in cooling?

What is the function of this coil in heating?

Locate the indoor metering device.

What type of metering device does this unit use for the indoor coil?

When does this device meter refrigerant?

●

What type of refrigerant storage does this system use?

Start at the compressor discharge and show the refrigerant path through the system until it returns to the suction of the compressor in cooling.

Describe the refrigerant changes through the cycle.

Start at the compressor discharge and show the refrigerant path through the system until it returns to the suction of the compressor in heating.

Describe the refrigerant changes through the cycle.

●

LAB 62.4 SPLIT SYSTEM HEAT PUMP REFRIGERATION CYCLE

LABORATORY OBJECTIVE
You will identify the refrigeration cycle components and describe their function in a split system heat pump.

LABORATORY NOTES
You will identify the refrigeration components on a split system heat pump and describe their function. You should be prepared to show each component to the instructor and describe what happens to the refrigerant going through each component.

FUNDAMENTALS OF HVACR TEXT REFERENCE
Unit 62 Heat Pump System Fundamentals

REQUIRED MATERIALS PROVIDED BY STUDENT
Safety Glasses
6-in-1 Screwdriver

REQUIRED MATERIALS PROVIDED BY SCHOOL
Split System Heat Pump

PROCEDURE

Examine the split system heat pump.

Locate the compressor.

Locate the reversing valve.

What does the reversing valve actually reverse?

Locate the outdoor coil.

What is the function of this coil in cooling?

What is the function of this coil in heating?

Locate the outdoor metering device.

What type of metering device does this unit use for the outdoor coil?

When does this device meter refrigerant?

What type of refrigerant storage does this system use?

Locate the indoor coil.

Where in the airstream is the coil in relation to the blower: Before of After?

What is the function of this coil in cooling?

What is the function of this coil in heating?

Locate the indoor metering device.

What type of metering device does this unit use for the indoor coil?

When does this device meter refrigerant?

Start at the compressor discharge and show the refrigerant path through the system until it returns to the suction of the compressor in cooling.

Describe the refrigerant changes through the cycle.

Start at the compressor discharge and show the refrigerant path through the system until it returns to the suction of the compressor in heating.

Describe the refrigerant changes through the cycle.

LAB 62.5 REVERSING VALVE FUNDAMENTALS

LABORATORY OBJECTIVE
You will describe the operation of a reversing valve and identify the parts of the valve including the common suction, common discharge, pilot tubes, pilot solenoid, and reversing valve coil.

LABORATORY NOTES
You will examine a reversing valve that has been cut open. You will identify the common suction line, common discharge line, pilot solenoid, pilot tubes, and reversing valve coil. Using the reversing valve, you will show the instructor how the reversing valve shifts.

***FUNDAMENTALS OF HVACR* TEXT REFERENCE**
Unit 62 Heat Pump System Fundamentals

REQUIRED MATERIALS PROVIDED BY STUDENT
Safety Glasses

REQUIRED MATERIALS PROVIDED BY SCHOOL
Reversing Valve that is cut open to observe the inside parts

PROCEDURE

Examine the reversing valve.

Identify the following parts of the valve:

- Common suction Line
- Common Discharge Line
- Pilot solenoid
- Pilot tubes
- Solenoid valve

Describe where each of the lines on the reversing valve connects to the unit.

Describe the how the valve shifts.

LAB 62.6 REVERSING VALVE OPERATION

LABORATORY OBJECTIVE
You will measure the temperature of the lines entering and leaving an operating reversing valve and observe their change when the valve shifts.

LABORATORY NOTES
You will operate the heat pump in cooling and heating, measure the temperature of the refrigerant lines entering and leaving the reversing valve, and complete the data sheet.

***FUNDAMENTALS OF HVACR* TEXT REFERENCE**
Unit 62 Heat Pump System Fundamentals

REQUIRED MATERIALS PROVIDED BY STUDENT
Safety Glasses
Thermocouple temperature tester
6-in-1 Screwdriver

REQUIRED MATERIALS PROVIDED BY SCHOOL
Operational Heat Pump

PROCEDURE

Locate the reversing valve in the heat pump.

Examine the diagram and determine when the reversing valve is energized: heating or cooling?

Operate the heat pump in cooling.

Measure the temperature of the four lines in this order:

- Common Discharge Line (small line by itself)
- Line to outdoor coil (one of the outside lines on the side with three large lines)
- Common Suction line (line in the middle on the side with three large lines)
- Line to indoor coil (should be the only line left)

Operate the heat pump in heating.

Measure the temperature of the four lines in this order:

- Common Discharge Line (small line by itself)
- Line to indoor coil (one of the outside lines on the side with three large lines)
- Common Suction line (line in the middle on the side with three large lines)
- Line to indoor coil (should be the only line left)

Operate the heat pump in the mode in which the reversing valve is energized (heating or cooling)

While the system is operating, carefully remove one wire from the reversing valve solenoid coil. The valve should shift.

Replace the wire and the valve should shift back

LAB 62.7 HEAT PUMP ELECTRICAL COMPONENTS

LABORATORY OBJECTIVE

You will identify common electrical heat pump components and explain their function.

FUNDAMENTALS OF HVACR TEXT REFERENCE

Unit 62 Heat Pump System Fundamentals

REQUIRED MATERIALS PROVIDED BY STUDENT

Safety Glasses

REQUIRED MATERIALS PROVIDED BY SCHOOL

Heat Pump

PROCEDURE

You will examine the heat pump assigned by the instructor and identify the listed components.

- Contactor

- Defrost Board

- High Pressure Switch

- Reversing Valve Solenoid

- Dual run capacitor

- Compressor Starting components

- Anti short cycling timer

- Outdoor fan motor

- Compressor

- Crankcase heater

LAB 63.1 IDENTIFYING DEFROST CONTROLS

LABORATORY OBJECTIVE
You will identify defrost controls as either an electronic time-temperature defrost control or a demand defrost control.

FUNDAMENTALS OF HVACR TEXT REFERENCE
Unit 63 Air Source Heat Pump Applications

REQUIRED MATERIALS PROVIDED BY STUDENT
Safety Glasses
6-in-1 screwdriver
Non-contact voltage probe
Multimeter

REQUIRED MATERIALS PROVIDED BY SCHOOL
Heat Pumps with Defrost Controls
(one electronic time temperature and one demand defrost)

PROCEDURE

Turn the power off to the unit.

Check with your non-contact voltage detector or multimeter to insure power is off.

Locate the defrost thermostat or coil temperature thermistor.

Which type does this unit use – defrost thermostat (switch) or coil temperature sensor (thermistor)?

Locate the defrost control board.

Does the board have a time jumper or pins?

Determine if this is a time temperature defrost board or a demand defrost board based on the presence or absence of time pins and the type of defrost thermostat (switch or sensor)

Time – temperature systems use round, physical defrost thermostat switches and have time period jumpers or switches.

Demand defrost boards use two thermistor temperature sensors: one for the coil and one for the outdoor ambient temperature.

Check the resistance of the defrost thermostat or the thermistor sensors.

The defrost thermostat should be open (OL) until it reaches its closing temperature, usually around 28°F.

The thermistor sensors should have a resistance in the thousands of ohms that vary with temperature.

LAB 63.2 WIRING DUAL FUEL HEAT PUMP

LABORATORY OBJECTIVE
You will wire the dual fuel heat pump assigned by the instructor. This includes control wiring.

FUNDAMENTALS OF HVACR TEXT REFERENCE
Unit 63 Air Source Heat Pump Applications

REQUIRED MATERIALS PROVIDED BY STUDENT
Safety Glasses
6-in-1 screwdriver
Wire cutters/crimpers/strippers
Non-contact voltage detector
Multimeter

REQUIRED MATERIALS PROVIDED BY SCHOOL
Split System Heat Pump
Gas furnace
Dual fuel panel or Dual Fuel capable thermostat
Wire
Wire connectors

PROCEDURE

This system should be a gas furnace with a heat [pump coil installed on it connected to a heat pump condensing unit. The coordination of the furnace and heat pump will be provided by either a dual fuel panel, or a dual fuel capable thermostat.

Turn the power off to both the furnace and the outdoor unit.

Check the power with your non-contact voltage detector or multimeter to make sure they are off.

Each piece will have its own power wire and disconnect switch.

If using a dual fuel panel:

> The control wiring should be run from the thermostat to the dual fuel panel, and then from the dual fuel panel to both the furnace and the outdoor unit.

If using a dual fuel capable thermostat:

> The control wiring should be run from the thermostat to the furnace, and then from the furnace to the outdoor unit.

> The thermostat may need to be configured to operate in dual fuel mode.

The thermostat wire does not need to be run in conduit.

The thermostat wire should NOT run in the same conduit as the power wire.

Mount the thermostat.

Wire the control wiring at the thermostat, the dual fuel panel, the furnace and the outdoor unit.

Have the instructor check your work BEFORE energizing.

636

Turn on the power to the unit.

Operate the unit in both the cooling, first stage heating, and second stage heating cycles.

LAB 63.3 ELECTRIC STRIP HEAT OPERATION

LABORATORY OBJECTIVE
You will operate electric strip heat and measure the key operating characteristics.

FUNDAMENTALS OF HVACR **TEXT REFERENCE**
Unit 61 Troubleshooting Electric Heat/ Unit 63 Air Source Heat Pump Applications

REQUIRED MATERIALS PROVIDED BY STUDENT
Safety Glasses
6-in-1 screwdriver
Wire cutters/crimpers/strippers
Non-contact voltage detector
Multimeter
Ammeter

REQUIRED MATERIALS PROVIDED BY SCHOOL
Heat Pump with Electric strip auxiliary heat or electric furnace

PROCEDURE

Turn the power off to the unit.

Check the voltage with you non-contact voltage detector or multimeter to insure it is off.

Disconnect the wires from one side of each strip heater, paying attention to how they are wired on the unit.

Measure the resistance of each strip heater.

Measure the resistance of the thermal overload limit switches and the fuse links.

Replace the wires, making sure they are properly connected and tight.

Turn the power back on.

Operate the electric furnace, measure and record the following information:

- Element voltage
- Element current (individual)
- Element current(total)
- Fan current

Note: there will normally be a delay before each strip heater is energized.

What controls the blower motor?

When does the blower motor turn on?

When does the blower shut off?

How many heat strips are there?

What is the kilowatt rating of each heater?

What is the rated amp draw of each heater?

What is the ACTUAL amp draw and kilowatt usage of each heater?

Describe the sequence in which the heat strips come on.

LAB 65.1 WIRING PACKAGED HEAT PUMP

LABORATORY OBJECTIVE
You will wire the packaged heat pump assigned by the instructor. This includes all power and control wiring.

FUNDAMENTALS OF HVACR TEXT REFERENCE
Unit 65 Heat Pump Installation

REQUIRED MATERIALS PROVIDED BY STUDENT
Safety Glasses
6-in-1 screwdriver
Wire cutters/crimpers/strippers
Non-contact voltage detector
Multimeter

REQUIRED MATERIALS PROVIDED BY SCHOOL
Packaged Heat Pump
Wire
Wire connectors

PROCEDURE

Turn the power off to the unit.

Check the power with your non-contact voltage detector or multimeter to make sure it is off.

Determine the correct size power wire using the minimum circuit ampacity listed on the unit

Size the power wire according to this minimum circuit ampacity.

Size the disconnect switch. It should be at least 115% of the minimum circuit ampacity.

Size the circuit breaker or fuse using the maximum over-current protection listed n the unit.

The power wire should be run in liquid tight conduit from the disconnect to the unit.

Make sure the unit is grounded.

The control wiring should be run from the thermostat to the unit.

The thermostat wire does not need to be run in conduit.

The thermostat wire should NOT run in the same conduit as the power wire.

Mount the thermostat.

Wire the control wiring at the thermostat and the unit.

Have the instructor check your work BEFORE energizing.

Turn on the power to the unit.

Operate the unit in both the cooling and heating cycles.

Note: Many units have a built in delay of several minutes the first time the unit is powered up. This will keep it from coming on immediately after the power has been turned off.

LAB 65.2 INSTALL SPLIT SYSTEM HEAT PUMP REFRIGERATION PIPING

LABORATORY OBJECTIVE
You will demonstrate your ability to install refrigeration piping for a split system heat pump and check the piping for leaks.

LABORATORY NOTES
You will run the refrigeration piping for a split system heat pump and pressure test it for leaks using nitrogen and soap bubbles.

FUNDAMENTALS OF HVACR TEXT REFERENCE
Unit 26 Refrigerant System Piping/ Unit 65 Heat Pump Installation

REQUIRED MATERIALS PROVIDED BY STUDENT
Refrigeration Gauges
Tubing cutter
Flaring tool

REQUIRED MATERIALS PROVIDED BY SCHOOL
Split System Air Conditioner
Refrigeration Copper
Oxy-Acetylene torch
Brazing material
Armaflex Insulation
Nitrogen cylinder, regulator, and pressure relief valve assembly
Soap bubbles or ultrasonic leak detector

PROCEDURE
Locate the condensing unit and indoor coil.

Determine the tubing size needed for the suction and liquid lines. . For residential air conditioning, this is normally done by measuring the pipe stub out on the condensing unit.

Roll out the tubing by holding the end of the tubing against the ground and straightening to form a long, straight length of tubing long enough to go between the condenser and the evaporator.

Note: To keep the tubing clean and dry, the caps or plugs on the ends of the tubing should remain in place until just before the tubing is connected.

If you are working with a line set, roll the entire length out.

If you are working with a roll of refrigeration copper, roll out just the amount you need plus a little extra for security.

If you are working with refrigeration copper, slide lengths of armaflex insulation over the suction tubing.

Pull a length of thermostat wire beside the piping, leaving plenty of wire on each end.

Tape the lines and wire together at regular intervals.

Pull the line set through the conduit, chase, or wall between the indoor and outdoor units.

Connect the lines to the outside pipe stubs.

If the lines are brazed, be sure to use nitrogen to minimize internal oxidation inside the tubing.

The armaflex insulation should be pushed back and taped to keep it clear of the flames during brazing.

Note: The insulation must be rated for use on heat pump systems. Because the gas line is hot during the heating cycle, not all pipe insulation is rated for heat pump application.

Connect the suction line to the inside pipe stub.

Brazing should be done after a nitrogen purge to prevent oxidation.

Work quickly when brazing the suction line.

If the evaporator has an internally installed TEV, overheating the suction line will destroy the TEV.

A filter drier should be installed in the liquid line.

The filter must be a bi-flow filter rated for heat pump use.

Connect the liquid line to the filter drier.

Connect a short line from the filter drier to the evaporator coil.

Pressurize the lines with nitrogen and check for leaks using soap bubbles or an ultra-sonic leak detector.

LAB 65.4 WIRE THERMOSTAT AND CONTROL WIRING

LABORATORY OBJECTIVE
The purpose of this lab is to learn how to wire a low voltage thermostat to a split system heat pump.

LABORATORY NOTES
You will wire a two stage heat pump thermostat to control a split system heat pump. You will need to wire the control wiring at the thermostat, blower coil control panel, and condensing unit.

FUNDAMENTALS OF HVACR TEXT REFERENCE
Unit 65 Heat Pump Installation

REQUIRED TOOLS AND MATERIALS
Low voltage heat pump thermostat
Split system heat pump
Multimeter
Electrical hand tools

PROCEDURE

Read the installation instructions for the thermostat and heat pump.

Turn off the power to the unit.

Check the voltage to the unit with a multimeter to be sure the power is off.

Mount the thermostat. The thermostat should be level.

If the thermostat is mounted on metal, be careful that electrical circuits are not contacting the metal.

Wire the thermostat wire to the thermostat sub-base.

Wire the thermostat wire to the unit terminal connections or pigtails (depending on the unit)

Have an instructor check your wiring.

Install thermostat on sub-base.

Turn the power on and operate the unit.

LAB 65.5 EVACUATE AND CHARGE A SPLIT SYSTEM HEAT PUMP

LABORATORY OBJECTIVE
You will demonstrate your ability to evacuate and charge a split system heat pump.

LABORATORY NOTES
You will pull a deep vacuum on a split system air conditioner down to 500 microns and weight in the correct refrigerant charge in liquid form into the high side of the system.

FUNDAMENTALS OF HVACR TEXT REFERENCE
Unit 30 Refrigeration System Evacuation/ Unit 31 Refrigeration System Charging/ Unit 65 Heat Pump Installation

REQUIRED MATERIALS PROVIDED BY STUDENT
Hand tools
Refrigeration Gauges
Temperature Tester

REQUIRED MATERIALS PROVIDED BY SCHOOL
Split System Heat Pump
Vacuum Pump
Extension cord
Vacuum Gauge
Refrigerant
Scale

PROCEDURE
Connect your gauges to the system.
If the system has a nitrogen holding charge, release the nitrogen to the air.
Do NOT release the charge if it is charged with refrigerant.
Connect the vacuum gauge and vacuum pump.
Evacuate the system down to 500 microns on the vacuum gauge.
Note: on a new split system you are only evacuating the lines and evaporator coil. With a leak free system, evacuating just the lines and coil should not take very long.
While the vacuum is pulling, calculate the total system charge.
The total charge should be the condensing unit factory charge + an amount of refrigerant for line length.
For most systems, the amount of refrigerant to add for line length is 0.6 ounces per foot over 15 feet of line.
Once the vacuum is achieved, close the refrigeration gauges and turn off the vacuum pump.
On new split systems, weigh in the amount of refrigerant needed for line length and then open the unit installation valves to let the refrigerant that is in the condenser into the rest of the system.
If the system is a dry system or an older system (not newly installed), you need to weigh in the full charge including the factory charge + the line length allowance.

LAB 65.6 HEAT PUMP START AND CHECK

LABORATORY OBJECTIVE

You will demonstrate your ability to perform an initial unit start and check procedure on a heat pump system.

LABORATORY NOTES

You will operate a packaged air conditioning unit, perform an initial unit start and check procedure, and record operational data on the worksheet.

FUNDAMENTALS OF HVACR TEXT REFERENCE

Unit 65 Heat Pump Installation/ Unit 89 Installation Techniques

REQUIRED MATERIALS PROVIDED BY STUDENT

Multimeter
Ammeter
Electrical hand tools
Refrigeration Gauges
Temperature Tester

REQUIRED MATERIALS PROVIDED BY SCHOOL

Heat Pump

PROCEDURE

Pre-start Inspection – Electrical

Record all electrical data from unit, including:

Minimum Supply Voltage, Maximum Supply Voltage, Compressor RLA, Outdoor Fan FLA, Indoor blower FLA

Unit MCA (Min Cir Amps) - Compare actual wire size in NEC to Unit MCA

Unit MOP (Maximum Overcurrent Protection) - Make sure actual fuse or circuit breaker does not exceed the maximum overcurrent protection

Make sure the unit is properly grounded.

Pre-start Inspection – Refrigeration

Measure and record the outdoor ambient temperature, the indoor temperature, and the indoor wet bulb temperature.

Check the type of refrigerant and measure the equalized refrigerant pressures.

Pre-start Inspection – Mechanical

Check to see that all packing materials and shipping supports have been removed.

Spin the fan blades by hand to verify that they are not hanging or seized.

Check to see that an air filter is installed.

Pour water in the evaporator condensate drain and verify that the drain work properly

Initial Power Up

Power should be applied to the unit for 24 hours before turning it on. (This requirement is waived for this lab.)

Check the incoming voltage to the unit with the thermostat set to off and the disconnect switch on.

The voltage should be between the minimum and maximum operating voltages stated on the unit data plate.

If the voltage is NOT within the acceptable operating range, turn the disconnect switch off and determine why.

If the outdoor temperature is above 60°F, set the thermostat to "Cool" and run the temperature down to call for cooling.

If the outdoor temperature is below 60°F, set the thermostat to "Heat" and run the temperature up to call for heat.

After the unit starts, recheck the voltage to the unit.

It should still be within the acceptable operating voltages listed on the unit data plate.

If the voltage is NOT within the acceptable operating range, the voltage drop through the power wire is unacceptably high – turn off the power!

Check the amp draw of the power wire feeding the unit and compare it to the unit data plate.

The amp draw should be lower than the minimum circuit ampacity stated on the data plate

Check the amp draw of the compressor, condenser fan motor, and evaporator fan motor and compare to the RLA and FLA ratings in the data plate.

The readings most likely will not be exactly what is on the data plate, but they should be in the same general range.

Operational Checks - General

You should check to see that the heat pump operates correctly in all stages, including:

Off – nothing should run. If the unit has been operating, there may be a delay before the unit shuts off.

Fan On- only the indoor blower should operate

Cooling – Compressor and both fans should run. Cool air should be leaving the ducts.

Heating, first stage – Compressor and both fans should run. Warm air leaving the ducts.

Defrost – compressor, indoor blower, and strips should operate. Outdoor fan should not operate. See 66.17 Forcing a Defrost in *Fundamentals of HVACR* for details on how to initiate the defrost cycle.

Heating Second Stage – Compressor, both fans, and heat strips should all operate. There may be a delay before all strips are energized. Hot air should be leaving the ducts.

Emergency Heat – only the indoor blower and strips should operate. Hot air should be leaving the ducts.

Operational Checks - Airflow

If the outdoor temperature is above 60°F, set the thermostat to "Cool" and run the temperature down to call for cooling.

If the outdoor temperature is below 60°F, set the thermostat to "Heat" and run the temperature up to call for heat.

Read the static pressure difference between the return and supply duct.

Compare this total external static reading to the manufacturer's specification.

If you do not have a specification, assume 0.5" wc to be "normal." Anything over 0.7" wc is normally too high.

Measure the system airflow using a flowhood or other airflow measuring instrument.

Compare the actual airflow reading to the manufacturer's specification.

If you do not have a manufacturer's specification, the airflow should be approximately 400 CFM per ton.

Operational Checks - Refrigeration

Operate the unit in cooling.

Note: if the outdoor temperature is below 50°, it may not be possible to check the system refrigerant pressures in cooling.

After the unit has operated long enough for the temperatures and pressures to stabilize, record the system suction and discharge pressures.

Compare the pressures to the manufacturer's specifications.

Measure the suction line temperature and calculate the suction superheat (suction line temperature – evaporator saturation temperature from gauges)

Compare the superheat to the manufacturer's specifications.

Measure the liquid line temperature and calculate the liquid line subcooling (Condenser saturation temperature from gauges – liquid line temperature)

Compare the liquid subcooling to the manufacturer's specifications.

Read the temperature of the discharge line leaving the compressor.

For most units, any temperature 200°F and higher indicates that the compressor is overheating.

Read the temperature of the return air and supply air.

Calculate the temperature difference across the coil (return air temperature – supply air temperature)

Compare the coil temperature difference to the manufacturer's specifications. This varies depending upon the conditions and airflow.

In cooling, normal temperature drops can be anywhere between 10°F and 20°F.

In heating, normal temperature rise can be anywhere between 10°F and 30°F.

Set the thermostat to "Off" and wait for the system to shut off.

Temporarily disable the heat strips by disconnecting the 24-volt wire control wire to the heat strips.

Operate the unit in heating and repeat all the measurements.

Note: If the outdoor temperature is above 75°, it may not be possible to check the system refrigerant pressures in heating.

HEAT PUMP START AND CHECK					
Pre-Start Check List					
Electrical Pre-Start Checks		**Refrigeration Pre-Start Checks**		**Mechanical Checks**	
Minimum Supply Voltage		Outdoor Ambient Temperature		Shipping Materials Removed	
Maximum Supply Voltage		Indoor Dry Bulb Temperature		Condenser Fan Spins	
Compressor RLA		Indoor Wet Bulb Temperature		Evaporator Fan Spins	
Condenser Fan FLA		Type of Refrigerant		Air Filter in Place	
Evaporator Fan FLA		Equalized Refrigerant Pressure		Evaporator Drain Works	
MCA (Min Cir Amps)					
Wire sized to MCA					
Max Fuse Size					
Actual Fuse Size Installed					

Initial Power-up – Cooling		**Initial Power-up – Emergency Heat**	
Voltage at outdoor unit off		Voltage at indoor unit with unit off	
Voltage at unit while running		Voltage at unit while running	
Amp draw on power wire		Amp draw on power wire	
Compressor amp draw		Amp Draw on Strips	
Condenser fan motor amp draw		Indoor Blower motor amp draw	

Refrigeration Operational Checks	Cool	Heat	**Airflow Operational Checks**	
Suction Pressure			Total External Static Specified	
Evaporator Saturation			Measured total external static measured	
Condenser Pressure			Airflow CFM Specified	
Condenser Saturation			Measured CFM	
Suction Line Temperature			Return air temperature	
Suction Superheat			Supply air temperature	
Liquid Line Temperature			Indoor Coil Temperature Difference	
Liquid Subcooling				
Discharge line temperature				

LAB 65.7 DETERMINING BALANCE POINT

LABORATORY OBJECTIVE

Given the unit specifications and house heat load, you will determine the system balance point.

***FUNDAMENTALS OF HVACR* TEXT REFERENCE**

Unit 63 Air Source Heat Pump Applications, Unit 65 Heat Pump Installation

SPECIFICATIONS

House heat load is 45,000 Btuh at 20°F outside, Heat pump capacity is 36,000 Btuh at 47°F and 18,000 Btuh at 17°F.

PROCEDURE

Construct a graph with the house load on one axis and the unit capacity on the other axis.

Plot the house line using the design load of 45,000 Btuh at 20°F and an assumed load of 0 Btuh 65°F.

Plot the unit capacity line using 36,000 Btuh at 47°F and 18,000 Btuh at 17°F.

The intersection is the balance point.

What is the balance point temperature?

What is the house heat load at the balance point?

What is the system capacity at the balance point?

How much supplemental heat (Btuh) does the system require at the balance point?

How many 5 kw electric strip heaters would be required ? (each produces approximately 17,000 Btuh)

LAB 66.1 PACKAGED HEAT PUMP CHARGING – COOLING CYCLE

LABORATORY OBJECTIVE
You will determine if the charge is correct on a packaged heat pump using a manufacturer's charging chart in the cooling cycle.

FUNDAMENTALS OF HVACR TEXT REFERENCE
Unit 31 Refrigerant System Charging/ Unit 65 Heat Pump Installation/ Unit 66 Troubleshooting Heat Pumps

REQUIRED MATERIALS PROVIDED BY STUDENT
Safety Glasses
Gloves
Refrigeration Gauges
6-in-1 Screwdriver
Temperature tester

REQUIRED MATERIALS PROVIDED BY SCHOOL
Packaged Heat Pump with a manufacturer's charging chart, which correlates suction pressure, liquid pressure, and outdoor ambient temperature.

PROCEDURE

Find the manufacturer's charging chart.

Read the instructions on the chart.

Pay attention to any specific operating conditions that must be met for the chart to be accurate.

Most charts assume that the return air is near comfort conditions.

Install your gauges on the assigned packaged heat pump.

Start the unit in cooling.

Verify that the system has good airflow across both the evaporator and condenser.

Measure the outdoor ambient temperature.

Allow the unit to run long enough for the pressures to stabilize.

Determine the intersection of the suction and liquid pressures on the chart.

If the intersection is below the ambient temperature line, the system is undercharged.

If the intersection is above the ambient temperature line, the system is overcharged.

Note: Some allowance for error must be made. If the intersection is within 3°F, the system is probably charged correctly.

LAB 66.2 HEAT PUMP CHARGING – COOLING CYCLE SUPERHEAT

LABORATORY OBJECTIVE
Using the superheat method, you will determine if the charge is correct on a heat pump with a fixed restriction metering device operating in cooling.

FUNDAMENTALS OF HVACR TEXT REFERENCE
Unit 31 Refrigerant System Charging/ Unit 65 Heat Pump Installation/ Unit 66 Troubleshooting Heat Pumps

REQUIRED MATERIALS PROVIDED BY STUDENT
Safety Glasses
Gloves
Refrigeration Gauges
6-in-1 Screwdriver
Temperature tester

REQUIRED MATERIALS PROVIDED BY SCHOOL
Heat Pump with fixed restriction metering device and a superheat charging chart.

PROCEDURE

You will be operating the unit in cooling.

Find the manufacturer's charging specifications.

Read the instructions on the chart.

Pay attention to any specific operating conditions that must be met for the chart to be accurate.

Many charts assume that the return air is near comfort conditions.

Install your gauges on the assigned heat pump.

Start the unit in cooling.

Verify that the system has good airflow across both the evaporator and condenser.

Measure the outdoor ambient temperature.

Read the suction pressure on the compound gauge.

Use a PT chart to convert the suction pressure to a saturated temperature.

Measure the suction line temperature.

The superheat is the temperature difference between the suction line temperature and the evaporator saturation temperature.

If the superheat is too low, the unit is overcharged.

If the superheat is too high, the unit is undercharged.

LAB 66.3 HEAT PUMP CHARGING – COOLING CYCLE SUBCOOLING

LABORATORY OBJECTIVE
Using the subcooling method, you will determine if the charge is correct on a heat pump with a fixed restriction metering device operating in cooling.

FUNDAMENTALS OF HVACR TEXT REFERENCE
Unit 31 Refrigerant System Charging/ Unit 65 Heat Pump Installations / Unit 66 Troubleshooting Heat Pumps

REQUIRED MATERIALS PROVIDED BY STUDENT
Safety Glasses
Gloves
Refrigeration Gauges
6-in-1 Screwdriver
Temperature tester

REQUIRED MATERIALS PROVIDED BY SCHOOL
Heat Pump with TEV metering device and subcooling charging chart.

PROCEDURE

You will be operating the unit in cooling.

Find the manufacturer's charging specifications.

Read the instructions on the chart.

Pay attention to any specific operating conditions that must be met for the chart to be

accurate.

Many charts assume that the return air is near comfort conditions.

Install your gauges on the assigned heat pump.

Start the unit in cooling.

Verify that the system has good airflow across both the evaporator and condenser.

Read the liquid pressure on the high pressure gauge.

Use a PT chart to convert the liquid pressure to a saturated temperature.

Measure the liquid line temperature.

The subcooling is the temperature difference between the liquid saturation temperature

and the liquid line temperature.

If the subcooling is too low, the unit is undercharged.

If the subcooling is too high, the unit is overcharged.

HEATING CYCLE CHARGING

Labs 66.4, 66.5, 66.6, 66.7

The fur really starts flying when discussing heat pump charging during the heating cycle. Many manufacturers now give heating charging charts that resemble their cool cycle charging charts. Heating superheat charts are also used which specify a superheat for a particular operating condition. A few manufacturers have developed unique approaches that involve liquid sub cooling or even compressor DISCHARGE LINE SUPERHEAT. Still, some manufacturers insist that you can only CHECK the charge in the heating cycle and can only weigh in a complete charge if the charge is low. Yet others state that you CANNOT CHECK OR CHARGE the refrigerant charge during the heating cycle. Needless to say, this all can lead to considerable confusion when attempting to charge a heat pump in the heating cycle. Again, the best advice is to follow whatever procedure the manufacturer recommends.

Heating cycle charging charts are similar to the cooling charging charts, requiring the same information, and yielding about the same results. The primary difference is that the unit is operating during the heating cycle.

A check chart is just that, a chart for CHECKING the charge during the heating cycle. It is not intended to be used for actually CHARGING the unit. A check chart normally specifies an outdoor temperature, a suction pressure, and a liquid pressure. It works similarly to a cooling charging chart; the difference is that it is only used to CHECK the charge. Why? Many manufacturers point out that the capacity of the heat pump is greatly diminished during the heating cycle at low outside ambient temperatures. This means that less refrigerant is being circulated in the refrigerant system and a large quantity is just sitting somewhere, usually in the accumulator. For this reason, a charge which is adequate to maintain correct operation and pressures at 25°F may not be adequate at 35°F. How? Suppose that the unit was undercharged according to the check chart. If you add just enough refrigerant to bring it up to the required pressures, you have added only enough refrigerant for correct operation at that one condition. When the outside temperature rises, the unit will be capable of pumping more refrigerant. However, since you didn't add any extra refrigerant, there won't be any more to pump. This is why some manufacturers say that the check chart cannot be used to CHARGE the system. Even more radical are the manufacturers who will not even supply a check chart. Their argument against check charts is similar to the argument against charging charts. So how do you know when to recover the charge and weigh in a new one if you don't even have a heating check chart? The most common answer from people against heating check charts is that you must run the unit in cooling and check the charge with the cooling check chart. Of course if it is 25°F outside, you will also need to do something to simulate a warmer ambient. Blocking the airflow across the condenser or possibly even disconnecting the condenser fan motor can do this. Most people subscribing to this method readily acknowledge that it is not very accurate, so they return in the spring to recheck the refrigerant charge.

One other method of checking AND charging heat pumps during the heating season is to operate the unit in cooling, block the condenser to obtain a head pressure of at least 250 psig, and check the liquid sub cooling. The amount of liquid sub cooling tells you how much liquid is sitting in the bottom of the condenser. Too much sub cooling indicates an overcharge, too little indicates an undercharge. This procedure is ONLY used on systems with thermostatic expansion valves. The idea is that once there is enough refrigerant in the system you will have the correct amount of subcooling. Manufacturers recommending this charging method list the subcooling on the panel, typically around 10°F - 15°F.

Another way to check the refrigerant charge during the heating season without a heating check chart is to measure the temperature rise across the indoor coil and compare it to data published for that unit. One problem with this method is that: the amount of temperature rise achieved is directly related to the amount of air moving across the coil. If the ACTUAL airflow is different from the manufacturer's stated conditions, the temperature rise will also vary. Heat pumps are rated by AHRI for heat output at two temperatures, 47°F and 17°F. Most heat pumps produce a heat output of close to their "tonnage rating" at the 47°F AHRI rating point and approximately half of their "tonnage rating" at 17°F. Most units operate correctly with an airflow somewhere in the range of 400 CFM per ton of capacity. Of course these are gross generalizations, given the numerous manufacturers and design available.

However, a general idea of temperature rise parameters can prove more useful than a general idea of operating pressures. Using these general assumptions, let's develop some temperature rise performance parameters. The basic formula to use is BTU = 1.08 x CFM x TEMP RISE. This can be rewritten as TEMP RISE = BTU / (1.08 x CFM). At 47°F the capacity per ton will be 12,000 BTU. Plugging this into the formula we get TEMP RISE = 12,000 / (1.08 x 400). This yields a projected temperature rise of 28°F at a 47°F ambient. On the other end of things, the assumed capacity per ton at 17°F ambient temperature is 6000 BTU. Plugging this in we get TEMP RISE = 6000 / (1.08 x 400). This yields a projected temperature rise of 14°F. You can readily see that the result of a 50% reduction in capacity yields a 50% reduction in temperature rise. Now let's test our assumptions against some REAL data taken from a manufacturer's performance specifications.

Model	ΔT @ 47°F	ΔT @ 17°F	BTUh @ 47°F	BTUh @ 17°F	CFM
ERHQ18	26°F	13°F	18,500	11,000	700
ERHQ24	25°F	12°F	24,400	12,500	800
ERHQ30	27°F	16°F	32,200	18,800	1055
ERHQ36	26°F	13°F	36,000	19,500	1310

The manufacturer states that a properly operating unit should be + or – 3°F of these typical values. You will note that our extrapolated values are all within the 3°F margin of error.

Another manufacturer publishes a temperature rise chart. They rate the temperature rise from 37°F to -8°F. For their PCB-036 unit operating at 17°F ambient temperature, the rise is published as 14°F. Note that this is EXACTLY what our mathematical extrapolations gave us. It is difficult to directly test our 47°F example against theirs since they don't publish a temperature rise for any temperature above 37°F. But this brings up an interesting point, what if the temperature is neither 47°F nor 17°F, what temperature rise do you look for then? Another assumption is in order here; we will assume that the capacity reduction due to ambient temperature drop is even. Since the temperature rise changed from 28°F to 14°F over a 30°F ambient change, this would represent approximately a 1°F change in temperature rise for every 2°F change in ambient temperature. In the above table, we see a temperature rise change of 13°F in 3 out of the 4 units listed. This gives similar results. In our assumed system, 28°F rise at 47°F ambient and a 14°F rise at 17°F ambient, a temperature rise of 23°F would be expected at 37°F ambient based on the relationship of 1°F rise per 2°F ambient temperature. Clearly, manufacturer's data for a specific unit is more accurate than extrapolated benchmarks; however, extrapolated temperature rise calculations are considerably more accurate than blocking the coil and pretending it's summer.

One of the most recent methods to gain acceptance with a few manufacturers is to compare the discharge line temperature to the ambient temperature during the heating cycle. For a correct charge, the discharge line should be 100°F – 110°F warmer than the outdoor temperature. Similar to this method is one that expects a 50°F to 60°F DISCHARGE SUPERHEAT. In other words, the discharge line is 50°F to 60°F warmer than the condensing temperature. To get this you convert the discharge pressure to temperature using a pressure temperature chart. You then subtract this condensing temperature from the measured discharge line temperature. The difference should be between 50°F and 60°F. A lower superheat indicates an overcharge while a higher superheat indicates an undercharge. For this method to work you need to be within 6" of the compressor. Reading the gas line leaving the unit will give incorrect results.

Remember: The best method for charging any unit is the one recommended by the people who made it!

LAB 66.4 SPLIT SYSTEM HEAT PUMP CHARGING – COOLING CYCLE: COLD AMBIENT TEMPERATURE

LABORATORY OBJECTIVE
You will determine if the charge is correct on a split system heat pump using a liquid line approach temperature method operating the heat pump in a cold ambient temperature.

FUNDAMENTALS OF HVACR TEXT REFERENCE
Attached Article **Heating Cycle Charging** / Unit 65 Heat Pump Installation / Unit 66 Troubleshooting Heat Pumps

REQUIRED MATERIALS PROVIDED BY STUDENT
Safety Glasses
Gloves
Refrigeration Gauges
6-in-1 Screwdriver
Temperature tester

REQUIRED MATERIALS PROVIDED BY SCHOOL
Split System Heat Pump with liquid line ambient temperature approach charging instructions

PROCEDURE
You will be operating the unit in cooling even though the outdoor ambient is cold.

Find the manufacturer's specification for the liquid line approach method.

Read the instructions on the chart.

Pay attention to any specific operating conditions that must be met for the chart to be accurate.

Some manufacturers require blocking the condenser to achieve a minimum liquid pressure.

Most charts assume that the return air is near comfort conditions.

Install your gauges on the assigned heat pump.

Start the unit in cooling.

Verify that the system has good airflow across both the evaporator and condenser.

Measure the outdoor ambient temperature.

Measure the liquid line temperature.

The approach is the temperature difference between the liquid line temperature and the ambient temperature.

If the liquid line is cooler than expected, the unit is overcharged.

If the liquid line is warmer than expected, the unit is undercharged.

LAB 66.5 PACKAGED HEAT PUMP CHARGING – HEATING CYCLE

LABORATORY OBJECTIVE
You will determine if the charge is correct on a packaged heat pump using a manufacturer's charging chart in the heating cycle.

FUNDAMENTALS OF HVACR TEXT REFERENCE
Attached Article **Heating Cycle Charging** / Unit 65 Heat Pump Installation / Unit 66 Troubleshooting Heat Pumps

REQUIRED MATERIALS PROVIDED BY STUDENT
Safety Glasses
Gloves
Refrigeration Gauges
6-in-1 Screwdriver
Temperature tester

REQUIRED MATERIALS PROVIDED BY SCHOOL
Packaged Heat Pump with a manufacturer's charging chart, which correlates suction pressure, liquid pressure, and outdoor ambient temperature.

PROCEDURE
Find the manufacturer's charging chart.

Read the instructions on the chart.

Pay attention to any specific operating conditions that must be met for the chart to be accurate.

Most charts assume that the return air is near comfort conditions.

Install your gauges on the assigned packaged heat pump.

Start the unit in heating.

Verify that the system has good airflow across both the evaporator and condenser.

Measure the return air temperature.

Allow the unit to run long enough for the pressures to stabilize.

Determine the intersection of the suction and liquid pressures on the chart.

If the intersection is below the ambient temperature line, the system is undercharged.

If the intersection is above the ambient temperature line, the system is overcharged.

Note: Some allowance for error must be made. If the intersection is within 3°F, the system is probably charged correctly.

LAB 66.6 HEAT PUMP CHARGING – HEATING CYCLE DISCHARGE LINE TEMPERATURE

LABORATORY OBJECTIVE
You will determine if the charge is correct on a heat pump operating in the heating cycle using the discharge line temperature method.

FUNDAMENTALS OF HVACR TEXT REFERENCE
Attached Article **Heating Cycle Charging** / Unit 65 Heat Pump Installation / Unit 66 Troubleshooting Heat Pumps

REQUIRED MATERIALS PROVIDED BY STUDENT
Safety Glasses
Gloves
Refrigeration Gauges
6-in-1 Screwdriver
Temperature tester

REQUIRED MATERIALS PROVIDED BY SCHOOL
Heat Pump with discharge line temperature specification.

PROCEDURE
You will be operating the unit in heating.

Find the manufacturer's discharge line temperature specifications.

Read the instructions on the chart.

Pay attention to any specific operating conditions that must be met for the chart to be accurate.

Read the ambient temperature.

Start the unit in heating.

Verify that the system has good airflow across both the evaporator and condenser.

Read the discharge line temperature within 6" of the compressor.

It should be 100°F –110°F warmer than the ambient temperature. (Use manufacturer's specific specification if available.)

If the temperature more than 100°F–110°F warmer than the ambient temperature, the unit is undercharged.

If the subcooling is less than 100°F–110°F warmer than the ambient temperature too high, the unit is overcharged.

LAB 66.7 HEAT PUMP CHARGING – HEATING CYCLE TEMPERATURE RISE

LABORATORY OBJECTIVE
You will determine if the charge is correct on a heat pump operating in the heating cycle using the temperature rise method.

FUNDAMENTALS OF HVACR TEXT REFERENCE
Attached Article **Heating Cycle Charging** / Unit 65 Heat Pump Installation / Unit 66 Troubleshooting Heat Pumps

REQUIRED MATERIALS PROVIDED BY STUDENT
Safety Glasses
Gloves
Refrigeration Gauges
6-in-1 Screwdriver
Temperature tester

REQUIRED MATERIALS PROVIDED BY SCHOOL
Heat Pump with indoor temperature rise specification

PROCEDURE
You will be operating the unit in heating.

Find the manufacturer's temperature rise specifications.

Read the instructions on the chart.

Pay attention to any specific operating conditions that must be met for the chart to be accurate.

Read the ambient temperature.

Start the unit in heating.

Verify that the system has good airflow across both the evaporator and condenser.

Read the return air temperature at the return air plenum.

Read the supply air temperature at the supply air plenum.

Determine the appropriate temperature rise for the outdoor ambient temperature.

The actual measured temperature difference should be + or – 3°F of the desired temperature rise.

If the temperature rise is less than specified, the unit is undercharged or the airflow is above specification.

If the temperature rise is more than specified, the airflow is lower than specification.

LAB 66.8 CHECKING DEFROST CONTROLS

LABORATORY OBJECTIVE
You will check an electronic time-temperature defrost control by manually putting it into defrost.

***FUNDAMENTALS OF HVACR* TEXT REFERENCE**
Unit 66 Troubleshooting Heat Pumps

REQUIRED MATERIALS PROVIDED BY STUDENT
Safety Glasses
6-in-1 screwdriver
Non-contact voltage probe
Multimeter

REQUIRED MATERIALS PROVIDED BY SCHOOL
Heat Pump with Defrost Control

PROCEDURE
First, you need to get the outdoor coil to freeze.

If the weather is cool, you can block the airflow to the outdoor coil using plastic trash bags.

If it is hot outside, you may need to disconnect the outdoor fan motor.

> Turn the power off to the unit.

> Check with your non-contact voltage detector or multimeter to insure power is off.

> Remove the fan wire from the defrost control and tape it up.

Operate the unit in heating until the coil has a good coating of frost around the defrost thermostat.

While the unit is still operating in heating, jump the test pins.

The location and labeling of the test pins will vary depending upon the board and manufacturer.

The unit may not react immediately – hold on the jumper for up to 2 minutes.

The unit should shift into cooling cycle.

Remove the jumper and wait for the unit to come out of defrost.

Turn the power off after the unit has come back out of defrost.

If you removed the outdoor fan wire, replace the fan wire.

LAB 66.9 TROUBLESHOOTING HEAT PUMPS, SCENARIO 1

LABORATORY OBJECTIVE
Given a heat pump with a problem, you will identify the problem, its root cause, and recommend corrective action. Labs 66.9 – 66.11 will all follow the same procedure, however each lab will present a different problem scenario based upon the unit and corrective action required.

FUNDAMENTALS OF HVACR TEXT REFERENCE
Unit 66 Troubleshooting Heat Pumps

REQUIRED MATERIALS PROVIDED BY STUDENT
Safety Glasses
Hand tools
Manometer
Multimeter

REQUIRED MATERIALS PROVIDED BY SCHOOL
Heat pump with a problem

PROCEDURE
Troubleshoot the gas furnace assigned by the instructor. Be sure to be complete in your description of the problem. You should include:

What is the furnace is doing wrong?

What component or condition is causing this?

What tests did you perform that told you this?

How would you correct this?

LAB 66.10 TROUBLESHOOTING HEAT PUMPS, SCENARIO 2

LABORATORY OBJECTIVE
Given a heat pump with a problem, you will identify the problem, its root cause, and recommend corrective action. Labs 66.9 – 66.11 will all follow the same procedure, however each lab will present a different problem scenario based upon the unit and corrective action required.

FUNDAMENTALS OF HVACR TEXT REFERENCE
Unit 66 Troubleshooting Heat Pumps

REQUIRED MATERIALS PROVIDED BY STUDENT
Safety Glasses
Hand tools
Manometer
Multimeter

REQUIRED MATERIALS PROVIDED BY SCHOOL
Heat pump with a problem

PROCEDURE
Troubleshoot the gas furnace assigned by the instructor. Be sure to be complete in your description of the problem. You should include:

What is the furnace is doing wrong?

What component or condition is causing this?

What tests did you perform that told you this?

How would you correct this?

LAB 66.11 TROUBLESHOOTING HEAT PUMPS, SCENARIO 3

LABORATORY OBJECTIVE
Given a heat pump with a problem, you will identify the problem, its root cause, and recommend corrective action. Labs 66.9 – 66.11 will all follow the same procedure, however each lab will present a different problem scenario based upon the unit and corrective action required.

FUNDAMENTALS OF HVACR **TEXT REFERENCE**
Unit 66 Troubleshooting Heat Pumps

REQUIRED MATERIALS PROVIDED BY STUDENT
Safety Glasses
Hand tools
Manometer
Multimeter

REQUIRED MATERIALS PROVIDED BY SCHOOL
Heat pump with a problem

PROCEDURE
Troubleshoot the gas furnace assigned by the instructor. Be sure to be complete in your description of the problem. You should include:

What is the furnace is doing wrong?

What component or condition is causing this?

What tests did you perform that told you this?

How would you correct this?

LAB 70.1 MANUAL J8 BLOCK LOAD

LABORATORY OBJECTIVE
You will demonstrate how to use the Manual J8ae Speed-Sheet to calculate a block load.

LABORATORY NOTES
The Manual J Speed-Sheet is an Excel file that automates Manual J calculations. You will use the Manual J8ae Speed-Sheet and tables from Manual J8 Abridged Edition to calculate the heat loss and gain of a house using the block load procedure.

FUNDAMENTALS OF HVACR TEXT REFERENCE
Unit 41 Fundamentals of Psychrometrics and Airflow / Unit 37 Electric Motors/ Unit 75 Fans and Air Handling Units

REQUIRED MATERIALS
Manual J8 abridged or tables from Manual J8 abridged
Computer with Microsoft Excel or Open Office
Manual J8ae Speed-Sheet
Plans for a small one story house

PROCEDURE
This Excel Sheet has tabs for different parts of the calculation.
Design Conditions
Start by clicking on the "J1 Form" tab at the bottom.
Enter the Project Name and Indoor design conditions in the white spaces at the top of the J1 Form.
Select the State and City by clicking in the white box and then clicking on the arrow to get a drop down list.
Most of the J1 Form is completed by the entries made on the Yellow tabs.
Doors
Click on the "Doors" tab .
Fill in the construction number and details in the left column and the U-value in the right column. You can find these data in manual J.
You do not need an entry for every door. Rather, you need an entry for every type of door. If a house has three exterior doors that are all identical in construction, you only need one entry.
Walls
Next, click on the Yellow "Walls" tab. (We will do the glass and windows last)
Fill in the construction number and details in the left column and the U-value in the middle column, and the Group number in the right column. You can find these data in manual J.
You do not need an entry for every wall. Rather, you need an entry for every type of wall. If a house has a basement , you would need up to three entries: the normal frame walls, the basement wall below grade, and the basement wall above grade.
Ceilings
Click on the yellow "Ceiling" tab.

Fill in the construction number and details in the left column and the U-value in the middle column, and the Cooling Temperature Difference (CLTD) in the right column. You can find these data in manual J.

You do not need an entry for every ceiling. Rather, you need an entry for every type of ceiling. If a house has a cathedral ceiling it would have two entries: one for the regular ceilings and another for the cathedral ceiling.

Floors

Click on the yellow "Floors" tab.

Fill in the construction number and details in the left column and the U-value in the right column. You can find these data in manual J.

You do not need an entry for every floor. Rather, you need an entry for every type of floor. Most houses will only have one type of floor.

Windows

Click on the yellow "Glass" tab.

Fill in the construction number and details in the left column, the U-value in the middle column, and the Cooling Heat Transfer Multiplier (HTM) on the right under the column for the direction the window faces. You can find these data in manual J.

You need an entry for every type of window and every facing. If a house has only one type of window with four facings, it needs four entries.

Click on the Green "Glass Schedule" tab.

This tab calculates the window shade line from the roof overhang.

You need an entry for every facing of each type of window in this table.

Click on the first yellow space on the left and choose the window type and direction from the drop down list.

Enter the overhang information in the white spaces beside it.

Repeat this for every window facing and type.

Net Area

The white columns labeled net area are where the area of different materials is entered.

The first white column labeled net area is for the entire house, or block load

Line 6A Windows Form J1

Click on the "J1 Form" green tab.

You should see the different window types and HTMs listed in the blue section in the top left.

Find line 6A and the white space under the "Net Area" column next to the block load column.

Enter the total ft2 area of the first window facing. This is not the area of a single window, but the total of all the similar windows facing that direction.

Repeat for each window facing. Most houses will have three window facings since east and west are the same.

Line 7 Doors Form J1

Go to line 7 for doors and click on the first yellow line to get a drop down box.

Select the type of door.

In the white box under "Net Area", enter the total area of all the doors that are similarly constructed.

If the house has more than one type of door, repeat on the second yellow line.

Line 8 & 9 Walls Form J1

Click on the first yellow box on line 8 and select the type of wall.

If the house has any below grade walls, repeat on line 9 "Below Grade Walls"

In the white box under "Net Area" enter the total ft2 area of the above grade walls.

If the house has a basement, enter the total area of the below grade walls in the white box on line 9.

666

Line 10 Ceiling Form J1

Click on the first yellow box on line 10 and select the type of ceiling.

In the white box under "Net Area" enter the total ft2 area of the ceiling.

Line 11 Floor Form J1

Click on the first yellow box on line 11 and select the type of floor.

In the white box under "Net Area" enter the total ft2 area of the floor.

Note: If the floor is a concrete slab on grade, the total perimeter distance is used instead of the square feet.

Line 12 Infiltration Form J1

On line 12, click to select the house tightness and the number of fireplaces.

In the large white box fill in the total heated and cooled floor area.

In the next white box enter the total above grade volume. You find that by multiplying the floor area by the ceiling height. Do not count basement floor area even if it is heated.

Line 13 Internal Gains Form J1

Click on the yellow box to select the appliance "Scenario"

You MUST enter enough BTUs in one more rooms to total exactly the BTUs shown in the Yellow box.

This number is entered in the white **Net Area** box in the Block Load column.

In the small white box under it enter the number of bedrooms.

The program will calculate the number of occupants based on the number of bedrooms and display the number in the blue box next to occupants. You MUST enter that exact number in the white box under "Net Area" or the program will refuse to total the calculations.

Line 15 Duct Loss Form J1

In the yellow boxes, select the duct type and location, the R-value of the duct insulation, and the duct leakage.

In the white boxes enter the installed square feet of duct surface area for the supply and return.

Line 15 Ventilation Form J1

Check the boxes for water heater or furnace if the house has a gas water heater or gas furnace.

Select the amount of ventilation in the yellow drop down box.

Line 19 Blower Heat Gain Form J1

Select whether or not the manufacturer considers blower heat gain in the cooling capacity

Line 20 shows the total sensible loss (heating) and gain (cooling)

Line 21 shows the total latent gain (cooling)

LAB 70.2 MANUAL J8 ROOM BY ROOM LOAD

LABORATORY OBJECTIVE
You will demonstrate how to use the Manual J8ae speed sheet to calculate a room by room load calculation.

LABORATORY NOTES
The Manual J Speed-Sheet is an Excel file that automates Manual J calculations. You will use the Manual J8ae Speed-Sheet and tables from Manual J8 Abridged Edition to calculate the heat loss and gain of a house using the room by room procedure.

FUNDAMENTALS OF HVACR TEXT REFERENCE
Unit 41 Fundamentals of Psychrometrics and Airflow / Unit 37 Electric Motors/ Unit 75 Fans and Air Handling Units

REQUIRED MATERIALS
Manual J8 abridged or tables from Manual J8 abridged
Computer with Microsoft Excel or Open Office
Manual J8ae Speed-Sheet
Plans for a small one story house

PROCEDURE
This Excel Sheet has tabs for different parts of the calculation.
The room by room procedure is the same as the block load, but the numbers entered under the "Net Area" are for each individual room. The summary tab shows the CFM for each room based on the heat loss and gain in each room. The sections labeled *Design Conditions, Doors, Walls, Ceilings, Floors,* and *Windows* are exactly the same for both the block load and the room by room load.
Design Conditions
Start by clicking on the "J1 Form" tab at the bottom.
Enter the Project Name and Indoor design conditions in the white spaces at the top of the J1 Form.
Select the State and City by clicking in the white box and then clicking on the arrow to get a drop down list.
Most of the J1 Form is completed by the entries made on the Yellow tabs.
Doors
Click on the "Doors" tab .
Fill in the construction number and details in the left column and the U-value in the right column. You can find these data in manual J.
You do not need an entry for every door. Rather, you need an entry for every type of door. If a house has three exterior doors that are all identical in construction, you only need one entry.
Walls
Next, click on the Yellow "Walls" tab. (We will do the glass and windows last)
Fill in the construction number and details in the left column and the U-value in the middle column, and the Group number in the right column. You can find these data in manual J.

You do not need an entry for every wall. Rather, you need an entry for every type of wall. If a house has a basement , you would need up to three entries: the normal frame walls, the basement wall below grade, and the basement wall above grade.

Ceilings

Click on the yellow "Ceiling" tab.

Fill in the construction number and details in the left column and the U-value in the middle column, and the Cooling Temperature Difference (CLTD) in the right column. You can find these data in manual J.

You do not need an entry for every ceiling. Rather, you need an entry for every type of ceiling. If a house has a cathedral ceiling it would have two entries: one for the regular ceilings and another for the cathedral ceiling.

Floors

Click on the yellow "Floors" tab.

Fill in the construction number and details in the left column and the U-value in the right column. You can find these data in manual J.

You do not need an entry for every floor. Rather, you need an entry for every type of floor. Most houses will only have one type of floor.

Windows

Click on the yellow "Glass" tab.

Fill in the construction number and details in the left column, the U-value in the middle column, and the Cooling Heat Transfer Multiplier (HTM) on the right under the column for the direction the window faces. You can find these data in manual J.

You need an entry for every type of window and every facing. If a house has only one type of window with four facings, it needs four entries.

Click on the Green "Glass Schedule" tab.

This tab calculates the window shade line from the roof overhang.

You need an entry for every facing of each type of window in this table.

Click on the first yellow space on the left and choose the window type and direction from the drop down list.

Enter the overhang information in the white spaces beside it.

Repeat this for every window facing and type.

Room Information

Type in the name of each room in the white boxes across the top on the right.

Note: Only name areas that might reasonably have a supply register. The area of closets and interior halls can be added to adjacent rooms.

To see all the room columns, use the right and left arrows at the bottom right of the spread sheet. This scrolls the rooms but leaves the main form in place.

Net Area

The white columns labeled *net area* are where the area of different materials are entered.

The first white column labeled *net area* is for the entire house, or block load.

You should **NOT** enter anything into this column in the room by room form. The spreadsheet will total the entries from the rooms in this column.

The other *net area* columns are for each individual room. This is where all your entries should be.

Form J1

Click on the "J1 Form" green tab.

Line 6A Windows

You should see the different window types and HTMs listed in the blue section in the top left.

For each room - enter the total ft^2 area of the first window facing in that room. This is not the area of a single window, but the total of all the similar windows facing that direction in each room.

Repeat for each room.

Line 7 Doors

Go to line 7 for doors and click on the first yellow line to get a drop down box.

Select the type of door.

For each room - in the white box under **Net Area**, enter the total area of all the **exterior** doors for that room.

All rooms will not have an exterior door. If a room does not have an exterior door, leave that space blank.

Line 8 & 9 Walls

Click on the first yellow box on line 8 and select the type of wall.

If the house has any below grade walls, repeat on line 9 "Below Grade Walls"

For each room - In the white box under **Net Area** enter the total ft^2 area of the above grade walls for each room.

Repeat for each room in the house.

For rooms below ground, enter the total area of the below grade walls in the white box on line 9.

Line 10 Ceiling

Click on the first yellow box on line 10 and select the type of ceiling.

For each room - In the white box under **Net Area** enter the total ft^2 area of the ceiling for that room.

Repeat for each room.

Line 11 Floor

Click on the first yellow box on line 11 and select the type of floor.

For each room - In the white box under **Net Area** enter the total ft^2 area of the floor for that room.

Note: If the floor is a concrete slab on grade, the total perimeter distance is used instead of the square feet.

Repeat for each room.

Line 12 Infiltration

On line 12, click to select the house tightness and the number of fireplaces.

In the large white box fill in the total heated and cooled floor area.

In the next white box enter the total above grade volume. You find that by multiplying the floor area by the ceiling height. Do not count basement floor area even if it is heated.

The Wall Area Ratio (WAR) is automatically calculated based on the ratio of each room's wall area compared to the total; wall area.

Line 13 Internal Gains

Click on the yellow box to select the appliance "Scenario"

You MUST enter enough BTUs in one more rooms to total exactly the BTUs shown in the Yellow box.

Do NOT enter the number in the block load net area, but in the net area for one or more rooms.

In the small white box under it enter the number of bedrooms.

The program will calculate the number of occupants based on the number of bedrooms and display the number in the blue box next to occupants.

You MUST enter enough people in one or more rooms to total that exact number.

Do NOT enter the number in the block load net area, but in the net area for one or more rooms.

Line 15 Duct Loss

In the yellow boxes, select the duct type and location, the R-value of the duct insulation, and the duct leakage.

In the white boxes enter the installed square feet of duct surface area for the supply and return.

Line 15 Ventilation

Check the boxes for water heater or furnace if the house has a gas water heater or gas furnace.

Select the amount of ventilation in the yellow drop down box.

Line 19 Blower Heat Gain

Select whether or not the manufacturer considers blower heat gain in the cooling capacity

Line 20 shows the total sensible loss (heating) and gain (cooling)

Line 21 shows the total latent gain (cooling)

Summary

Click on the Green *Summary* tab

At the top right are two white boxes: Sensible heat ratio and Design CFM

The sensible heat ratio is the percentage of equipment cooling capacity that actually cools the air.

This is affected by equipment design, airflow, and operating conditions.

In humid areas, you should be between 0.6 to 0.8.

Lower ratios yield more water removal because more capacity is going into latent cooling.

The Design CFM is the airflow the equipment moves.

400 CFM per ton of total cooling capacity (sensible + latent) works on most equipment if you don't have a particular [piece of equipment selected.

LAB 72.1 WIRING ZONE CONTROL SYSTEM

LABORATORY OBJECTIVE
You will wire a complete zone control system including:

Zone panel

Zone thermostats

Zone dampers

Furnace control wiring

Air Conditioner control wiring

LABORATORY NOTES
You will wire a complete zone control system. You will wire the transformer, zone thermostats, and zone dampers to the zone control panel. The zone control panel will wire to the furnace and air conditioner.

FUNDAMENTALS OF HVACR TEXT REFERENCE
Unit 38 Diagrams / Unit 39 Control Systems / Unit 72 Zone Control Systems

REQUIRED MATERIALS PROVIDED BY STUDENT
Safety Glasses

Gloves

Electrical hand tools

Multimeter

Amp Meter

PROCEDURE

The zone panel requires its own transformer. The transformer primary can be wired to the line voltage for the furnace. The transformer secondary wires to the zone panel. The zone panel needs to be powered 24-7.

The zone thermostats each wire to the zone panel at the terminals designated for the thermostats. The zone dampers wire to the panel at the terminals designated for them.

The system control wires that normally go to the thermostat are wired to the zone panel instead. As far as the system is concerned, the zone panel is the thermostat.

Some zone panels use an outdoor temperature sensor and a supply plenum sensor. They are wired to the terminals on the zone panel designated for them.

LAB 73.1 ROTATING VANE ANEMOMETER

LABORATORY OBJECTIVE
Demonstrate your ability to use a rotating vane anemometer to measure air flow velocity and air flow volume.

FUNDAMENTALS OF HVACR TEXT REFERENCE

Unit 73 Testing and Balancing Air Systems

REQUIRED MATERIALS PROVIDED BY STUDENT
Ruler
Calculator

REQUIRED MATERIALS PROVIDED BY SCHOOL
Rotating vane anemometer
Register or duct

LABORATORY NOTES
You will measure the average air velocity in feet per minute (FPM) leaving a duct. You will use that velocity to calculate the volume of air in cubic feet per minute (CFM). You calculate the air volume in CFM by multiplying the velocity in FPM by the duct cross sectional area in square feet.

PROCEDURE

Turn on the blower.

Place the anemometer in the airstream and turn it on.

Set the anemometer to average the reading and continue to hold it in the airstream.

Move the anemometer slowly across the face of the duct or grille.

Make sure to read the whole face of the duct or grille.

Push the hold button and take the average reading in FPM.

Calculate the area of the duct or grille in square feet.

Multiply the FPM by the square feet to arrive at the volume in cubic feet per minute CFM.

LAB 73.2 FLOW HOOD

LABORATORY OBJECTIVE
Demonstrate your ability to use a flow hood to measure air volume.

LABORATORY NOTES
You will measure air volume in cubic feet per minute (CPM) at supply registers and return grilles using the flowhood.

***FUNDAMENTALS OF HVACR* TEXT REFERENCE**
Unit 73 Testing and Balancing Air Systems

REQUIRED MATERIALS PROVIDED BY SCHOOL
Flow hood
Duct system, with supply registers and return air grilles

PROCEDURE
Turn on the blower.

Assemble the hood on the flow hood.

Set the flowhood to read supply air on the highest scale.

Place the flowhood over the supply register and look at the reading.

If the reading is low, adjust the setting until you have a measureable reading.

Record the reading.

Repeat on another supply register.

Set the flowhood to measure return air on the highest scale.

Place the flowhood over the return grille and look at the reading.

If the reading is low, adjust the setting until you have a measureable reading.

Record the reading.

LAB 73.3 USING MAGNEHELIC GAUGE

LABORATORY OBJECTIVE
Demonstrate your ability to use a magnehelic gauge to measure total pressure, static pressure, and velocity pressure in a duct system.

LABORATORY NOTES
You will measure the total pressure, static pressure, and velocity pressure in a duct system. You will calculate the velocity using the velocity pressure, and then you will calculate the air volume in CFM using the calculated velocity and duct cross sectional area.

***FUNDAMENTALS OF HVACR* TEXT REFERENCE**
Unit 41 Psychrometrics and Airflow / Unit 73 Testing and Balancing Air Systems

REQUIRED MATERIALS PROVIDED BY STUDENT
Ruler
Calculator

REQUIRED MATERIALS PROVIDED BY SCHOOL
Magnehelic Gauge
Duct system
Static Pressure pickup tube
Pitot tube
Drill
¼" drill bit

PROCEDURE

Drill a hole in the duct large enough to allow the static pressure tube and the pitot tube to pass through – approximately ¼".

Insert the static pressure tube into the duct with the tip facing the airflow.

Position the magnehelic horizontally so that the needle rests at 0.

Connect the rubber hose from the high pressure connection on the magnehelic to the static pressure tube.

Turn on the blower and red the static pressure in inches of water column – wc.

Replace the static pressure tip with a pitot tube.

Connect a rubber hose from the end of the pitot tube to the magnehelic high pressure connection.

Read the total pressure in inches of water column – wc.

Connect another rubber hose from the side of the pitot tube to the magnehelic low pressure connection.

Read the velocity pressure in inches of water column – wc.

Convert velocity pressure to velocity

Velocity = 4005√ velocity pressure

Calculate air volume by multiplying the air velocity by the cross sectional area.

Cross sectional area in square feet = (length inches x width inches) / 144

CFM = FPM velocity x Area ft^2

676

LAB 75.1 BELT DRIVES

LABORATORY OBJECTIVE
You will learn to adjust and correctly tension a belt drive.

FUNDAMENTALS OF HVACR **TEXT REFERENCE**
Unit 75 Fans and Air Handling Units / Unit 90 Planned Maintenance

REQUIRED MATERIALS PROVIDED BY STUDENT
Amp meter
Electrical hand tools

REQUIRED MATERIALS PROVIDED BY SCHOOL
Belt drive blower with adjustable pulley.

PROCEDURE

1. Operate the belt drive blower assigned by the instructor.

2. Use a digital photo-tachometer to measure the motor RPM and the blower RPM.

3. Measure the motor amp draw.

4. Loosen and remove the belt.

5. Adjust the pulley out counterclockwise ½ round. Make sure to position the set screw over a flat on the pulley before tightening the set screw.

6. Reinstall and re-tension the belt.

7. Operate the blower and measure the motor amp draw.

8. Use a digital photo-tachometer to measure the motor RPM and the blower RPM.

9. Loosen and remove the belt.

10. Adjust the pulley back in clockwise ½ round. Make sure to position the set screw over a flat on the pulley before tightening the set screw.

11. Reinstall and re-tension the belt.

12. Operate the blower and measure the motor amp draw.

13. Use a digital photo-tachometer to measure the motor RPM and the blower RPM.

LAB 75.2 DIRECT DRIVE MOTOR APPLICATIONS

LABORATORY OBJECTIVE
You will learn to identify the types of direct drive electric motor applications in HVACR.

FUNDAMENTALS OF HVACR TEXT REFERENCE
Unit 75 Fans and Air Handling Units / Unit 90 Planned Maintenance

REQUIRED MATERIALS PROVIDED BY STUDENT
Amp meter
Electrical hand tools

REQUIRED MATERIALS PROVIDED BY SCHOOL
Selection of Direct Drive Motor Applications

PROCEDURE
Find three different direct drive electric motor applications in the shop. In a direct drive application the motor is connected directly to the device it is operating.

Application	Connection: Hub, Spring Coupling, Rubber Coupling

LAB 75.3 AC INDUCTION MOTOR BLOWER PROPERTIES

LABORATORY OBJECTIVE

You will demonstrate the effect airflow restriction has on a forward curved centrifugal blower with an AC induction motor.

LABORATORY NOTES

You will operate the blower and read the motor amp draw. You will then restrict first the intake, and then the exhaust and measure the change in the blower motor current.

***FUNDAMENTALS OF HVACR* TEXT REFERENCE**

Unit 41 Fundamentals of Psychrometrics and Airflow / Unit 75 Fans and Air Handling Units

REQUIRED MATERIAL PROVIDED BY STUDENT

Clamp Ammeter

REQUIRED MATERIAL PROVIDED BY SCHOOL

PSC centrifugal blower
Piece of cardboard, wood, or metal to block airflow
 irflow hood

PROCEDURE

Operate the blower with no restriction and measure the blower motor amp draw.
Measure the blower CFM with a flowhood.
Restrict the blower intake, operate the blower and measure the amp draw.
Measure the blower CFM with a flowhood.
What happens to the blower motor amp draw when the intake is blocked?
What happens to the blower CFM when the intake is blocked?
Restrict the blower exhaust, operate the blower and measure the amp draw.
Measure the blower CFM with a flowhood.
What happens to the blower CFM when the exhaust is blocked?
What happens to the blower motor amp draw when the exhaust is blocked?

PSC BLOWER DATA		
CONDITION	CFM	AMPS
Normal		
Intake Restricted		
Exhaust Restricted		

LAB 75.4 ECM BLOWER PROPERTIES

LABORATORY OBJECTIVE
You will demonstrate the effect airflow restriction has on a forward curved centrifugal blower with an electronically commutated motor (ECM).

LABORATORY NOTES
You will operate the blower and read the motor amp draw. You will then restrict first the intake, and then the exhaust and measure the change in the blower motor current.

FUNDAMENTALS OF HVACR TEXT REFERENCE
Unit 41 Fundamentals of Psychrometrics and Airflow / Unit 37 Electric Motors/ Unit 75 Fans and Air Handling Units

REQUIRED MATERIAL PROVIDED BY STUDENT
Clamp Ammeter

REQUIRED MATERIAL PROVIDED BY SCHOOL
ECM centrifugal blower
Piece of cardboard, wood, or metal to block airflow
Airflow low hood

PROCEDURE
Operate the blower with no restriction and measure the blower motor amp draw.
Measure the blower CFM with a flowhood.
Restrict the blower intake, operate the blower and measure the amp draw.
Measure the blower CFM with a flowhood.
What happens to the blower motor amp draw when the intake is blocked?
What happens to the blower CFM when the intake is blocked?
Restrict the blower exhaust, operate the blower and measure the amp draw.
Measure the blower CFM with a flowhood.
What happens to the blower CFM when the exhaust is blocked?
What happens to the blower motor amp draw when the exhaust is blocked?

PSC BLOWER DATA		
CONDITION	CFM	AMPS
Normal		
Intake Restricted		
Exhaust Restricted		

LAB 80.1 WATER CIRCULATING PUMP COMPLETE SERVICE

LABORATORY OBJECTIVE
The purpose of this lab is to demonstrate your ability to conduct the proper maintenance service procedure for a water circulating pump.

LABORATORY NOTES
Hot water heating boilers are dependent on water pumps to distribute the hot water they produce. The pumps are primarily driven by electric motors. The electric motor requires typical motor service, cleaning, lubrication, proper mounting, insulation resistance and amperage checks. The pump is usually connected to the pump with a spider type coupler or flexible spring connected coupler. Improper alignment is the main reason for coupler problems. Other pump maintenance issues fall into two major categories: impeller problems and shaft seal problems. On smaller pumps it is typical to exchange the entire pump or bearing assembly for a new or rebuilt replacement. Larger pumps can generally be brought in to be rebuilt by the local factory representative. If the pump is to be rebuilt on site, it is recommended that you have attended factory service demonstrations and have been trained in this area.

FUNDAMENTALS OF HVACR TEXT REFERENCE
Unit 80 Hydronic Heating Systems/ Unit 81 Boilers and Related Equipment

REQUIRED TOOLS AND EQUIPMENT
Circulating pump
Tool kit
T-handle Allen wrench
Strap wrench
Compressed air
Cleaner - degreaser
Rubber hammer
Thin chisel

SAFETY REQUIREMENTS
A. Check all circuits for voltage before doing any service work.

B. Be careful to avoid lubricant contact with exposed skin.

C. Always wear safety glasses whenever using compressed air or CO_2 for blowing out dust and dirt.

PROCEDURE
Step 1

Collect the pump data and fill in the chart.

Blower Unit Data

Circulating Pump Make:		Model Number:	
Motor Data:	Horsepower:		Type and Size of Coupler:

Step 2

Complete an initial inspection. Complete the Initial Inspection & Cleaning Check List and make sure to check each step off in the appropriate box as you finish it. This will help you to keep track of your progress.

Initial Inspection & Cleaning Check List

STEP	PROCEDURE	CHECK
1	Make sure that the power to the circulating pump motor is secured. Verify this with a multi-meter voltage test. Refer to Unit 35 in *Fundamentals of HVACR*.	
2	Use CO_2 or compressed air pressure to blow out and clean all open air passages of the motor and bearing assembly.	
3	Use a non-detergent or recommended lubricant for the motor and bearing assembly.	
4	Using a pump sprayer, apply a cleaner degreaser to the motor and pump assembly.	

STEP	PROCEDURE	CHECK
5	Use cleaning rags to wipe off all excess oil and any accumulated dirt, dust, grease, etc.	
6	Inspect the bearing assembly for any water dripping, metal shavings, scraping or grinding noise, etc.	
7	Inspect motor mounts. Is the motor centered within the rubber mount or is the motor sagging in the rubber?	
8	If the motor is sagging or out of center in the rubber mount, the motor mounts must be replaced.	

Step 3

After finishing the initial circulating pump inspection and cleaning, make sure all power is off. Lock and tag the power panel before removing any parts. Complete the Motor Mount Replacement Check List.

Motor Mount Replacement Check List

STEP	PROCEDURE	CHECK
1	Make sure that the power to the blower motor is secured. Verify this with a multi-meter voltage test. Refer to Unit 35 in *Fundamentals of HVACR*.	
2	Using a thin open end wrench, loosen the two top motor mount bolts from inside the bearing assembly.	

STEP	PROCEDURE	CHECK
3	Loosen and remove the two bottom motor mount bolts.	
4	Use a long T-handle Allen wrench to remove the motor end of the spring coupler.	
5	Supporting the motor in one hand, remove the previously loosened top two motor mount bolts.	
6	Remove the motor from its cradle and the two machine screws connected to the motor mount straps.	
7	Using an old screwdriver or a thin chisel, pry of both rubber motor mounts from the motor.	
8	Use a small metal rubber or plastic hammer to tap in place two new rubber motor mounts.	

Step 4

If the bearings drag and need to be replaced then complete the Bearing Replacement Check List.

Bearing Replacement Check List

STEP	PROCEDURE	CHECK
1	Make sure that the power to the blower motor is secured. Verify this with a multi-meter voltage test. Refer to Unit 35 in *Fundamentals of HVACR*.	

2	Turn off the water isolation valves for the pump and drain off any remaining water.	
3	Remove the motor as in Step # 3 from the Motor Mount Replacement Check List.	
4	Remove the bolts holding the bearing assembly to the pump housing.	
5	Remove the bearing assembly along with the pump impeller.	
6	Inspect and clear any debris from the pump housing.	
7	Hold the impeller with a strap wrench and use a socket wrench to remove the bolt holding the impeller to the pump shaft.	
8	Inspect the impeller and replace it if there are any signs of wear or deterioration.	
9	Install the impeller on the new bearing assembly.	
10	Remove any old gaskets and carefully scrape and clean the surface metal. Always install new gaskets.	

11	After installing new gasket, slip the bearing assembly into the pump housing.	
12	Install and tighten the bolts holding the bearing assembly to the pump housing.	
13	Install the coupler to the bearing assembly shaft.	
14	Hold the motor in position and slide the coupler on to the motor shaft. Be sure to tighten the set screws into the hollow spot on both the pump and motor shafts.	
15	While still holding the motor, start the two top motor mount bolts first, then the bottom two bolts.	
16	Tighten the motor mount bolts.	
17	Open the water isolation valves and obtain normal pressure on the system. Purge any air from the system.	
18	Start up the pump and obtain normal operation.	
19	Lubricate the bearing assembly and motor as required.	
20	Inspect the bearing assembly for any water leaks.	

Name _____

Date _____

Instructor's OK ☐

LAB 88.1 TESTING POTENTIAL START RELAYS

LABORATORY OBJECTIVE

The student will demonstrate how to properly test a potential starting relay.

LABORATORY NOTES

A potential starting relay assists in starting the motor by allowing current flow to the starting winding through the start capacitor and the normally closed contacts in the relay.

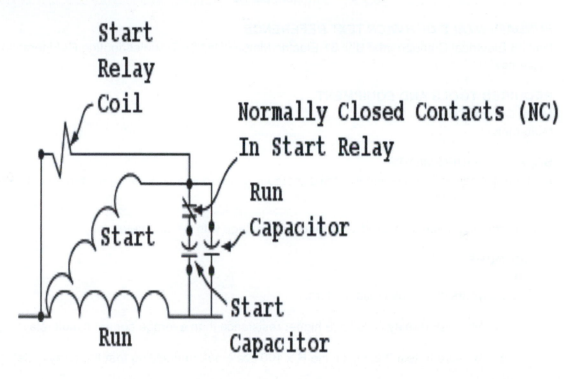

Figure 88-1-1

The potential starting relay coil is energized by the counter electromotive force developed by the motor. The faster the motor turns, the higher the counter electromotive force. When the motor comes up to speed the start relay coil energizes and acting as an electromagnet it will open the normally closed contacts in the start relay. The coil is wired in parallel to the start winding. The contacts are wired in series with the start winding and the start capacitor.

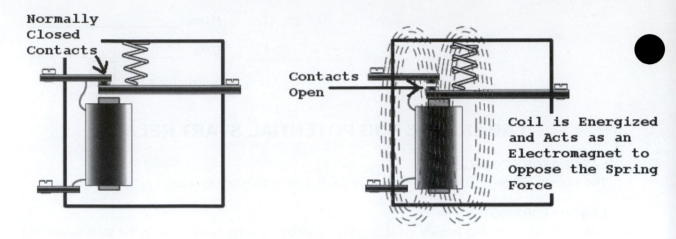

Normally
Closed
Contacts

Contacts
Open

Coil is Energized
and Acts as an
Electromagnet to
Oppose the Spring
Force

Figure 88-1-2

FUNDAMENTALS OF HVACR TEXT REFERENCE
Unit 36 Electrical Components/ Unit 37 Electric Motors/ Unit 88 Troubleshooting Refrigeration Systems

REQUIRED TOOLS AND EQUIPMENT
Motor Control Circuit
Multi-Meter

SAFETY REQUIREMENTS
A. Turn the power off if the winding to be tested is installed in an operating system. Lock and tag out the power supply.

B. Confirm the power is secured by testing for 0 voltage with a multi-meter.

PROCEDURE
Step 1
Familiarize yourself with electrical meters.

    A. Many start relay coils have higher resistance than average control circuit relays.

    B. Be sure to test the coil on the R x 100 scale before deciding that the relay is defective.

    C. The pull-in/drop-out voltage of the relay is unique. Do not attempt to replace it with an ordinary, similar voltage relay.

Step 2
Test the relay coil as follows:

    A. Start with the resistance reading on the highest scale.

    B. Record the resistance reading _____Ohms.

    C. Compare this reading to the manufacturer's data to see if the coil is satisfactory. A resistance exceeding 5000 ohms is common for most potential start relays.

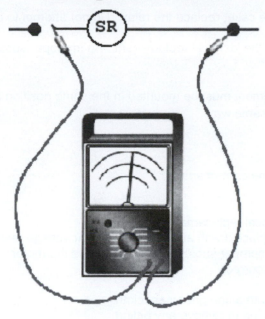

Figure 88-1-3

Step 3
Test the relay contacts as follows:

A. When testing start relay contacts, the contacts should be closed and the ohmmeter will read zero resistance.

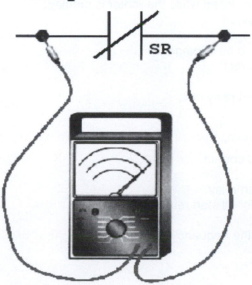

Figure 88-1-4

B. Sometimes while in operation the contacts stick closed and become badly burned. This will make for a poor connection which will be indicated by a positive resistance reading.

C. If such is the case, replace the relay. Do not attempt to clean the contacts.

D. Always use the identical replacement. An improper substitution can damage the motor.

E. The replacement must be mounted in the same position as the original and connected the same way.

QUESTIONS

(Circle the letter that indicates the correct answer.)

1. When replacing a start relay:
   A. always use an identical replacement.
   B. do not attempt to replace it with and ordinary, similar voltage relay.
   C. remember that an improper substitution can damage the motor.
   D. All of the above are correct.
2. Burned start relay contacts:
   A. should be cleaned with a fine grit emery cloth.
   B. should be carefully filed to remove any pitting.
   C. should be cleaned with the proper solvent.
   D. should be replaced.
3. Good start relay contacts should have:
   A. an infinite resistance when closed.
   B. a fairly high resistance when closed.
   C. no resistance when closed.
   D. a resistance of exactly 100 mega ohms.
4. Start relay coils have lower resistance than average control circuit relays:
   A. True.
   B. False.
5. If the start relay contacts stick closed while the motor is running:
   A. the motor will over speed.
   B. the motor will stall.
   C. the start winding could be damaged.
   D. Both A and C are correct.

6. An infinite resistance in a start relay coil:
   A. indicates a short.
   B. indicates an open.
   C. is normal.
   D. None of the above is correct.
7. The pull-in/drop-out voltage:
   A. is unique to each start relay.
   B. is always the same for all start relays.
   C. is easily adjusted.
   D. is directly related to the frequency of the circuit.

LAB 88.2 OHMING SINGLE PHASE HERMETIC COMPRESSOR MOTORS

LABORATORY OBJECTIVE
The purpose of this lab is to learn how to check a hermetic compressor motors for faults using an ohm meter.

LABORATORY NOTES
We will use an ohm meter to test the windings of hermetic compressors. You will determine if the motor is good, open, shorted, or grounded. If the motor is good, you will use the ohm readings to identify the Common, Start, and Run terminals.

FUNDAMENTALS OF HVACR TEXT REFERENCE
Unit 37 Motors / Unit 88 Troubleshooting Refrigeration Systems

REQUIRED MATERIALS PROVIDED BY STUDENT
Volt-Ohm meter
Electrical hand tools

REQUIRED MATERIALS PROVIDED BY SCHOOL
Hermetic compressors.

PROCEDURE
1. Take ohm readings between each combination of two terminals – there should be three readings
2. Write the readings down.
3. If any of the three readings is infinite, the motor winding is open.
4. If any of the three readings is less than 0.3 ohms, the motor is most likely shorted.
   Note: Some very large compressors may have windings with resistances this low.
5. Read the resistance between each terminal and one of the copper lines on the compressor.
6. Any reading other than infinite indicates a grounded motor.
   Note: A reading of 4 million ohms or higher may indicate contamination, not a grounded motor.
7. If the motor is not open, shorted, or grounded you can determine the C, S, and R terminals.
8. The highest reading will be between the start and run terminals. Therefore, the terminal that is NOT involved in the highest reading is common.
9. The lowest reading is between Common and Run. The terminal that is involved in the lowest reading with the Common terminal is Run.
10. The start terminal will be the terminal left over after identifying both the common and run terminals.
11. Complete the table on the next page and be prepared to explain your results to the instructor.

The three most typical terminal arrangements are shown in the boxes below.

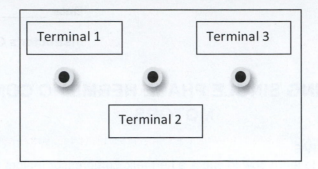

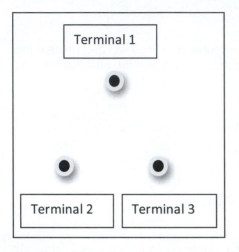

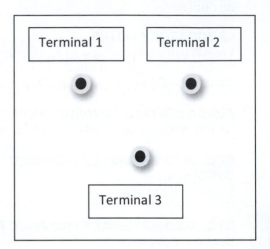

Compressor Motors Resistance Readings							
	TERMINAL READINGS						**CONDITION**
	Term 1 to Term 2	Term 1 to Term 3	Term 2 to Term 3	Terminal 1 to Ground	Terminal 2 to Ground	Terminal 3 to Ground	Good Shorted Open Grounded
Comp 1							
Comp 2							
Comp 3							

Determining Common Start and Run Terminals			
	Terminal 1	Terminal 2	Terminal 3
Comp 1			
Comp 2			
Comp 3			
Comp 4			

LAB 88.3 WIRE AND OPERATE HERMETIC PSC MOTORS

LABORATORY OBJECTIVE
You will learn how to wire and operate hermetic PSC motors used in compressors.

LABORATORY NOTES
You will wire PSC compressors, operate them, and measure their amp draw.

***FUNDAMENTALS OF HVACR* TEXT REFERENCE**
Unit 37 Motors / Unit 88 Troubleshooting Refrigeration Systems

REQUIRED MATERIALS PROVIDED BY STUDENT
Volt-Ohm meter
amp meter
Electrical hand tools

REQUIRED MATERIALS PROVIDED BY SCHOOL
PSC compressors
Run capacitor

PROCEDURE

1. Use the wiring diagram on the motor and/or the diagrams in the book to wire the PSC Compressor motors assigned by the instructor.

2. Be sure to check the motor nameplate voltage and compare it to the voltage you are connecting to the motor.

3. Operate the motor.

4. Use a clamp on amp meter to read the motor current through the common wire.

5. Write down your reading in the chart below.

6. Discuss your results with the instructor.

7. Put up all the motors.

Voltage	Nameplate LRA	Nameplate RLA (if available)	Running Amps

LAB 88.4 TROUBLESHOOTING HERMETIC COMPRESSOR MOTORS

LABORATORY OBJECTIVE
The purpose of this lab is to learn how to troubleshoot PSC hermetic compressor motors for faults using volt, amp, and ohm meters.

LABORATORY NOTES
You will use volt, amp, and ohm meters to identify problems in hermetic PSC compressors. You will determine if the motor is good, open, shorted, grounded, locked, or has a defective run capacitor.

***FUNDAMENTALS OF HVACR* TEXT REFERENCE**
Unit 37 Motors / Unit 88 Troubleshooting Refrigeration Systems

REQUIRED MATERIALS PROVIDED BY STUDENT
Safety glasses
Gloves
Multi- meter with volts, ohms, and microfarads scales.
Electric hand tools.

REQUIRED MATERIALS PROVIDED BY SCHOOL
Hermetic compressors with PSC motors.

PROCEDURE
The procedure varies depending upon the compressor reaction when you try to energize it
The Compressor does not try to start.
1. You need to determine if the compressor is receiving the correct voltage.
2. Check the voltage on the load side of the compressor contactor.
3. If there is no voltage, the problem is in the circuit to the compressor. Problems could include
    a. Tripped breaker or blown fuse
    b. Defective contactor
    c. Loss of control voltage to contactor due to
        i. open safety – high pressure, low pressure, overload
        ii. incorrect control wiring
        iii. bad control transformer
        iv. bad thermostat
        v. bad control board in unit
4. If there is correct voltage on the load side of the contactor, turn power off.
5. Discharge the run capacitor.
6. Remove the compressor terminal cover and check the wire connections.
7. Check the wires against the manufacturer's diagram.
8. Note where the wires are connected and remove them.
9. Use the procedure from Lab 21 to ohm the compressor motor out.
10. If the compressor ohms out, check the capacitor microfarads with the capacitor test function of the multi-meter.

The Compressor tries to start but just hums.

Are the pressures equalized?

PSC compressors cannot start against system pressures, the pressures must be equalized.

1. Turn the power off and install gauges to check the system pressures.
2. If the pressures are not equalized, wait for them to equalize and then turn the power back on.
3. Check the amp draw on the common wire while the compressor is trying to start.
    a. If the amp draw spikes and then drops to around the rated load amperage (RLA), the compressor is running. Check the gauges to see if the high side pressure is rising and the low side pressure is dropping.
    b. If the amp draw is considerably lower than the compressor rated load amps (RLA), the compressor may actually be operating. Check the gauges. If it is drawing a low amp draw and not pumping, it has a mechanical failure.
    c. If the amp draw is high, somewhere around locked rotor amps, the compressor is locked – turn off the power.

Is the compressor receiving the correct voltage?

You need to determine if the compressor is receiving the correct voltage.

Since it is humming, you know it has voltage, but it may not be correct.

1. You are going to briefly turn the power back on, make four voltage checks on the contactor, and turn the power back off. The voltages to check are
    a. Line side voltage
    b. Load side voltage
    c. Voltage across each of the sets of contacts (L1 to T1 and L2 to T2)
2. A voltage reading between L1 and T1 or L2 and T2 indicates a defective contactor.
3. If the voltage on the load side of the contactor is correct
    a. turn off the disconnect
    b. Discharge the run capacitor
    c. Note where the wires are connected to the capacitor and then remove them.
    d. Check the capacitor microfarads with the capacitor test function of the multi-meter.
4. If the capacitor checks OK, remove the compressor terminal cover and check the wire connections.
    a. Check the wires against the manufacturer's diagram.
    b. Note where the wires are and remove them.
    c. Use the procedure from Lab 21 to ohm the compressor motor out.
    d. If the compressor ohms out, the compressor is locked down.
5. If the voltage on the line side of the contactor is lower than the voltage stated on the equipment:
    a. Check connections at the contactor
    b. Check the circuit breaker
    c. Check connections at the disconnect
    d. Check the wiring from the disconnect to the unit
    e. Check the wiring from the power panel to the disconnect

LAB 88.5 IDENTIFYING TYPES OF STARTING RELAYS

LABORATORY OBJECTIVE
You will learn to identify different types of hermetic motor starting relays.

LABORATORY NOTES
You will use an ohm meter to identify and check different types of hermetic motor starting relays.

FUNDAMENTALS OF HVACR TEXT REFERENCE
Unit 37 Motors / Unit 88 Troubleshooting Refrigeration Systems

REQUIRED MATERIALS PROVIDED BY STUDENT
Volt-Ohm meter
Electrical hand tools

REQUIRED MATERIALS PROVIDED BY SCHOOL
Current relays
Potential relays
Solid state relays

PROCEDURE
Use the photos and diagrams in the book to identify each relay.

Use the ohm meter to check each of the relays.

Current Relays	Coil Resistance	Contact Resistance Upright	Contact Resistance Upside Down
Potential Relays	Coil Resistance (2 to 5)	Contact Resistance (1 to 2)	
Solid State Relays	Line to Run Resistance	Line to Start Resistance	

LAB 88.6 WIRE AND OPERATE HERMETIC SPLIT PHASE AND CAPACITOR START COMPRESSORS

LABORATORY OBJECTIVE
You will learn how to wire and operate hermetic split phase and capacitor start compressor motors used in HVACR.

LABORATORY NOTES
You will wire split phase compressor motors, starting relay, and overload. You will then operate them, and measure their amp draw.

FUNDAMENTALS OF HVACR TEXT REFERENCE
Unit 37 Motors / Unit 88 Troubleshooting Refrigeration Systems

REQUIRED MATERIALS PROVIDED BY STUDENT
Volt-Ohm meter
amp meter
Electrical hand tools

REQUIRED MATERIALS PROVIDED BY SCHOOL
Split phase or capacitor start hermetic compressors
Current starting relay
Solid state starting relay

PROCEDURE

1. Use the wiring diagrams in the book to wire the compressor, starting relay, and overload.

2. Be sure to check the compressor nameplate voltage and compare it to the voltage you are connecting to the motor.

3. Operate the motor.

4. Use a clamp on amp meter to read the motor current.

5. Write down your reading in the chart below.

6. Discuss your results with the instructor.

Put up all the motors. **Voltage**	**Nameplate LRA**	**Nameplate RLA** (if available)	**Running Amps**

LAB 88.7 WIRE AND OPERATE CSR MOTORS

LABORATORY OBJECTIVE
You will learn how to wire and operate CSR motors used in compressors.

LABORATORY NOTES
You will wire CSR compressors, operate them, and measure their amp draw.

***FUNDAMENTALS OF HVACR* TEXT REFERENCE**
Unit 37 Motors / Unit 88 Troubleshooting Refrigeration Systems

REQUIRED MATERIALS PROVIDED BY STUDENT
Volt-Ohm meter
amp meter
Electrical hand tools

REQUIRED MATERIALS PROVIDED BY SCHOOL
CSR compressors
Run capacitor
Potential relay
Start capacitor

PROCEDURE

1. Use the wiring diagram on the motor and/or the diagrams in the book to wire the CSR Compressor motors assigned by the instructor.

2. Be sure to check the motor nameplate voltage and compare it to the voltage you are connecting to the motor.

3. Operate the motor.

4. Use a clamp on amp meter to read the current through the start capacitor. There should be a measurable current spike when the motor starts, and then the current should drop to 0 after startup.

5. Use a clamp on amp meter to read the current through the common wire.

6. Write down your reading in the chart below.

7. Discuss your results with the instructor.

8. Put up all the motors.

Voltage	Nameplate LRA	Nameplate RLA (if available)	Running Amps

LAB 88.8 INSTALL HARD START KIT

LABORATORY OBJECTIVE
You will learn how to wire and operate CSR motors used in compressors.

LABORATORY NOTES
You will wire CSR compressors, operate them, and measure their amp draw.

***FUNDAMENTALS OF HVACR* TEXT REFERENCE**
Unit 37 Motors / Unit 88 Troubleshooting Refrigeration Systems

REQUIRED MATERIALS PROVIDED BY STUDENT
Volt-Ohm meter
amp meter
Electrical hand tools

REQUIRED MATERIALS PROVIDED BY SCHOOL
CSR compressors
Run capacitor
Potential relay
Start capacitor

PROCEDURE

1. Use the wiring diagram on the motor and/or the diagrams in the book to wire a hard start kit to the CSR Compressor motors assigned by the instructor.

2. Be sure to check the motor nameplate voltage and compare it to the voltage you are connecting to the motor.

3. Operate the motor.

4. Use a clamp on amp meter to read the current through the start capacitor. There should be a measurable current spike when the motor starts, and then the current should drop to 0 after startup.

5. Use a clamp on amp meter to read the current through the common wire.

6. Write down your reading in the chart below.

7. Discuss your results with the instructor.

8. Put up all the motors.

Voltage	Nameplate LRA	Nameplate RLA (if available)	Running Amps

LAB 89.1 WIRE DISCONNECT SWITCH AND POWER SUPPLY

LABORATORY OBJECTIVE
You will size and install the power wiring and disconnect switch to an air conditioning system.

LABORATORY NOTES
You will use the unit minimum circuit ampacity to size the conductor and disconnect switch. You will use the maximum overcurrent protection to size the fuse or circuit breaker protection. You will then install the disconnect switch, run conduit between the disconnect and the unit, and run power wire from the disconnect to the unit.

FUNDAMENTALS OF HVACR TEXT REFERENCE
Unit 34 Alternating Current Fundamentals / Unit 89 Installation

REQUIRED MATERIALS PROVIDED BY STUDENT
Safety glasses   Wire cutter
Multi-meter   Wire crimper/stripper
Ammeter   6 in 1 screwdriver

REQUIRED MATERIALS PROVIDED BY SCHOOL
Wire   Disconnect switch
Conduit   Air conditioning unit
Conduit connections

PROCEDURE
Find the minimum circuit ampacity and maximum overcurrent protection on the data plate. Use the National Electric Code to size the wire and conduit.

The circuit breaker or fuse should be no larger than the maximum overcurrent protection. The disconnect switch must be rated for no less than 115% of the minimum circuit ampacity. In most cases, a disconnect switch that will hold the maximum allowable fuse size is adequate.

Mount the disconnect switch.

Run conduit between the disconnect switch and the unit.

Pull the power wire between the disconnect switch and the unit.

Have the instructor check your work.

With the disconnect off but power to it, check the voltage into the top of the disconnect.

Compare the voltage to the minimum and maximum allowable voltages printed on the nameplate.

If the voltage is outside of the allowable voltage range LEAVE THE DISCONNECT SWITCH OFF!

If the voltage is within the allowable voltage range, turn on the disconnect switch.

Check voltage to the unit.

If the voltage to the unit is within the allowable voltage range, turn the unit on.

Check the voltage to the unit with the unit operating.

The voltage should still be within the allowable voltage range.

The difference between the voltage at the unit with the unit off and the voltage to the unit with the unit operating is the voltage drop through the circuit.

LAB 89.2 WIRING A PACKAGED UNIT

LABORATORY OBJECTIVE
You will wire a packaged air conditioning unit, including:
>Sizing the power wire
>Sizing the Disconnect switch
>Sizing the Circuit Breaker
>Connecting the power wire
>Installing the thermostat
>Connecting the control wiring

LABORATORY NOTES
You will size the power wire and disconnect based on the Unit Minimum circuit Ampacity on the Unit data plate. The overcurrent protection will be sized by the Maximum Overcurrent Protection on the Unit Data plate. You will wire the power wire from the disconnect to the unit and the control wire from the thermostat to the unit. After completing the installation, you will operate the unit and check the voltage and amp draw.

***FUNDAMENTALS OF HVACR* TEXT REFERENCE**
Unit 34 Alternating Current Fundamentals / Unit 76 Single Zone Rooftop Unit Installation / Unit 89 Installation

REQUIRED MATERIALS PROVIDED BY STUDENT
Safety Glasses
Gloves
Electrical hand tools
Multimeter
Amp Meter

REQUIRED MATERIALS PROVIDED BY SCHOOL
Packaged Air Conditioning Unit
Low voltage thermostat
Power Wire
Control Wire

PROCEDURE
Find the minimum circuit ampacity (MCA) and the maximum overcurrent protection on the unit data plate and record both.

Use the National Electrical Code Wire sizing charts or the manufacturer's installation literature to look up the correct wire size.

Size the disconnect by multiplying 1.15 x MCA

The fuse or circuit breaker should be no larger than the Maximum Overcurrent Protection

Mount the disconnect switch within sight of the unit. It can be mounted on the unit, but it should not be mounted on a service access panel.

Run weatherproof conduit between the disconnect switch and the unit.

Run the power wire between the disconnect switch and the unit.

Single-phase systems require 3 wires – 2 power wires and a ground.

Three phase systems require four wires – 3 power wires and a ground.

Mount the thermostat 5 feet off the floor on an inside wall out of drafts and sunlight.

Run 18-4 solid thermostat wire between the thermostat and the unit.

Do not use wire smaller than 18 gauge. Do not use stranded wire.

Follow the unit diagram to wire the control wires.

Have the instructor check your work.

Before starting the unit, check the voltage at the disconnect and compare it to the unit nameplate to make sure it is the correct voltage for the unit.

Start the unit and check its operating current.

LAB 89.3 WIRING A SPLIT SYSTEM

LABORATORY OBJECTIVE
You will wire a split system air conditioning unit, including:

Sizing the power wire to blower coil
Sizing the power wire to condenser
Sizing the Disconnect switch to blower
Sizing the Disconnect switch to condenser
Sizing the Circuit Breaker to blower
Sizing the Circuit Breaker to condenser
Connecting the power wire
Installing the thermostat
Connecting the control wiring

LABORATORY NOTES
A split system consists of two parts – an indoor blower coil and an outdoor condensing unit. Each has its own disconnect and power supply. You will need to size the power wire, disconnect, and breaker size for each unit separately. You will size the power wire and disconnect for both the blower and condenser based on the Unit Minimum circuit Ampacity on the Unit data plates. The overcurrent protection will be sized by the Maximum Overcurrent Protection on the Unit Data plates. You will wire the power wire from each disconnect to each unit. You will wire the control wire from the thermostat to the blower and from the blower to the condenser. After completing the installation, you will operate the unit and check the voltage and amp draw.

FUNDAMENTALS OF HVACR TEXT REFERENCE
Unit 34 Alternating Current Fundamentals / Unit 45 Residential Split-System Air Conditioning Installations / Unit 76 Single Zone Rooftop Unit Installation / Unit 89 Installation

REQUIRED MATERIALS PROVIDED BY STUDENT
Safety Glasses
Gloves
Electrical hand tools
Multimeter
Amp Meter

REQUIRED MATERIALS PROVIDED BY SCHOOL
Split System Air Conditioning Unit
Low voltage thermostat
Power Wire
Control Wire

PROCEDURE
Find and record the minimum circuit ampacity (MCA) and the maximum overcurrent protection on the unit data plates of both units.

Use the National Electrical Code Wire sizing charts or the manufacturer's installation literature to look up the correct wire size for both units.

Size the disconnect swtich for each unit by multiplying 1.15 x MCA

The fuse or circuit breaker for each unit should be no larger than the Maximum Overcurrent Protection.

Mount each disconnect switch within sight of the unit it controls. The disconnect switches can be mounted on the units, but they should not be mounted on service access panels.

Run weatherproof conduit between each disconnect switch and the unit.

Run the power wire between each disconnect switch and the unit.

Single-phase systems require 3 wires – 2 power wires and a ground.

Three phase systems require four wires – 3 power wires and a ground.

Mount the thermostat 5 feet off the floor on an inside wall out of drafts and sunlight.

Run 18-4 solid thermostat wire between the thermostat and the blower coil.

Run 18-2 solid thermostat wire between the blower coil and the condensing unit.

Do not use wire smaller than 18 gauge. Do not use stranded wire.

Follow the unit diagram to wire the control wires.

Before starting the unit, check the voltage at each disconnect switch and compare it to the unit nameplate to make sure that the correct voltage is supplied to each unit..

The voltage requirements may not be the same for both units. Furnaces normally operate on 120 volts while condensing units operate on 240 volts. Be careful to put the correct voltage to each unit.

Have the instructor check your work.

Start the unit and check its operating current.

LAB 90.1 BELT DRIVE BLOWER COMPLETE SERVICE

LABORATORY OBJECTIVE

The purpose of this lab is to demonstrate your ability to conduct the proper maintenance service procedure for a belt driven blower.

LABORATORY NOTES

Some older residential and most commercial air handlers use belt drive blowers to deliver air through the duct system. The advantage of belt drive blowers over direct drive is the flexibility in choosing blower speed and the ease of manufacturing a blower wheel to withstand the RPM. It is too difficult to match all the factors, RPM, CFM, motor HP, etc. in all commercial applications. Most new residential systems have gone over to direct drive blowers but in commercial systems, belt drives will be around for may years to come. Occasionally a belt drive blower needs a complete overhaul including any or all of the following: motor and bearings, motor pulley, belt, blower pulley, blower bearings, blower shaft, cleaning, balancing, or rebalance.

***FUNDAMENTALS OF HVACR* TEXT REFERENCE**

Unit 75 Fans and Air Handling Units/ Unit 90 Planned Maintenance

REQUIRED TOOLS AND EQUIPMENT

Belt driven blower assembly
Tool kit
Multi-meters
Clamp-on Ammeter

SAFETY REQUIREMENTS

A. Check all circuits for voltage before doing any service work.

B. Stand on dry nonconductive surfaces when working on live circuits.

C. Never bypass any electrical protective devices.

PROCEDURE
Step 1

Collect the blower data and fill in the chart.

Blower Unit Data

Unit Name:		Model Number:	Blower Motor Amperage Rating:	
Unit Type:				
Blower Motor Data:	Voltage:	Horsepower:	Locked Rotor Amps:	Rated Load Amps:
				Full Load Amps:
Blower Wheel Data:	Width:	Diameter:	Shaft Size OD:	
Pulley Size:	Motor Pulley OD:	Motor Pulley ID:	Blower Shaft OD:	Blower Shaft ID:
Belt Size:		**Belt Number:**		

Lubrication required for blower bearings: **(circle one)**	Grease	Oil
Lubrication required for motor bearings: **(circle one)**	Grease	Oil
Inspect greased motors for grease relief fitting at bottom of bearing below grease opening:	None _____	

706

Step 2

Complete an initial inspection. Complete the Initial Inspection Check List and make sure to check each step off in the appropriate box as you finish it. This will help you to keep track of your progress.

Initial Inspection Check List

STEP	PROCEDURE	CHECK
1	Spin the blower wheel slowly by hand.	
2	Notice any drag noise or pulley wobble.	
3	Install the clamp-on ammeter to the L1 terminal of the motor.	
4	Place the blower door in the normal position for normal air flow and load on the motor.	
5	Obtain the fan only operation and record the motor amps. Amps = _____	
6	Remove blower door for further inspection.	
7	Observe motor begin to turn and watch for belt slipping during startup.	
8	Note any running noise, bearing, grinding or any noise other than normal airflow noise.	

Step 3

After finishing the initial blower inspection, make sure all power is off. Lock and tag the power panel before removing any parts. Complete the Blower Removal And Cleaning Check List.

Blower Removal And Cleaning Check List

STEP	PROCEDURE	CHECK
1	Make sure that the power to the blower motor is secured. Verify this with a multi-meter voltage test. Refer to Unit 35 in *Fundamentals of HVACR*.	
2	Remove and store carefully all panels and covers.	
3	Loosen tension on belt for removal.	
4	Remove belt and inspect for any cracks ore severe shine (caused by slipping).	
5	Inspect pulleys for wear and grooving.	
6	Inspect blower wheel blades and scrape with a screwdriver or thin tool for any evidence of accumulated debris.	
7	Clean blower wheel with water, compressed air, or CO_2, whichever is most available and appropriate.	
8	Blow motor air passages with compressed air or CO_2.	
9	Wipe down the motor and nameplate with a clean rag.	

708

After cleaning the blower assembly can be disassembled. Complete the Blower Disassembly Check List.

Blower Disassembly Check List

STEP	PROCEDURE	CHECK
1	Loosen setscrew from the blower pulley and remove the pulley.	
2	Support the blower wheels to keep them in position.	
3	Loosen the screws on the set collars and pull the blower shaft out of the blower.	
4	Record the length and diameter of the blower shaft. Length = _____ OD = _____	
5	List any special features of blower keyways, flat sides, etc. for replacement.	
6	Make notes on the blower shaft condition such as wear, grooving, rust, etc.	
7	Roll the shaft on a flat surface to be sure it is straight and true.	

8	Remove and inspect blower wheel bearings and replace if necessary.	
9	Install the original or a replacement blower shaft.	
10	Install the original or replacement shaft pulley.	
11	Install the original or replacement shaft pulley.	
12	Align or check alignment of pulleys with a straight edge on the pulley outer edge. The straight edge should touch and outer edge of both pulleys in at least four different places.	
13	Lubricate the motor and shaft bearings.	
14	Install new belt of the correct size.	
15	Adjust the motor tension assembly for the correct belt tension. Consult the manufacturer recommendation for correct belt tension (belt play is typically 2 to 3 inches).	

Step 5

After reassembling the blower you can check for correct operation with a final test run. Complete the Blower Test Run Check List.

Blower Test Run Check List

STEP	PROCEDURE	CHECK
1	Spin the blower wheel slowly by hand.	
2	Notice any drag noise or pulley wobble.	
3	Install the clamp-on ammeter to the L1 terminal of the motor.	
4	Turn on the main power.	
5	Place the blower door in the normal position for normal air flow and load on the motor.	
6	Obtain the fan only operation and record the motor amps. Amps = _____	
7	Compare measured amps with rated motor amps. Actual amps is: (circle one) Higher Lower	

8	If the running amperage is incorrect then loosen the motor belt tension and remove belt.	
9	Loosen and adjust the motor pulley or replace with a pulley of larger or smaller size as required. Move the pulley sheave closer to increase the size.	
10	Install the belt and adjust the belt tension.	
11	Run the blower and once again measure the amperage. Amps = _____	
12	Repeat steps 8 through 11 to obtain rated full load amps and maximum CFM.	

LAB 90.2 CLEANING COILS

LABORATORY OBJECTIVE
You will demonstrate your ability to safely clean an air conditioning coil.

LABORATORY NOTES
You will clean a coil using both mechanical and chemical means. You will turn off power to the unit and remove any access panels in the way. Next, you will use a stiff brush and vacuum cleaner to remove debris on the surface of the coil. Then you will apply a chemical coil cleaner according to the manufacturer's instructions.

***FUNDAMENTALS OF HVACR* TEXT REFERENCE**
Unit 16 Condensers/ Unit 90 Planned Maintenance

REQUIRED MATERIALS PROVIDED BY STUDENT
Safety Glasses
Gloves
Flat blade screwdriver
Phillips Screwdriver
¼" nut driver
5/16" nut driver
OR
6 in 1 with all the above
Adjustable jaw wrench

REQUIRED MATERIALS PROVIDED BY SCHOOL
Air Conditioning Units with Dirty Coils
Brush
Shop Vacuum
Bug Sprayer with Coil Cleaner
Bug sprayer with water.

PROCEDURE

Safety Glasses are required!

Turn off power to the unit.

Some units may require panel removal to get to the coil.

Use a stiff brush and/or shop vacuum to remove debris accumulated on the face of the coil.

CAUTION: The coil fins are like thousands of razor blades ready to slice you up. Wear gloves and avoid contacting the fins while cleaning the coil.

Be careful not to bend the coil fins. They are very thin and easily damaged. The brush and/or vacuum should move in the same direction as the fins, not perpendicular to the fins.

Read the instructions for the coil cleaner.

Apply the coil cleaner according to the manufacturer's instructions.

LAB 90.3 TYPES OF MOTOR BEARINGS

LABORATORY OBJECTIVE
You will learn to identify the types of bearing used in electric motors.

FUNDAMENTALS OF HVACR TEXT REFERENCE
Unit 90 Planned Maintenance

REQUIRED MATERIALS PROVIDED BY STUDENT
Electrical hand tools

REQUIRED MATERIALS PROVIDED BY SCHOOL
Selection of Electric Motors

PROCEDURE

Select five motors in the shop and record their model numbers in the chart below. Determine the type of bearing (ball or sleeve) and record. Show method of lubrication used.

Motor	Lubrication Method	Shaft Bearing	End Bell Bearing

LAB 90.4 BELT DRIVES

LABORATORY OBJECTIVE
You will learn to adjust and correctly tension a belt drive.

***FUNDAMENTALS OF HVACR* TEXT REFERENCE**
Unit 75 Fans and Air Handling Units / Unit 90 Planned Maintenance

REQUIRED MATERIALS PROVIDED BY STUDENT
Amp meter
Electrical hand tools

REQUIRED MATERIALS PROVIDED BY SCHOOL
Belt drive blower with adjustable pulley.

PROCEDURE

1. Operate the belt drive blower assigned by the instructor.

2. Use a digital photo-tachometer to measure the motor RPM and the blower RPM.

3. Measure the motor amp draw.

4. Loosen and remove the belt.

5. Adjust the pulley out counterclockwise ½ round. Make sure to position the set screw over a flat on the pulley before tightening the set screw.

6. Reinstall and re-tension the belt.

7. Operate the blower and measure the motor amp draw.

8. Use a digital photo-tachometer to measure the motor RPM and the blower RPM.

9. Loosen and remove the belt.

10. Adjust the pulley back in clockwise ½ round. Make sure to position the set screw over a flat on the pulley before tightening the set screw.

11. Reinstall and re-tension the belt.

12. Operate the blower and measure the motor amp draw.

13. Use a digital photo-tachometer to measure the motor RPM and the blower RPM.

LAB 90.5 DIRECT DRIVE MOTOR APPLICATIONS

LABORATORY OBJECTIVE
You will learn to identify the types of direct drive electric motor applications in HVAC/R.

***FUNDAMENTALS OF HVACR* TEXT REFERENCE**
Unit 75 Fans and Air Handling Units / Unit 90 Planned Maintenance

REQUIRED MATERIALS PROVIDED BY STUDENT
Amp meter
Electrical hand tools

REQUIRED MATERIALS PROVIDED BY SCHOOL
Selection of Direct Drive Motor Applications

PROCEDURE
Find three different direct drive electric motor applications in the shop. In a direct drive application the motor is connected directly to the device it is operating.

Application	Connection: Hub, Spring Coupling, Rubber Coupling

LAB 92.1 WATERFLOW EFFECT ON SYSTEM PERFORMANCE

LABORATORY OBJECTIVE
You will demonstrate the effect evaporator and condenser waterflow have on system performance.

LABORATORY NOTES
You will operate a water source heat pump in cooling, adjust the condenser waterflow and record the effect. You will them operate the unit in heating, adjust the evaporator waterflow, and record the effect.

FUNDAMENTALS OF HVACR TEXT REFERENCE
Unit 16 Condensers / Unit 18 Evaporators / Unit 92 Troubleshooting

REQUIRED MATERIALS PROVIDED BY STUDENT
Safety Glasses
Gloves
6 in 1
Temperature Tester

REQUIRED MATERIALS PROVIDED BY SCHOOL
Operating Water Source Heat Pump

PROCEDURE

Condenser Waterflow

Operate water-cooled system in cooling with 3 GPM waterflow.

Check for proper operation, charge and adjust as necessary.

Record pressures and temperatures on data sheet.

Increase water flow to 5 GPM and observe for 5 minutes.

Record pressures/temperatures.

What effect does increased condenser water flow have on the system pressures/temperatures?

Reduce the water flow to 2 GPM and operate for 5 minutes.

Record pressures/temperatures.

What effect does low condenser water flow have on system pressures/temperatures?

Turn water flow off and observe for 1 minute.

Do not let the head pressure climb over 300 psig.

Turn off the system if head pressure should start to climb over 300 psig.

What effect does no condenser water flow have on system pressures/temperatures? Record pressures/temperatures.

Summarize the effect of condenser waterflow on total system operation.

	Cond Pres	Change	Cond Temp	Change	Liquid Temp	Change	Subcooling	Change
	CONDENSER WATERFLOW							
Normal								
5 GPM								
3 GPM								
2 GPM								

Evaporator Waterflow

Operate water source heat pump in heat with 3 GPM waterflow.

Check for proper operation, charge and adjust as necessary.

Record pressures and temperatures on data sheet.

Increase water flow to 5 GPM and observe for 5 minutes.

Record pressures/temperatures.

What effect does increased condenser water flow have on the system pressures and temperatures?

Reduce the water flow to 2 GPM and operate for 5 minutes.

Record pressures/temperatures.

What effect does low water flow have on system pressures and temperatures?

Summarize the effect of evaporator waterflow on total system operation.

	Evap Pres	Change	Evap Temp	Change	Suction Temp	Change	Superheat	Change
Normal								
5 GPM								
3GPM								
2 GPM								

LAB 92.2 EFFECT OF AIRFLOW ON SYSTEM PERFORMANCE

LABORATORY OBJECTIVE
You will demonstrate the effect evaporator and condenser airflow have on system performance.

LABORATORY NOTES
You will operate a refrigeration trainer, adjust the condenser airflow and record the effect. You will them operate the trainer, adjust the evaporator airflow, and record the effect.

***FUNDAMENTALS OF HVACR* TEXT REFERENCE**
Unit 16 Condensers / Unit 18 Evaporators / Unit 92 Troubleshooting

REQUIRED MATERIALS PROVIDED BY STUDENT
Safety Glasses

REQUIRED MATERIALS PROVIDED BY SCHOOL
Refrigeration trainer with variable speed blower controls.

PROCEDURE

Condenser Airflow

Operate refrigeration trainer with both the evaporator and condenser fans set to medium.

Check for proper operation, charge and adjust as necessary.

Record pressures and temperatures on data sheet.

Turn condenser fan speed to high and observe for 5 minutes. What effect does increased condenser air have on the system pressures/temperatures? Record pressures/temperatures.

Turn the condenser fan speed to low and observe for 5 minutes.

What effect does low condenser air has on system pressures/temperatures? Record pressures/temperatures.

Turn condenser fans speed to off and observe for 1 minute. Do not let the head pressure climb over 250 psig. Turn fan speed to high if head pressure should start to climb over 200 psig.

What effect does no condenser air flow have on system pressures/temperatures? Record pressures/temperatures.

Summarize the effect of condenser airflow on total system operation.

LAB 92.2 EFFECT OF AIRFLOW ON SYSTEM PERFORMANCE								
	Cond Pres	Change	Cond Temp	Change	Liquid Temp	Change	Subcooling	Change
Normal								
High Fan Speed								
Low Fan Speed								
Fan Off								

Evaporator Airflow

1. Operate refrigeration trainer with both the evaporator and condenser fans set to medium.

2. Check for proper operation, charge and adjust as necessary.

3. Record pressures and temperatures on data sheet.

4. Turn evaporator fan speed to high and observe for 5 minutes.
    Record pressures/temperatures.

5. What effect did increased air have on the system operation?

6. Turn evaporator fan speed to low and observe for 5 minutes. Record pressures, temperatures pressures/temperatures.

7. What effect did decreased air have on system operation?

8. Turn evaporator fan speed to off and observe for 5 minutes. Record pressures/temperatures.

9. What effect did no air have on system operation?

10. Summarize the effect of evaporator airflow on total system operation. Be sure to include changes in superheat, sub-cooling, high side pressure and low side pressure.

	Evap Pres	Change	Evap Temp	Change	Suction Temp	Change	Superheat	Change
Normal								
High Fan Speed								
Low Fan Speed								
Fan Off								

LAB 92.3 TROUBLESHOOTING SCENARIO 1

LABORATORY OBJECTIVE

Given a system with a problem, you will identify the problem, its root cause, and recommend corrective action. Labs 92.3 through 92.5 will all follow the same procedure, however each lab will present a different problem scenario based upon the type of furnace and corrective action required.

***FUNDAMENTALS OF HVACR* TEXT REFERENCE**

Unit 92 Troubleshooting

REQUIRED MATERIALS PROVIDED BY STUDENT

Safety Glasses

Hand tools

Manometer

Multimeter

Refrigeration Gauges

REQUIRED MATERIALS PROVIDED BY SCHOOL

System with a problem

PROCEDURE

Troubleshoot the system assigned by the instructor. Be sure to be complete in your description of the problem. You should include:

What is the unit is doing wrong?

What component or condition is causing this?

What tests did you perform that told you this?

How would you correct this?

LAB 92.4 TROUBLESHOOTING SCENARIO 2

LABORATORY OBJECTIVE
Given a system with a problem, you will identify the problem, its root cause, and recommend corrective action. Labs 92.3 through 92.5 will all follow the same procedure, however each lab will present a different problem scenario based upon the type of furnace and corrective action required.

FUNDAMENTALS OF HVACR TEXT REFERENCE
Unit 92 Troubleshooting

REQUIRED MATERIALS PROVIDED BY STUDENT
Safety Glasses
Hand tools
Manometer
Multimeter
Refrigeration Gauges

REQUIRED MATERIALS PROVIDED BY SCHOOL
System with a problem

PROCEDURE
Troubleshoot the system assigned by the instructor. Be sure to be complete in your description of the problem. You should include:

What is the unit is doing wrong?

What component or condition is causing this?

What tests did you perform that told you this?

How would you correct this?

LAB 92.5 TROUBLESHOOTING SCENARIO 3

LABORATORY OBJECTIVE
Given a system with a problem, you will identify the problem, its root cause, and recommend corrective action. Labs 92.3 through 92.5 will all follow the same procedure, however each lab will present a different problem scenario based upon the type of furnace and corrective action required.

FUNDAMENTALS OF HVACR TEXT REFERENCE
Unit 92 Troubleshooting

REQUIRED MATERIALS PROVIDED BY STUDENT
Safety Glasses
Hand tools
Manometer
Multimeter
Refrigeration Gauges

REQUIRED MATERIALS PROVIDED BY SCHOOL
System with a problem

PROCEDURE
Troubleshoot the system assigned by the instructor. Be sure to be complete in your description of the problem. You should include:

What is the unit is doing wrong?

What component or condition is causing this?

What tests did you perform that told you this?

How would you correct this?